Microbial Biopolymers: Trends in Synthesis, Modification, and Applications

Microbial Biopolymers: Trends in Synthesis, Modification, and Applications

Editor

Shashi Kant Bhatia

MDPI • Basel • Beijing • Wuhan • Barcelona • Belgrade • Manchester • Tokyo • Cluj • Tianjin

Editor
Shashi Kant Bhatia
Dept. of Biological
Engineering
Konkuk University
Seoul
Korea, South

Editorial Office
MDPI
St. Alban-Anlage 66
4052 Basel, Switzerland

This is a reprint of articles from the Special Issue published online in the open access journal *Polymers* (ISSN 2073-4360) (available at: www.mdpi.com/journal/polymers/special_issues/Microbial_Biopolym).

For citation purposes, cite each article independently as indicated on the article page online and as indicated below:

LastName, A.A.; LastName, B.B.; LastName, C.C. Article Title. *Journal Name* **Year**, *Volume Number*, Page Range.

ISBN 978-3-0365-7453-0 (Hbk)
ISBN 978-3-0365-7452-3 (PDF)

Contents

About the Editor

Shashi Kant Bhatia

Dr. Shashi Kant Bhatia works as an associate professor in the Department of Biological Engineering, Konkuk University, Seoul, South Korea. He received his M.Sc. degree and a Ph.D. degree in biotechnology from Himachal Pradesh University (India). Previously, Dr. Bhatia worked as a Brain Pool Postdoc Fellow at Konkuk University (2014–2016), South Korea. His major research areas include environmental biotechnology; biochar production and applications in bioenergy and emerging contaminant removal; resource recovery from waste; valorization of biowaste into biomaterials, bioenergy, and biochemicals; biotransformation and biocatalysis; fermentation; enzyme production; and immobilizations. He has published more than 190 research and review articles, has edited 4 books, and holds 12 international patents. His current H-index is 46 with an i10-index of 146, and he has a total number of citations of 7000. He is a life member of the Biotech Research Society of India (BRSI) and the Association of Microbiologists of India (AMI). He is serving as associate editor of *Microbial Cell Factories*, *Frontiers in Microbiology*, *3 Biotech*, and *PLOS One*, and as an editorial board member of *Biomass Conversion and Biorefineries*, *Bioprocess and Biosystems Engineering*, *Sustainability*, *Biomass*, and *Energies*. He is also an active and regular reviewer of international journals published by Elsevier, ACS, Springer, Wiley, and Taylor & Francis.

Preface to "Microbial Biopolymers: Trends in Synthesis, Modification, and Applications"

This is a reprint of the Special Issue, "Microbial Biopolymers: Trends in Synthesis, Modification, and Applications". The reprint provides information on recent technological advancements in the area of the production, functionalization, and applications of polyhydroxyalkanoates (PHAs). There are ten chapters in this reprint that discuss the production of polyhydroxyalkanoates, feedstock studies, functionalization, and their applications. Chapter 1 is an introduction to microbial polyhydroxyalkanoates and current research scenarios around the globe, and it also provides information about the various chapters published in this reprint. Chapter 2 provides information about cheap and alternate carbon sources for PHA production and strategies to search for microbes for galactose-enriched algal biomass utilization. Chapter 3 is about the utilization of *Bacillus mycoides* for PHA production using cardboard as a feedstock. Chapter 4 discusses the possibility of alkanes as a source for PHA copolymer production using *Pseudomonas* sp. Chapter 5 reports on PHA recovery and downstream processing using a novel detergent-based PHA extraction method. Chapter 6 discusses a technology for PHA and hydrogen production from spent coffee grounds using *Pseudomonas resinovorans* and *Clostridium butyricum*. Chapter 7 provides information about improving the properties of polyhydroxyalkanoates by preparing copolymers or mixing and blending them with other natural and synthetic polymers. Chapter 8 provides information about polyhydroxyalkanoate applications in the packaging industry. Chapter 9 discusses the preparation of the poly(3-hydroxybutyrate-co-3-hydroxyvalerate) P(3HB-co-3HV) scaffold and its application in periodontal tissue engineering. Chapter 10 is about PHA blending with other plasticizers and its effect on the biodegradability of *Microbulbifer* sp. SOL66. Chapter 11 is about exopolysaccharide production by the *Bacillus amyloliquefaciens* RT7 strain isolated from an extreme acidic environment using a combination of glucose–Tween 80 as feedstock. Chapter 12 provides information about the identification of a thermostable levansucrase from *Pseudomonas orientalis* and its application in levan production. Chapter 13 discusses *Phormidium ambiguum* and *Leptolyngbya ohadii* exopolysaccharide production under low water availability. Chapter 14 provides information about microbial exopolysaccharide production, the preparation of their composite materials, and their applications in the health sector.

Shashi Kant Bhatia
Editor

polymers

Editorial

Microbial Biopolymers: Trends in Synthesis, Modification, and Applications

Shashi Kant Bhatia [1,2]

[1] Department of Biological Engineering, College of Engineering, Konkuk University, Seoul 05029, Republic of Korea; shashibiotechhpu@gmail.com
[2] Institute for Ubiquitous Information Technology and Applications, Konkuk University, Seoul 05029, Republic of Korea

Citation: Bhatia, S.K. Microbial Biopolymers: Trends in Synthesis, Modification, and Applications. *Polymers* **2023**, *15*, 1364. https://doi.org/10.3390/polym15061364

Received: 7 March 2023
Accepted: 8 March 2023
Published: 9 March 2023

Microbes can act as a factory for the conversion of a variety of carbon and nitrogen sources into diverse kinds of intracellular and extracellular biopolymers, including polyhydroxyalkanoates (PHA) and exopolysaccharides (EPS), under different stress conditions [1,2]. Polyhydroxyalkanoates are intracellularly stored and serve as carbon and energy storage reserves. Almost 150 types of monomeric units have been reported and involved in the regulation of the physicochemical properties of the PHA [3]. A variety of microbes have recently been reported to produce PHA, such as *Bacillus mycoides*, *Halomonas cerina*, *Pseudomonas resinovorans*, etc. [4–6]. These biopolymers are biocompatible, biodegradable, and have different chemical and morphological properties that make them suitable for drug delivery, tissue engineering, packaging industry, and environmental applications [7]. Recent advances in molecular biology, transcriptomics, and metabolomics techniques have improved the understanding related to mechanisms and regulations involved in biopolymer synthesis [8,9]. A microbial system can be easily engineered and cultured under controlled conditions to produce different polymers. Biopolymers produced by microbial systems are rich in various functional groups that can be exploited further to modify the polymers for a variety of applications [10]. The production cost of biopolymers is the main challenge for their applicability on a commercial scale. Researchers are working on the utilization of diverse kinds of organic wastes, such as lignocellulosic waste, municipal waste, whey, paper, pulp industry waste, etc., as feedstocks for microbial fermentation. Biopolymer production from a microbial system is a clean and green approach that has recently become a popular topic worldwide, and it is considered a possible way to deal with plastic-based wastes, with extensive applications in the biotechnology sector. According to the Scopus database, approximately 1849 research and review articles have been published in the last few years (2018–2023). The top five countries involved in microbial biopolymer research are India (382), China (274), the United States (214), South Korea (122), and Brazil (135) (Figure 1).

To make biopolymer production economic, searching for microbes able to utilize cheap and abundantly available raw material is an important step. Galactose is a sugar found in marine algal biomass. Jung et al. screened 16 different *Halomonas* strains and reported a high PHB production (5.2 g/L) in *Halomonas cerina* with *Eucheuma spinosum* hydrolysate as a carbon source. *H. cerina* can survive in high-saline conditions and is able to accumulate PHA up to 72.41% w/w under unsterilized conditions [6]. In another study, researchers used cardboard hydrolysate as a carbon source for *Bacillus mycoides* and reported 56% w/w PHA accumulation [5]. Polyhydroxybutyrate (PHB) polymers have limited applications due to their brittle nature. To overcome this issue, researchers are working on the production of various copolymers of PHB. *Pseudomonas* sp. has the ability to produce medium-chain-length PHA (mcl-PHA) with improved properties. Jeon et al. used three different alkanes—n-octane, n-decane, and n-dodecane—as carbon sources, and the process resulted in mcl-PHA productionof 0.48 g/L, 0.27 g/L, or 0.07 g/L, respectively.

The optimization of cultural conditions using statistical design, and the use of mixed alkane under optimized conditions at a 7 L fermentation scale produces 2.1 g/L mcl-PHA [4]. PHA recovery and downstream processing are also expensive processes. To make the PHA recovery process more economic, Novakova et al. developed a detergent-based PHA extraction method. This method includes the exposition of PHA-accumulating microbial biomass to hypotonic conditions at elevated temperatures in the presence of sodium dodecyl sulfate (SDS). This process resulted in a high PHA recovery with 99% purity [11]. The use of SDS is able to remove hydrophobic impurities and the further removal of SDS can be easily achieved using KCl as a precipitation method. To efficiently utilize feedstocks, Kang et al. developed a technology for PHA and hydrogen production, where oil was first extracted from spent coffee grounds and used as a feedstock for *Pseudomonas resinovorans* to produce PHA (1.6 g/L). Further, oil extracted residual spent coffee biomass was hydrolyzed to produce sugars and used as a carbon source for *Clostridium butyricum* DSM10702 to produce hydrogen (181.19 mL) [12]. Polyhydroxyalkanoates properties can be further improved by preparing copolymers or mixing and blending them with other natural and synthetic polymers. Jo et al. reported that copolymers with a 4HB content greater than 16% are non-crystalline in nature, while P(3HB-*co*-4HB) mixtures with the same 4HB content are crystalline. Due to this effect, the mixture has a higher melt viscosity and a lower tangent with better melt processing properties [13]. Polyhydroxyalkanoates have applications in the packaging industry, and PHA must be prepared as a foamed material with a porous structure. Zhang et al. used supercritical CO_2 to prepare the foaming material of P(3HB-*co*-4HB) and reported that 4HB concentration plays an important role in foaming, while no foaming occurs beyond 50% 4HB content [14]. Phuegyod et al. prepared a poly(3-hydroxybutyrate-*co*-3-hydroxyvalerate) P(3HB-*co*-3HV) scaffold and compared its properties with other polymers for use in periodontal tissue engineering. The human gingival fibroblasts and periodontal ligament stem cells had a higher proliferation, healthy morphology, and better compatibility when cultured with P(3HB-*co*-3HV) as compared to other polymers [15]. To improve the mechanical strength of PHA, these polymers are blended with other materials such as plasticizers, which affect their biodegradability. Cho et al. prepared PHB blends with various plasticizers and studied the degradation of the blended materials. It was reported that a PHB blend with 10-20% tributyl citrate improves its mechanical properties as well as speeds up its degradation by *Microbulbifer* sp. SOL66 [16].

As the editor of this Special Issue, I have observed that the utilization of waste materials as feedstocks, the development of eco-friendly methods for PHA recovery, and the improvement of PHA properties by mixing are the current areas of interest among the published research and review articles. I am sure that this Special Issue will pique the interest of researchers in this area and provide readers with a broad and updated overview of this topic.

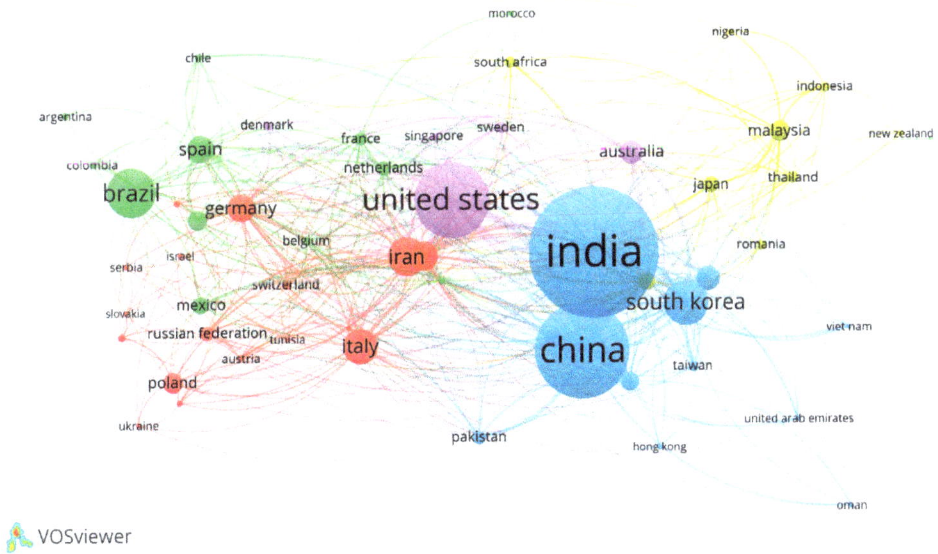

Figure 1. Co-occurrence mapping of publications related to microbial biopolymers (min. number of occurrence 5).

Acknowledgments: I acknowledge all the authors and reviewers who have contributed to achieving this Special Issue. In addition, I would like to thank the technical support team at MDPI for their assistance in preparing this Issue. The author acknowledge the KU Research Professor Program of Konkuk University, Seoul, South Korea.

Conflicts of Interest: The author declares no conflict of interest.

References

1. Bhatia, S.; Gurav, R.; Choi, Y.-K.; Choi, T.-R.; Kim, H.-j.; Song, H.-S.; Mi Lee, S.; Lee Park, S.; Soo Lee, H.; Kim, Y.-G.; et al. Bioprospecting of exopolysaccharide from marine *Sphingobium yanoikuyae* BBL01: Production, characterization, and metal chelation activity. *Bioresour. Technol.* **2021**, *324*, 124674. [CrossRef] [PubMed]
2. Jung, H.-R.; Choi, T.-R.; Han, Y.H.; Park, Y.-L.; Park, J.Y.; Song, H.-S.; Yang, S.-Y.; Bhatia, S.K.; Gurav, R.; Park, H.; et al. Production of blue-colored polyhydroxybutyrate (PHB) by one-pot production and coextraction of indigo and PHB from recombinant *Escherichia coli*. *Dyes. Pigments.* **2020**, *173*, 107889. [CrossRef]
3. Vu, D.H.; Wainaina, S.; Taherzadeh, M.J.; Åkesson, D.; Ferreira, J.A. Production of polyhydroxyalkanoates (PHAs) by *Bacillus megaterium* using food waste acidogenic fermentation-derived volatile fatty acids. *Bioengineered* **2021**, *12*, 2480–2498. [CrossRef] [PubMed]
4. Jeon, J.-M.; Park, S.-J.; Son, Y.-S.; Yang, Y.-H.; Yoon, J.-J. Bioconversion of Mixed Alkanes to Polyhydroxyalkanoate by *Pseudomonas resinovornas*: Upcycling of Pyrolysis Oil from Waste-Plastic. *Polymers* **2022**, *14*, 2624. [CrossRef] [PubMed]
5. Abdelmalek, F.; Steinbüchel, A.; Rofeal, M. The Hyperproduction of Polyhydroxybutyrate Using *Bacillus mycoides* ICRI89 through Enzymatic Hydrolysis of Affordable Cardboard. *Polymers* **2022**, *14*, 2810. [CrossRef] [PubMed]
6. Jung, H.J.; Kim, S.H.; Cho, D.H.; Kim, B.C.; Bhatia, S.K.; Lee, J.; Jeon, J.-M.; Yoon, J.-J.; Yang, Y.-H. Finding of Novel Galactose Utilizing *Halomonas* sp. YK44 for Polyhydroxybutyrate (PHB) Production. *Polymers* **2022**, *14*, 5407. [CrossRef] [PubMed]
7. Bhatia, S.K.; Otari, S.V.; Jeon, J.-M.; Gurav, R.; Choi, Y.-K.; Bhatia, R.K.; Pugazhendhi, A.; Kumar, V.; Rajesh Banu, J.; Yoon, J.-J.; et al. Biowaste-to-bioplastic (polyhydroxyalkanoates): Conversion technologies, strategies, challenges, and perspective. *Bioresour. Technol.* **2021**, *326*, 124733. [CrossRef] [PubMed]
8. Lee, H.S.; Lee, S.M.; Park, S.L.; Choi, T.-R.; Song, H.-S.; Kim, H.-J.; Bhatia, S.K.; Gurav, R.; Kim, Y.-G.; Kim, J.-H.; et al. Tung Oil-Based Production of High 3-Hydroxyhexanoate-Containing Terpolymer Poly(3-Hydroxybutyrate-co-3-Hydroxyvalerate-co-3-Hydroxyhexanoate) Using Engineered *Ralstonia eutropha*. *Polymers* **2021**, *13*, 1084. [CrossRef] [PubMed]
9. Jung, H.-R.; Yang, S.-Y.; Moon, Y.-M.; Choi, T.-R.; Song, H.-S.; Bhatia, S.K.; Gurav, R.; Kim, E.-J.; Kim, B.-G.; Yang, Y.-H. Construction of Efficient Platform Escherichia coli Strains for Polyhydroxyalkanoate Production by Engineering Branched Pathway. *Polymers* **2019**, *11*, 509. [CrossRef] [PubMed]

10. Emaimo, A.J.; Olkhov, A.A.; Iordanskii, A.L.; Vetcher, A.A. Polyhydroxyalkanoates Composites and Blends: Improved Properties and New Applications. *J. Compos. Sci.* **2022**, *6*, 206. [CrossRef]
11. Novackova, I.; Kourilova, X.; Mrazova, K.; Sedlacek, P.; Kalina, M.; Krzyzanek, V.; Koller, M.; Obruca, S. Combination of Hypotonic Lysis and Application of Detergent for Isolation of Polyhydroxyalkanoates from Extremophiles. *Polymers* **2022**, *14*, 1761. [CrossRef] [PubMed]
12. Kang, B.-J.; Jeon, J.-M.; Bhatia, S.K.; Kim, D.-H.; Yang, Y.-H.; Jung, S.; Yoon, J.-J. Two-Stage Bio-Hydrogen and Polyhydroxyalkanoate Production: Upcycling of Spent Coffee Grounds. *Polymers* **2023**, *15*, 681. [CrossRef] [PubMed]
13. Jo, M.; Jang, Y.; Lee, E.; Shin, S.; Kang, H.-J. The Modification of Poly(3-hydroxybutyrate-co-4-hydroxybutyrate) by Melt Blending. *Polymers* **2022**, *14*, 1725. [CrossRef] [PubMed]
14. Zhang, T.; Jang, Y.; Lee, E.; Shin, S.; Kang, H.-J. Supercritical CO_2 Foaming of Poly(3-hydroxybutyrate-co-4-hydroxybutyrate). *Polymers* **2022**, *14*, 2018. [CrossRef] [PubMed]
15. Phuegyod, S.; Pramual, S.; Wattanavichean, N.; Assawajaruwan, S.; Amornsakchai, T.; Sukho, P.; Svasti, J.; Surarit, R.; Niamsiri, N. Microbial Poly(hydroxybutyrate-co-hydroxyvalerate) Scaffold for Periodontal Tissue Engineering. *Polymers* **2023**, *15*, 855. [CrossRef] [PubMed]
16. Cho, J.Y.; Kim, S.H.; Jung, H.J.; Cho, D.H.; Kim, B.C.; Bhatia, S.K.; Ahn, J.; Jeon, J.-M.; Yoon, J.-J.; Lee, J.; et al. Finding a Benign Plasticizer to Enhance the Microbial Degradation of Polyhydroxybutyrate (PHB) Evaluated by PHB Degrader *Microbulbifer* sp. SOL66. *Polymers* **2022**, *14*, 3625. [CrossRef] [PubMed]

polymers

Article

Finding of Novel Galactose Utilizing *Halomonas* sp. YK44 for Polyhydroxybutyrate (PHB) Production

Hee Ju Jung [1], Su Hyun Kim [1], Do Hyun Cho [1], Byung Chan Kim [1], Shashi Kant Bhatia [1,2], Jongbok Lee [3], Jong-Min Jeon [4], Jeong-Jun Yoon [4] and Yung-Hun Yang [1,2,*]

1 Department of Biological Engineering, College of Engineering, Konkuk University, Seoul 05029, Republic of Korea
2 Institute for Ubiquitous Information Technology and Applications, Konkuk University, Seoul 05029, Republic of Korea
3 Department of Biological and Chemical Engineering, Hongik University, Sejong 30016, Republic of Korea
4 Green & Sustainable Materials R&D Department, Research Institute of Clean Manufacturing System, Korea Institute of Industrial Technology (KITECH), Cheonan 31056, Republic of Korea
* Correspondence: seokor@konkuk.ac.kr; Tel.: +82-2-450-2-3936

Abstract: Polyhydroxybutyrate (PHB) is a biodegradable bioplastic with potential applications as an alternative to petroleum-based plastics. However, efficient PHB production remains difficult. The main cost of PHB production is attributed to carbon sources; hence, finding inexpensive sources is important. Galactose is a possible substrate for polyhydroxyalkanoate production as it is abundant in marine environments. Marine bacteria that produce PHB from galactose could be an effective resource that can be used for efficient PHB production. In this study, to identify a galactose utilizing PHB producer, we examined 16 *Halomonas* strains. We demonstrated that *Halomonas cerina* (*Halomonas* sp. YK44) has the highest growth and PHB production using a culture media containing 2% galactose, final 4% NaCl, and 0.1% yeast extract. These culture conditions yielded 8.98 g/L PHB (78.1% PHB content (w/w)). When galactose-containing red algae (*Eucheuma spinosum*) hydrolysates were used as a carbon source, 5.2 g/L PHB was produced with 1.425% galactose after treatment with activated carbon. Since high salt conditions can be used to avoid sterilization, we examined whether *Halomonas* sp. YK44 could produce PHB in non-sterilized conditions. Culture media in these conditions yielded 72.41% PHB content. Thus, *Halomonas* sp. YK44 is robust against contamination, allowing for long-term culture and economical PHB production.

Keywords: polyhydroxybutyrate 1; *Halomonas* sp. YK44 2; galactose 3

Citation: Jung, H.J.; Kim, S.H.; Cho, D.H.; Kim, B.C.; Bhatia, S.K.; Lee, J.; Jeon, J.-M.; Yoon, J.-J.; Yang, Y.-H. Finding of Novel Galactose Utilizing *Halomonas* sp. YK44 for Polyhydroxybutyrate (PHB) Production. *Polymers* **2022**, *14*, 5407. https://doi.org/10.3390/polym14245407

Academic Editor: Luigi Botta

Received: 3 November 2022
Accepted: 8 December 2022
Published: 10 December 2022

1. Introduction

Plastic products are convenient in various fields such as food packaging, sterile medical uses, and construction. However, widespread plastic use leads to a huge amount of waste, which adversely affects the environment and humanity worldwide [1–3]. In particular, petroleum-based, non-degradable plastics cause environmental pollution, thus demonstrating the need for degradable plastic alternatives [4,5].

Polyhydroxyalkanoates (PHAs) are bioplastics that are potential alternatives to artificially synthesized plastics [6]. PHAs can be used for multiple applications such as biomedical devices, packaging, and coating materials. Poly(3-hydroxybutyrate) (PHB) is a PHA with properties similar to synthetic polymers [7]. PHB is produced by many microorganisms (*Halomonas* sp., *Ralstonia eutropha*, *Bacillus* sp., and *Pseudomonas* sp., etc.) [8,9], and several PHAs are produced by bacteria using commercial sugars [10,11]. PHB has a higher production cost than petrochemical plastics; hence, several researchers are studying low-cost PHA production by investigating readily available alternative and inexpensive carbon sources [12–14].

Marine biomasses such as large green algae, red algae, and brown macroalgae are promising because they present fast growth, large growth areas, and have low land, fertilizer, and water requirements [15]. In addition, these algae contain a large number of polysaccharides; hence, abundant monosaccharides can be derived from these species [16,17]. Among these, *Eucheuma spinosum*, red algae, grows in Southeast area and is used to produce agar and carrageenan [18]. Agar is a viscous complex polysaccharide constituting the cell wall of red algae and is a galactose polymer. Carrageenan is a linear polysaccharide composed of galactan with connected galactose residues [19]; 3,6-anhydrogalactose and sulfate ester groups are connected to the galactose unit, which cause structural deformation during acid treatment or heat treatment [20]. The ratio of 3,6-anhydrogalactose in *E. spinosum* is 56.2% galactose and 43.8% 3,6-anhydrogalactose [21–24].

The marine environment contains several strains that use galactose, suggesting that these could be used to generate galactose for PHB production [25–28]. *Halomonas* strains are present in salt-rich environments such as salt lakes, salt sand, salt soil, saline wells, solar salterns, saline wetlands, and marine environments [29,30]. *Halomonas* spp. can survive under high salinity [31,32], and their PHB production utilizing galactose has been organized into a table for comparison (Table 1) [25–28].

Table 1. PHA synthesis using galactose.

Microorganism	Carbon Source	DCW (g/L)	PHB Content (wt%)	PHB (g/L)	Culture	Reference
Halomonas halophila	Galactose	4.22 ± 0.10	80.7 ± 2.0	3.41 ± 0.12	Batch	[27]
Halomonas salina	Galactose	0.97 ± 0.03	12.3 ± 0.48	0.12 ± 0.01	Batch	[25]
Halomonas organivorans	Galactose	5.80 ± 0.22	90.55 ± 4.08	5.61 ± 0.01	Batch	[25]
Bacillus sp. 112A	Galactose	1.02	35.50	0.879	Batch	[26]
Halomonas sp. SF2003	Galactose	3.16	39	1.23	Batch	[28]

Many researchers examined PHB production using galactose in many marine strains (Table 1) [25–28], however, previous strains did not give significant titer for PHB production. As a result we tried to find out PHB producers with higher production capabilities from galactose and hypothesized that halotolerant bacteria could be applied to seaweed hydrolysates that have high salinity.

To realize this in this study, we identified *Halomonas* sp. YK44 as a potential PHB-producing species through utilizing galactose. Then, we examined the optimal culture media conditions for *Halomonas* sp. YK44 and tested its viability during PHB production. In addition, the *E.spinosum* hydrolysates were used to see if *Halomonas* could produce PHB using red algae that is abundant in marine environments [18].

2. Materials and Methods

2.1. Chemical Reagents

All chemicals used for cell culture were purchased from BD Difco Laboratories (Becton-Dickinson Franklin Lakes, NJ, USA). Reagents used for GC (Gas chromatography) and HPLC (High Performance Liquid Chromatography) analysis (chloroform, methanol, and other derivatization reagents) were purchased from Sigma-Aldrich (St. Louis, MO, USA). Red seaweed hydrolysate was prepared from *E. spinosum*, which was procured from the Korea Institute of Industrial Technology (Cheonan, KITECH, Korea). The lysate was prepared using dilute H_2SO_4 and an acid catalyst.

2.2. Sample Collection, Strain Isolation, and Phylogenetic Analysis

All sixteen strains were obtained from the National Marine Biodiversity Institute of Korea (Seochun, NMBIK, Korea). Each stock sample was suspended with distilled water and was spread on marine agar (MA; Difco Laboratories, Detroit, MI, USA) for isolation. The plates were incubated for two days at 30 °C. Colonies were isolated from each plate and cultured for one day in marine broth (MB; Difco Laboratories, Detroit, MI, USA). Stocks were prepared with 20% (w/v) glycerol and were stored at −81 °C until use.

The strain was identified at the species level using 16S rRNA. 16s rRNA was acquired from NMBIK, compared to those in the GenBank database of the NCBI using BLASTN tools, and used for making a phylogenetic tree utilizing MEGA 11 software, Molecular Evolutionary Genetics Analysis version 11 (Tamura, Stecher, and Kumar 2021).

2.3. Plate Assay for Strain Characterization

Plate assays were used to test the characteristics of the isolated strains [33]. An antibiotic susceptibility test was also performed via plate assay. The isolated strains were cultured for 2 days on MA with each antibiotic (100 μg/mL ampicillin, 25 μg/mL gentamicin, 50 μg/mL kanamycin, 100 μg/mL spectinomycin, and 25 μg/mL chloramphenicol). Colonies present after selection were considered resistant to the antibiotic.

2.4. Culture Conditions for PHA Synthesis

To identify which of the 16 *Halomonas* strains produces PHB, we cultivated 16 strains using MB containing 2% fructose. To optimize PHB production, *Halomonas* sp. YK44 was cultivated in the presence of various carbon and nitrogen sources. To determine which carbon source *Halomonas* uses, we added 2% glucose, xylose, galactose, glycerol, sucrose, fructose, or lactose to MB. Then, 0.1% yeast extract, NH_4NO_3, NH_4CL, $(NH_4)HSO_4$, NH_4NO_3, and $(NH_4)_2SO_4$ were used to determine the optimal nitrogen source. In addition, galactose (1–5% w/v) and yeast extract (0.1–1% w/v) were added to MB to optimize galactose and yeast extract concentrations. All experiments were performed in a shaking incubator at 30 °C (200 rpm) for 48 h. To determine optimal growth conditions, *Halomonas* sp. YK44 was cultured in MB with different concentrations of final NaCl (2–10% w/v) with 2% galactose and cultivated at 16 °C, 20 °C, 25 °C, 30 °C, 37 °C, and 42 °C. To confirm the optimal cultivation time for PHA production, *Halomonas* sp. YK44 was cultured in MB containing 2% galactose, final 4% NaCl, and 0.1% yeast extract at 25 °C for 96 h. The residual carbon source concentration was evaluated by HPLC using a PerkinElmer system equipped with a refractive index detector (Waltham, MA, USA) and an Aminex HPX-87H column (300 × 7.8 mm internal diameter)(Dublin, Ireland). The flow rate of the mobile phase containing 0.008 N sulfuric acid was constantly maintained at 0.6 mL/min. The oven temperature was set at 60 °C [34].

2.5. Analytical Methods

Using GC, PHB were quantified and characterized as previously described [34,35]. After culturing, the broth was centrifuged to collect cell pellets, which were subsequently washed twice with deionized water. After adjusting the amount of water to match the culture volume, 1 mL samples were collected in a glass vial for lyophilization. The dry weight of the lyophilized pellet was measured. Then, 1 mL chloroform and 1 mL 15% (v/v) H_2SO_4/85% methanol solution were added to the dried cell pellet. Methanolysis was conducted while heating for 2 h at 100 °C. After cooling to room temperature (25 °C), 1 mL of deionized water was added to the methyl ester solution, which was vortexed for approximately 5 s. The chloroform layer was carefully extracted and transferred to a microtube containing sodium sulfate anhydrous to remove any remaining water. Samples were filtered through a 0.22-μm Millex-GP syringe filter unit and 1 μL aliquots were injected into a gas chromatograph with split mode (1/10) (Young-lin 6500, Seoul, Korea), equipped with a fused silica capillary column (Agilent HP-FFAP, 30 m × 0.32 mm, i.d. 0.25 μm film) and a flame ionization detector (FID). The inlet temperature was 210 °C and helium carrier gas was supplied at 3 mL/min. The oven temperature was controlled following a gradient program of 0–5 min at 80 °C and 12–17 min at 220 °C. The FID temperature was maintained at 230 °C throughout the operation [34,36,37].

2.6. TEM Analysis

To conduct TEM analysis, 1 mL of cultured sample was collected, centrifuged, and mixed with Karnovsky's fixative solution containing 2% glutaraldehyde. The sample was fixed with 1% osmium tetroxide in 0.05 M sodium cacodylate buffer. The cells were gradually dehydrated using an ascending ethanol gradient (50%, 70%, 95%, and 100% *v*/*v*). Propylene was used for the transition step. The sample was settled in different ratios of propylene oxide and Spurr's resin (*v*/*v*) (1:1 and 1:2). After mixing with 100% Spurr's resin, the sample solidified overnight at 70 °C in a dry oven. Then, the samples were cut with an ultramicrotome (LEICA, EM UC7, Wetzlar, Germany). The slices were placed on a grid for EF-TEM (Carl Zeiss, LIBRA 120, Oberkochen, Germany) with an accelerating voltage of 120 kV [34].

2.7. Polymer Extraction and Characterization

Halomonas sp. YK44 cells were cultivated in 40 mL MB medium with 2% galactose for 48 h at 25 °C. The cells were pelleted, washed twice with deionized water, and lyophilized. Approximately 20 mL chloroform was added to the lyophilized cells and PHB was extracted for 16 h at 60 °C. The chloroform layer with dissolved polymer was collected by centrifugation and filtered using Whatman No. 1 filter paper to remove cell debris. The chloroform was evaporated at room temperature, to make a PHB film. The film was analyzed by differential scanning calorimetry (DSC) and gel permeation chromatography (GPC) [35,38].

DSC was used to analyze the extracted PHB film. Analysis was conducted with a Discovery DSC instrument (TA Instrument, Delaware, USA) from −60 °C to 180 °C. The heating and cooling rate was 10 °C/min in an N_2 atmosphere. PHB powder (Sigma-Aldrich) was used as a standard.

GPC analysis was performed using a GPC instrument (YL Chromass, Anyang, Korea) with a loop injector (Rheodyne 7725i), dual-head isocratic pump (YL9112), column oven (YL9131), column (Shodex, K-805, 8.0 I.D. × 300 mm), and RI detector (YL9170). For GPC analysis, 0.1 g PHB film was dissolved in chloroform and passed through a 0.2-μm syringe filter (Chromdisc, Hwaseong, Korea). Here, 60 μL of this solution (without air bubbles) was used as the injection volume. The flow rate of the chloroform mobile phase was 1 mL/min and was maintained at 35 °C. The molecular masses of 5000–2,000,000 Da were compared to polystyrene standards using YL-Clarity software (YL Chromass, Anyang, Korea).

2.8. PHB Production in E. spinosum Hydrolysate

Halomonas sp. YK44 was cultured with *E. spinosum* hydrolysate in MB. To determine the optimal culture growth conditions, *Halomonas* sp. YK44 was cultured in MB with different concentrations of *E. spinosum* hydrolysate (final galactose concentration: 0.475–1.9% *w*/*v*) and pH 6–10. DCW and PHB were measured after 2 days of cultivation at 30 °C while shaking at 200 rpm. The salinity of *E. spinosum* hydrolysate was analyzed using a digital salinity tester (ATAGO, Tokyo, Japan).

The *E. spinosum* hydrolysate was treated with 1 g/L activated carbon. Activated carbon treatment was performed by shaking for 5 min and centrifuging at 3700 rpm for 15 min. The treated hydrolysate was filtered using a 0.2-μm syringe filter (Chromdisc, Hwaseong, Korea).

2.9. PHB Production Reusing Non-Sterilized and Sterilized Medium for Comparison

With or without sterilization, *Halomonas* sp. YK44 was cultured in a high salinity MB medium with final 6% NaCl, 2% galactose and 0.1% yeast extract for 2 days at 25 °C and 200 rpm for comparing non-sterilized and sterilized media.

Culture medium reuse experiments were conducted by comparing culture media in sterilized medium with that in non-sterilized media. After centrifugation, the supernatant was collected and reused. We added 2% galactose and 0.1% yeast extract to the reused culture medium each cycle. It was done 10 cycles in total.

3. Results

3.1. Isolation and Characterization of Marine Strains for PHA Production

16 *Halomonas* strains were cultured in MB containing 2% fructose and analyzed for PHB production using GC. Among them, we found that *Halomonas* sp. YK44 produces the highest amount of PHB (11.5 g/L of DCW, 6.8 g/L *w/w* PHB; 58.8% PHB content) (Table 2, Figure 1A). Strain YK44 showed high 16s rRNA sequence similarity (97.93%) with *H. cerina* SP4 (red box) (Figure 1B) according to phylogenetic analysis. Given the high PHB peak, *Halomonas* sp. YK44 was selected for further study.

Table 2. *Halomonas* strains used in this study.

	Similar Species	Strain	Isolated Temperature	Resource Number
1	*Halomonas alkalicola*	DH-10	25 °C	MABIK MI00000003
2	*Halomonas cerina*	YK44	20 °C	MABIK MI00000284
3	*Halomonas sulfidaeris*	J05-14M-11R	20 °C	MABIK MI00000306
4	*Halomonas fontilapidosi*	O12	25 °C	MABIK MI00000338
5	*Halomonas gomseomensis*	CJCa107	25 °C	MABIK MI00000370
6	*Halomonas arcis*	CJCbj078	25 °C	MABIK MI00000396
7	*Halomonas janggokensis*	CJCbj082	25 °C	MABIK MI00000412
8	*Halomonas salicampi*	CJCbj041	25 °C	MABIK MI00000417
9	*Halomonas lutescens*	CJCbj058	25 °C	MABIK MI00000427
10	*Halomonas fontilapidosi*	MEBiC12169	25 °C	MABIK MI00005505
11	*Halomonas campaniensis*	S510	25 °C	MABIK MI00005561
12	*Halomonas saccharevitans*	MJ005	25 °C	MABIK MI00005664
13	*Halomonas shengliensis*	MEBiC12098	25 °C	MABIK MI00005446
14	*Halomonas denitrificans*	MEBiC13328	25 °C	MABIK MI00005839
15	*Halomonas aestuarii*	MEBiC13369	25 °C	MABIK MI00005871
16	*Halomonas lutea*	15A021	27 °C	MABIK MI00005926

Halomonas sp. YK44 was assessed for hydrolase activity, antibiotic resistance, carbon and nitrogen source utility, growth temperature, and salt resistance. *Halomonas* sp. YK44 did not show any hydrolase activity. To evaluate antibiotic resistance, the isolated strain was cultured in MB containing each 100 µg/mL ampicillin, 25 µg/mL gentamicin, 50 µg/mL kanamycin, 100 µg/mL spectinomycin, and 25 µg/mL chloramphenicol. Strain YK44 showed the highest resistance on the ampicillin plate. The strain was also resistant to spectinomycin (Supplementary Table S1). When YK44 was tested for poly(3-hydroxybutyrate-co-3-hydroxyvalerate) (P(3HB-co-3HV) production from propionate, it produced only 0.43% 3HV.

To visualize PHA accumulation in *Halomonas* sp. YK44, the strain was cultured for 48 h at 25 °C in MB with 2% galactose, final 4% NaCl, and 0.1% yeast extract. Then, TEM analysis was conducted. TEM showed PHB granule accumulation in the cells after 48 h cultivation (Figure 1C). The PHB granules completely filled the cytoplasm showing *Halomonas* sp. YK44 produces PHB.

3.2. Evaluating Carbon and Nitrogen Sources for PHB Production

Among the tested carbon sources (2% glucose, xylose, fructose, galactose, sucrose, glycerol), *Halomonas* sp. YK44 showed the best growth with sucrose but showed higher PHB production in the presence of galactose (Figure 2A; Table 1). To optimize the galactose concentration, we compared PHB production with different galactose concentrations (1–5% *w/v*). Growth and PHB production were the highest at 2% galactose; hence, this percentage was used for further analysis (Figure 2B).

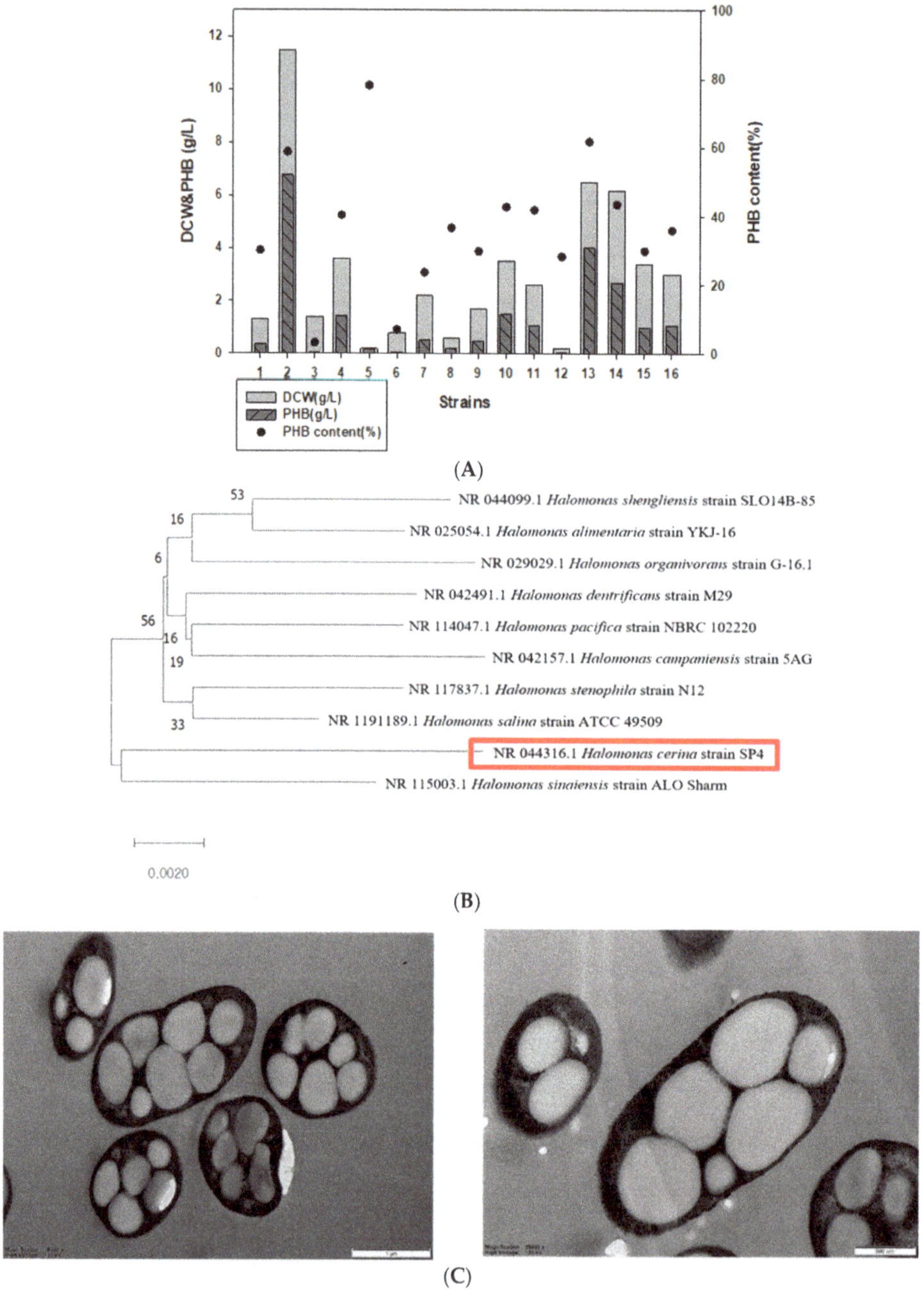

Figure 1. Identification of PHB-producing strains that produce PHB. (**A**) GC analysis comparison of DCW and PHB production in 16 *Halomonas* strains. (**B**) Phylogenetic analysis using 16S rRNA sequencing of *Halomonas* sp. YK44. (**C**) TEM image of *Halomonas* sp. YK44 after being cultured for 48 h at 25 °C in MB medium with 2% galactose, final 4% NaCl, and 0.1% yeast extract (Magnification: Right—16,300×, Left—25,000×).

Multiple nitrogen sources, such as 0.1% yeast extract, NH_4NO_3, NH_4CL, $(NH_4)HSO_4$, NH_4NO_3, and $(NH_4)_2SO_4$, were added to MB to determine the optimal nitrogen source. *Halomonas* sp. YK44 showed the most growth and PHB production with 0.1% yeast extract, followed by $(NH_4)_2SO_4$ (Figure 2C). To optimize the concentration of yeast extract, we compared PHB production in MB with different yeast extract concentrations (0.1–1% *w*/*v*). However, additional yeast extract did not change PHB production. Since PHB production was similar in MB containing 0.1% and 1% yeast extract, we decided to use 0.1% yeast extract to facilitate economical PHB production (Figure 2D).

3.3. Examination of Culture Temperature, Time, and NaCl Concentration

To determine the optimum culture conditions, we first examined different culture temperatures. When cultured at 16, 20, 25, 30, 37, and 42 °C, YK44 showed the best growth and PHB production at 25 °C (Figure 3A). Other studies mainly cultured *Halomonas* strains at 30 °C and 37 °C [39–41]. YK44 can grow and produce PHB at 25 °C similar to room temperature.

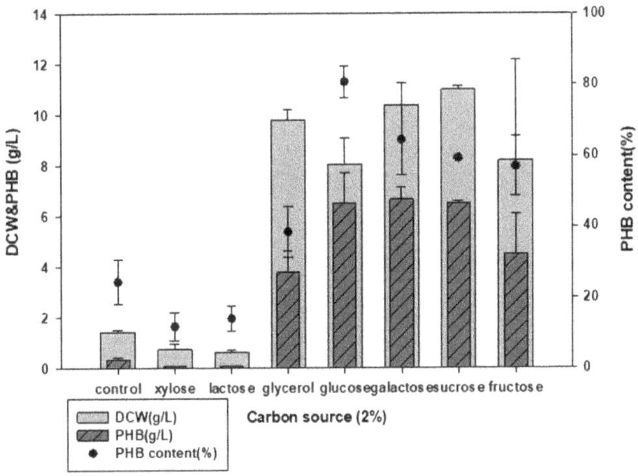

(A)

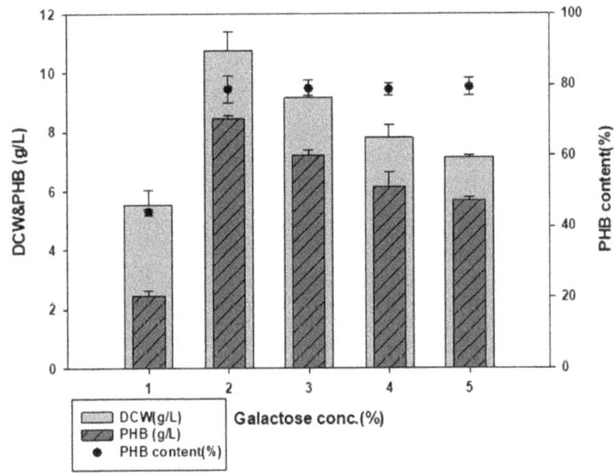

(B)

Figure 2. *Cont.*

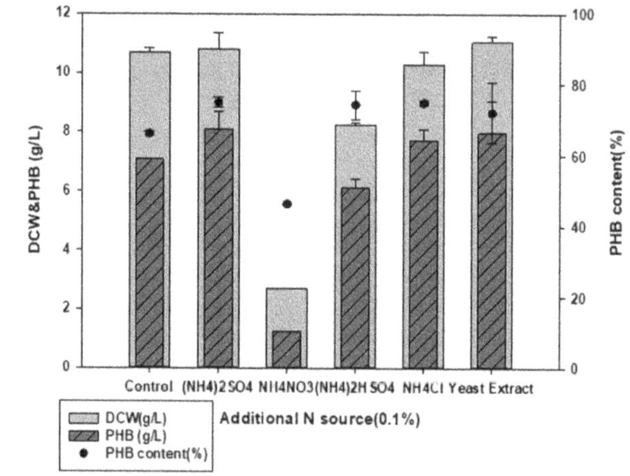

(C)

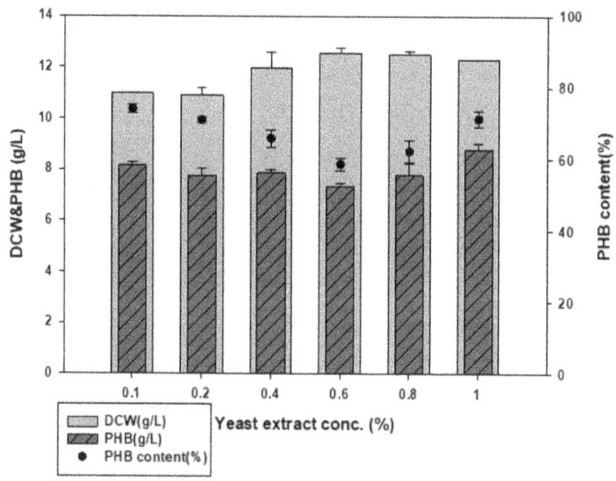

(D)

Figure 2. Optimized carbon and nitrogen sources for PHB production. (**A**) Comparison of PHB production in MB medium with 2% (*w*/*v*) additional carbon sources. (**B**) Effect of galactose concentration on PHB production. (**C**) Comparison of PHB production in MB medium with 0.1% (*w*/*v*) additional nitrogen sources. (**D**) Effect of yeast extract concentration on PHB production in MB.

Since *Halomonas* sp. YK44 is halotolerant and can grow under high NaCl conditions, we examined how various salt concentrations affected PHB production (Figure 3B). Growth and PHB production were highest under final 4% NaCl, though there was no difference of more than 2 g/L from 2% to final 8% NaCl.

To determine the optimal *Halomonas* sp. YK44 culture time, growth in the optimized medium was monitored every 24 h for 4 days. The highest growth and PHB production were observed after 48 h (Figure 3C), with PHB production reaching 8.98 g/L PHB (75.11% PHB content *w*/*w*) and 11.6 g/L DCW. Culturing for more than 48 h did not increase PHB production, likely because most of the sugar was consumed. Finally, 20 g/L of galactose was completely consumed after 72 h.

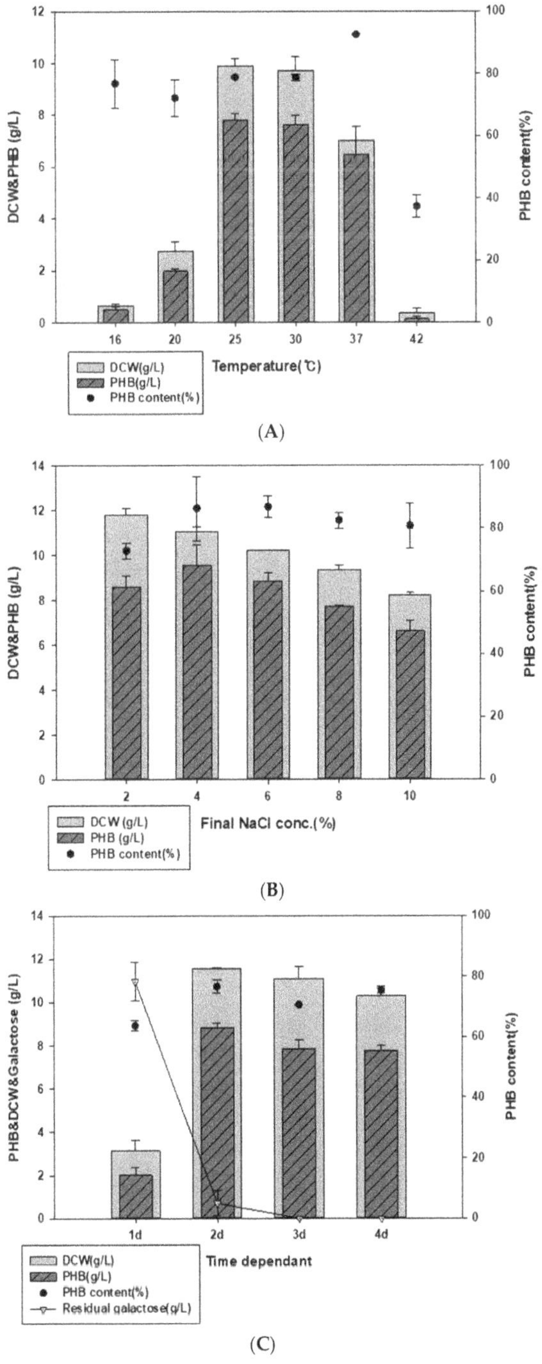

Figure 3. Optimized culture conditions for PHB production. (**A**) Effect of NaCl on PHB production in MB containing 2% (*w*/*v*) galactose. (**B**) Effect of temperature on PHB production in MB with 2% (*w*/*v*) galactose. (**C**) Optimized cultivation time for PHB production at 25 °C in MB with 2% galactose, final 4% NaCl, and 0.1% yeast extract.

3.4. Chemical Properties and Molecular Weight of PHB from *Halomonas* sp. YK44

PHB film was extracted from 40 mL of optimized cultivation media after 48 h cultivation at optimal culture conditions (25 °C in MB containing 2% galactose, final 4% NaCl, and 0.1% yeast extract) (Supplementary Figure S2). The melting and crystallization temperatures of the extracted PHB film were T_m = 170.57 °C and T_c = 126.60 °C, respectively. These values were similar to those of authentic PHB film: T_m = 176.55 °C and T_c = 126.29 °C (Supplementary Figure S1). Further, the T_m and T_c values were similar to those of authentic films and PHB films derived from other PHB producer strains [42–44].

To characterize the number average molecular weight (M_w), the weight average molecular weight (M_n), and the polydispersity index (PDI = M_w/M_n) of PHB extracted from *Halomonas* sp. YK44, we performed a GPC analysis. The M_w, M_n, and PDI of extracted PHB were 7.92×10^5, 6.90×10^5, and 1.48, respectively. GPC analysis revealed that strain YK44 produced higher molecular weight polymers than authentic PHB films ($M_w = 4.29 \times 10^5$, $M_n = 3.19 \times 10^5$, PDI = 1.34) (Supplementary Table S2). A high molecular weight polymer indicates high PHB quality, which is advantageous for industrial applications. Reduced molar masses during extraction are likely caused by damage to the granules. Compared with authentic PHB, intracellular PHB has low polydispersity; PDI is narrower than authentic PHB film. The narrow dispersity means homogeneous polymers [45,46].

3.5. PHB Production using *E. spinosum* Hydrolysates

To determine possible applications of *Halomonas* sp. YK44, cells were cultured in MB containing *E. spinosum* hydrolysate. The salinity of *E. spinosum* hydrolysates was approximately 2.6% and contained 3.8% galactose. To optimize the pH of *E. spinosum* hydrolysate, we compared PHB production capacity in the MB medium at different pH values. Growth and PHB production were the highest at pH 10 (Figure 4A). Further, growth was suppressed at galactose concentrations higher than 0.95%, possibly due to the influence of inhibitors. The highest cell growth and PHB production were obtained with 0.95% galactose concentration from *E. spinosum* hydrolysate in MB. Culturing in these conditions returned 4.9 g/L DCW and 27.2% PHB content (Figure 4B).

Activated carbon was used to remove impurities from the hydrolysate as activated carbon treatment removes toxic substances that may be released during hydrolysis [47,48]. *Halomonas* sp. YK44 showed optimal growth with 0.95% galactose but activated carbon treatment led to growth in media containing 1.9% galactose. Moreover, 5.2 g/L PHB was produced under 1.425% galactose culture condition (Figure 4C), which was 3.92-fold higher than the culture without activated carbon. These results suggest that activated carbon treatment increases hydrolysate utilization.

3.6. PHB Production in Reusing High Saline Non-Sterilized Medium and Sterilized Medium

To examine possible applications of *Halomonas* sp. YK44, we investigated whether this strain could be cultured in non-sterilized high-salt media without contamination and whether these culture conditions would change PHB production [49,50]. In sterilized or non-sterilized MB with final 6% NaCl and 2% galactose, *Halomonas* sp. YK44 produced PHB with or without sterilization. In non-sterilized samples, the highest amount of PHB was 7.24 g/L from 10.00 g/L DCW (72.41% PHB content *w/w*) in non-sterilized samples. This was comparable with results for sterilized media, which yielded 7.13 g/L PHB from 9.50 g/L DCW (75.02% PHB content *w/w*) (Figure 5). Overall, these results indicate that *Halomonas* sp. YK44 produces PHB under non-sterilized conditions like sterilized conditions.

When comparing sterilized and non-sterile media for reperirive use of media, there is no significant difference in growth or PHB production, indicating that *Halomonas* sp. YK44 could be cultured in non-sterilized medium without contamination. Accumulated data over 10 cycles indicated that PHB production showed a difference of approximately 1.8 g/L, but this difference was not significant considering that it is an accumulation of PHB production (Figure 5). Thus, *Halomonas* sp. YK44 shows high robustness to contamination, which will facilitate PHB production for industrial applications.

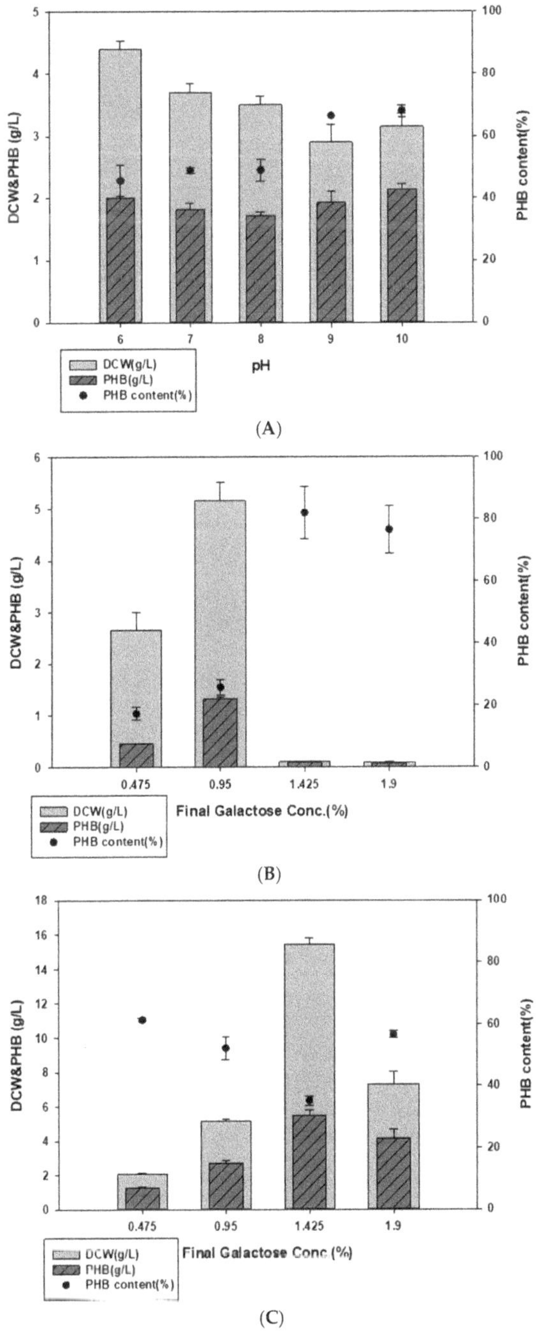

Figure 4. PHB production with *E. spinosum* hydrolysate and optimal pH. (**A**) Effect of pH on PHB production in MB with 1% galactose in *E. spinosum* hydrolysate. (**B**) DCW and PHB production after culture with *E. spinosum* hydrolysate (**C**) DCW and PHB production after culture with *E. spinosum* hydrolysate and activated carbon.

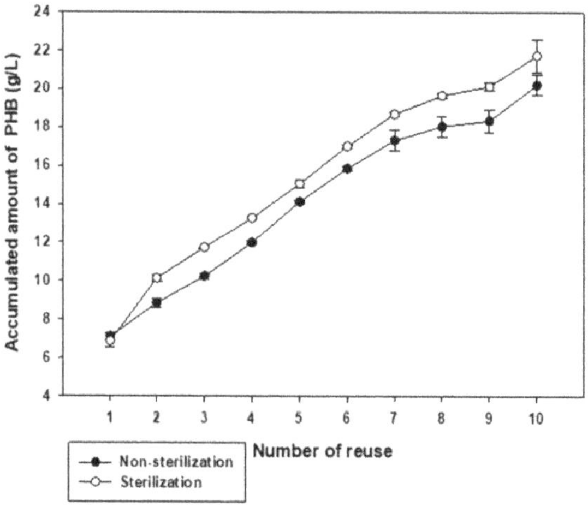

Figure 5. Effect of reused sterilized and non-sterilized high salinity medium on total PHB production. Total PHB production in non-sterilized high salinity medium with 2% galactose was compared to production in sterilized high salinity medium with 2% galactose.

4. Discussion

As *Halomonas* spp. can usually survive and produce PHB at high salinity, it is possible to utilize open non-sterile fermentation in a high salinity medium without antibiotics treatment or sterilization [31,32]. In addition, as this *Halomonas* strain is a benign native PHB producer, it is free of any environmental issues such as the law of genetically modified objects or pathogen issues. Since *Halomonas* sp. YK44 is salt-tolerant and uses galactose, this strain can be used to produce PHB by utilizing seaweed hydrolysate like *Euchenoma spinosum* hydrolysate resulting in reduced PHB production cost. Our results suggest that not only can red algae be used to produce PHB, but also that it allows microorganisms to grow better when hydrolysates are treated with activated carbon, as shown in other papers [51,52].

In conclusion, we determined that *Halomonas* sp. YK44 strain produces more PHB using galactose as a carbon source among 16 *Halomonas* strains. According to Figure 3B, as *Halomonas* sp. YK44 is salt-tolerant and uses galactose, it can be used to produce PHB utilizing seaweed hydrolysate. PHB was produced using *E. spinosum* hydrolysates that have abundant galactose with a salinity of 2.6% and production increased after activated carbon treatment. Furthermore, we reused the media to compare PHB productivity with sterilized and non-sterilized high-salt media to determine that *Halomonas* is resistant to contamination in high salinity conditions. Total PHB production showed no significant difference between cultures in sterilized or non-sterilized high-salinity culture media. Thus, cost-effective application will be possible even if PHB is produced using by-products without sterilization.

In *Halomonas* sp. YK44, PHB was effectively produced using various sugar components, especially galactose. We optimized culture media conditions to produce PHB in MB media with 2% galactose, 0.1% yeast extract, and final 4% NaCl concentration under 25 °C for 48 h. Under optimized PHB production conditions, the strain produced 8.98 g/L PHB from 11.6 g/L DCW (78.11% PHB content). *Halomonas* sp. YK44 showed about 50% more production than the maximum amount of the previously reported paper [25–28]. Similar PHB content was produced in MB medium containing final 2–8% NaCl. This suggests that *Halomonas* sp. YK44 is salt-tolerant similar to other *Halomonas* spp. [53,54]. Since PHB was produced at a low temperature (25 °C), it may be possible to produce only by shaking at room temperature. For PHB production and cultivation using *E. spinosum* hydrolysate, 1.425 g/L galactose, and activated carbon, the yield was 5.2 g/L PHB, which was the largest amount we produced using *E. spinosum* hydrolysates. In MB media with final 6% NaCl and 2% galactose without antibiotics and sterilization, PHB production was 4.2 g/L PHB from 7.3 g/L DCW.

Supplementary Materials: The following supporting information can be downloaded at: https://www.mdpi.com/article/10.3390/polym14245407/s1, Figure S1: DSC analysis of PHB from *Halomonas* sp. YK44; Table S1: Species identification using 16S rRNA sequencing and antibiotic resistance the identified strain.; Figure S2: GPC analysis of PHB from *Halomonas* sp. YK44; Table S2: Comparison of GPC analysis result *Halomonas* sp. YK44 film and Authentic PHB film; Figure S2; PHB film extracted from *Halomonas* sp. YK44.

Author Contributions: Conceptualization, H.J.J., S.K.B. and Y.-H.Y.; methodology, H.J.J., S.K.B. and Y.-H.Y.; validation, S.H.K., D.H.C. and B.C.K.; resources, J.-M.J. and J.-J.Y.; writing—original draft preparation, H.J.J.; writing—review and editing, S.K.B., J.L. and Y.-H.Y.; supervision, Y.-H.Y.; project administration, S.K.B. and Y.-H.Y. All authors have read and agreed to the published version of the manuscript.

Funding: This research was supported by Research Program to solve social issues of the National Research Foundation of Korea (NRF)s funded by the Ministry of Science and ICT, South Korea [grant number 2017M3A9E4077234] and [NRF-2022R1A2C2003138, NRF-2020R1A2C2102381]. This study was also performed with the support of the R&D Program for Forest Science Technology [grant number 2020261C10-2222-AC02] provided by Korea Forest Service (Korea Forestry Promotion Institute) and by the R&D Program of MOTIE/KEIT (20018072).

Institutional Review Board Statement: Not applicable.

Data Availability Statement: All data related to the study are provided with the manuscript and its associated files.

Conflicts of Interest: The authors declare that they have no known competing financial interest or personal relationships that could have influenced the work reported in this paper.

References

1. Barnes, D.K.A.; Galgani, F.; Thompson, R.C.; Barlaz, M. Accumulation and Fragmentation of Plastic Debris in Global Environments. *Philos. Trans. R. Soc. B Biol. Sci.* **2009**, *364*, 1985–1998. [CrossRef] [PubMed]
2. Chamas, A.; Moon, H.; Zheng, J.; Qiu, Y.; Tabassum, T.; Jang, J.H.; Abu-Omar, M.; Scott, S.L.; Suh, S. Degradation Rates of Plastics in the Environment. *ACS Sustain. Chem. Eng.* **2020**, *8*, 3494–3511. [CrossRef]
3. Patrício Silva, A.L.; Prata, J.C.; Walker, T.R.; Duarte, A.C.; Ouyang, W.; Barcelò, D.; Rocha-Santos, T. Increased Plastic Pollution Due to COVID-19 Pandemic: Challenges and Recommendations. *Chem. Eng. J.* **2021**, *405*, 126683. [CrossRef] [PubMed]
4. Vieira, M.G.A.; da Silva, M.A.; dos Santos, L.O.; Beppu, M.M. Natural-Based Plasticizers and Biopolymer Films: A Review. *Eur. Polym. J.* **2011**, *47*, 254–263. [CrossRef]
5. Kabir, E.; Kaur, R.; Lee, J.; Kim, K.H.; Kwon, E.E. Prospects of Biopolymer Technology as an Alternative Option for Non-Degradable Plastics and Sustainable Management of Plastic Wastes. *J. Clean. Prod.* **2020**, *258*, 120536. [CrossRef]
6. Costa, S.S.; Miranda, A.L.; de Morais, M.G.; Costa, J.A.V.; Druzian, J.I. Microalgae as Source of Polyhydroxyalkanoates (PHAs)—A Review. *Int. J. Biol. Macromol.* **2019**, *131*, 536–547. [CrossRef]
7. Khosravi-Darani, K.; Mokhtari, Z.B.; Amai, T.; Tanaka, K. Microbial Production of Poly(Hydroxybutyrate) from C1 Carbon Sources. *Appl. Microbiol. Biotechnol.* **2013**, *97*, 1407–1424. [CrossRef]
8. Venkateswar Reddy, M.; Mawatari, Y.; Yajima, Y.; Seki, C.; Hoshino, T.; Chang, Y.C. Poly-3-Hydroxybutyrate (PHB) Production from Alkylphenols, Mono and Poly-Aromatic Hydrocarbons Using *Bacillus* sp. CYR1: A New Strategy for Wealth from Waste. *Bioresour. Technol.* **2015**, *192*, 711–717. [CrossRef]
9. Tsuge, T.; Hyakutake, M.; Mizuno, K. Class IV Polyhydroxyalkanoate (PHA) Synthases and PHA-Producing *Bacillus*. *Appl. Microbiol. Biotechnol.* **2015**, *99*, 6231–6240. [CrossRef]
10. Hempel, F.; Bozarth, A.S.; Lindenkamp, N.; Klingl, A.; Zauner, S.; Linne, U.; Steinbüchel, A.; Maier, U.G. Microalgae as Bioreactors for Bioplastic Production. *Microb. Cell Factories* **2011**, *10*, 81. [CrossRef]
11. Verlinden, R.A.J.; Hill, D.J.; Kenward, M.A.; Williams, C.D.; Radecka, I. Bacterial Synthesis of Biodegradable Polyhydroxyalkanoates. *J. Appl. Microbiol.* **2007**, *102*, 1437–1449. [CrossRef] [PubMed]
12. Cui, Y.W.; Zhang, H.Y.; Lu, P.F.; Peng, Y.Z. Effects of Carbon Sources on the Enrichment of Halophilic Polyhydroxyalkanoate-Storing Mixed Microbial Culture in an Aerobic Dynamic Feeding Process. *Sci. Rep.* **2016**, *6*, 30766. [CrossRef] [PubMed]
13. Getachew, A.; Woldesenbet, F. Production of Biodegradable Plastic by Polyhydroxybutyrate (PHB) Accumulating Bacteria Using Low Cost Agricultural Waste Material. *BMC Res. Notes* **2016**, *9*, 509. [CrossRef] [PubMed]

14. Saratale, R.G.; Saratale, G.D.; Cho, S.K.; Kim, D.S.; Ghodake, G.S.; Kadam, A.; Kumar, G.; Bharagava, R.N.; Banu, R.; Shin, H.S. Pretreatment of Kenaf (*Hibiscus cannabinus* L.) Biomass Feedstock for Polyhydroxybutyrate (PHB) Production and Characterization. *Bioresour. Technol.* **2019**, *282*, 75–80. [CrossRef]
15. Sawant, S.S.; Salunke, B.K.; Kim, B.S. Consolidated Bioprocessing for Production of Polyhydroxyalkanotes from Red Algae *Gelidium amansii*. *Int. J. Biol. Macromol.* **2018**, *109*, 1012–1018. [CrossRef]
16. Jung, K.A.; Lim, S.R.; Kim, Y.; Park, J.M. Potentials of Macroalgae as Feedstocks for Biorefinery. *Bioresour. Technol.* **2013**, *135*, 182–190. [CrossRef]
17. Wei, N.; Quarterman, J.; Jin, Y.S. Marine Macroalgae: An Untapped Resource for Producing Fuels and Chemicals. *Trends Biotechnol.* **2013**, *31*, 70–77. [CrossRef]
18. Rhein-Knudsen, N.; Ale, M.T.; Meyer, A.S. Marine Drugs Seaweed Hydrocolloid Production: An Update on Enzyme Assisted Extraction and Modification Technologies. *Mar. Drugs* **2015**, *13*, 3340–3359. [CrossRef]
19. Estevez, J.M.; Ciancia, M.; Cerezo, A.S. The System of Sulfated Galactans from the Red Seaweed *Gymnogongrus torulosus* (Phyllophoraceae, Rhodophyta): Location and Structural Analysis. *Carbohydr. Polym.* **2008**, *73*, 594–605. [CrossRef]
20. Campo, V.L.; Kawano, D.F.; da Silva, D.B.; Carvalho, I. Carrageenans: Biological Properties, Chemical Modifications and Structural Analysis—A Review. *Carbohydr. Polym.* **2009**, *77*, 167–180. [CrossRef]
21. Kim, M.-J.; Kim, J.-S.; Ra, C.H.; Kim, S.-K. Bioethanol Production from *Eucheuma spinosum* Using Various Yeasts. *KSBB J.* **2013**, *28*, 315–318. [CrossRef]
22. Ra, C.H.; Jung, J.H.; Sunwoo, I.Y.; Kang, C.H.; Jeong, G.T.; Kim, S.K. Detoxification of *Eucheuma Spinosum* Hydrolysates with Activated Carbon for Ethanol Production by the Salt-Tolerant Yeast *Candida tropicalis*. *J. Microbiol. Biotechnol.* **2015**, *25*, 856–862. [CrossRef] [PubMed]
23. Adnan, H.; Porse, H. Culture of *Eucheuma cottonii* and *Eucheuma spinosum* in Indonesia. *Hydrobiologia* **1987**, *151–152*, 355–358. [CrossRef]
24. Andersen, N.S.; Dolan, T.C.S.; Rees, D.A. Carrageenans. Part VII. Polysaccharides from *Eucheuma spinosum* and *Eucheuma cottonii*. The Covalent Structure of L-Carrageenan. *J. Chem. Soc. Perkin 1* **1973**, *19*, 2173–2176. [CrossRef] [PubMed]
25. Pernicova, I.; Kucera, D.; Novackova, I.; Vodicka, J.; Kovalcik, A.; Obruca, S. Extremophiles—Platform Strains for Sustainable Production of Polyhydroxyalkanoates. In *Materials Science Forum*; Trans Tech Publications Ltd.: Wollerau, Switzerland, 2019; Volume 955, pp. 74–79.
26. Thirumala, M.; Reddy, S.V.; Mahmood, S.K. Production and Characterization of PHB from Two Novel Strains of *Bacillus* Spp. Isolated from Soil and Activated Sludge. *J. Ind. Microbiol. Biotechnol.* **2010**, *37*, 271–278. [CrossRef]
27. Kucera, D.; Pernicová, I.; Kovalcik, A.; Koller, M.; Mullerova, L.; Sedlacek, P.; Mravec, F.; Nebesarova, J.; Kalina, M.; Marova, I.; et al. Characterization of the Promising Poly(3-Hydroxybutyrate) Producing Halophilic Bacterium *Halomonas halophila*. *Bioresour. Technol.* **2018**, *256*, 552–556. [CrossRef] [PubMed]
28. Thomas, T.; Sudesh, K.; Bazire, A.; Elain, A.; Tan, H.T.; Lim, H.; Bruzaud, S. PHA Production and Pha Synthases of the Halophilic Bacterium *Halomonas* sp. SF2003. *Bioengineering* **2020**, *7*, 29. [CrossRef] [PubMed]
29. Poli, A.; Nicolaus, B.; Denizci, A.A.; Yavuzturk, B.; Kazan, D. *Halomonas Smyrnensis* sp. Nov., a Moderately Halophilic, Exopolysaccharide-Producing Bacterium. *Int. J. Syst. Evol. Microbiol.* **2013**, *63*, 10–18. [CrossRef]
30. Xu, X.W.; Wu, Y.H.; Zhou, Z.; Wang, C.S.; Zhou, Y.G.; Zhang, H.B.; Wang, Y.; Wu, M. *Halomonas saccharevitans* sp. Nov., *Halomonas arcis* sp. Nov. and *Halomonas subterranea* sp. Nov., Halophilic Bacteria Isolated from Hypersaline Environments of China. *Int. J. Syst. Evol. Microbiol.* **2007**, *57*, 1619–1624. [CrossRef]
31. Ling, C.; Qiao, G.Q.; Shuai, B.W.; Song, K.N.; Yao, W.X.; Jiang, X.R.; Chen, G.Q. Engineering Self-Flocculating *Halomonas campaniensis* for Wastewaterless Open and Continuous Fermentation. *Biotechnol. Bioeng.* **2019**, *116*, 805–815. [CrossRef]
32. Fu, X.Z.; Tan, D.; Aibaidula, G.; Wu, Q.; Chen, J.C.; Chen, G.Q. Development of *Halomonas* TD01 as a Host for Open Production of Chemicals. *Metab. Eng.* **2014**, *23*, 78–91. [CrossRef] [PubMed]
33. Lee, S.M.; Lee, H.J.; Kim, S.H.; Suh, M.J.; Cho, J.Y.; Ham, S.; Jeon, J.M.; Yoon, J.J.; Bhatia, S.K.; Gurav, R.; et al. Screening of the Strictly Xylose-Utilizing *Bacillus* sp. SM01 for Polyhydroxybutyrate and Its Co-Culture with *Cupriavidus necator* NCIMB 11599 for Enhanced Production of PHB. *Int. J. Biol. Macromol.* **2021**, *181*, 410–417. [CrossRef] [PubMed]
34. Park, Y.L.; Bhatia, S.K.; Gurav, R.; Choi, T.R.; Kim, H.J.; Song, H.S.; Park, J.Y.; Han, Y.H.; Lee, S.M.; Park, S.L.; et al. Fructose Based Hyper Production of Poly-3-Hydroxybutyrate from *Halomonas* sp. YLGW01 and Impact of Carbon Sources on Bacteria Morphologies. *Int. J. Biol. Macromol.* **2020**, *154*, 929–936. [CrossRef] [PubMed]
35. Jung, H.R.; Choi, T.R.; Han, Y.H.; Park, Y.L.; Park, J.Y.; Song, H.S.; Yang, S.Y.; Bhatia, S.K.; Gurav, R.; Park, H.A.; et al. Production of Blue-Colored Polyhydroxybutyrate (PHB) by One-Pot Production and Coextraction of Indigo and PHB from Recombinant *Escherichia coli*. *Dye. Pigment.* **2020**, *173*, 107889. [CrossRef]
36. Hong, J.W.; Song, H.S.; Moon, Y.M.; Hong, Y.G.; Bhatia, S.K.; Jung, H.R.; Choi, T.R.; Yang, S.Y.; Park, H.Y.; Choi, Y.K.; et al. Polyhydroxybutyrate Production in Halophilic Marine Bacteria *Vibrio Proteolyticus* Isolated from the Korean Peninsula. *Bioprocess. Biosyst. Eng.* **2019**, *42*, 603–610. [CrossRef]
37. Bhatia, S.K.; Gurav, R.; Choi, T.R.; Jung, H.R.; Yang, S.Y.; Moon, Y.M.; Song, H.S.; Jeon, J.M.; Choi, K.Y.; Yang, Y.H. Bioconversion of Plant Biomass Hydrolysate into Bioplastic (Polyhydroxyalkanoates) Using *Ralstonia eutropha* 5119. *Bioresour. Technol.* **2019**, *271*, 306–315. [CrossRef]

38. Bhatia, S.K.; Wadhwa, P.; Hong, J.W.; Hong, Y.G.; Jeon, J.M.; Lee, E.S.; Yang, Y.H. Lipase Mediated Functionalization of Poly(3-Hydroxybutyrate-Co-3-Hydroxyvalerate) with Ascorbic Acid into an Antioxidant Active Biomaterial. *Int. J. Biol. Macromol.* **2019**, *123*, 117–123. [CrossRef]
39. Tohme, S.; Hacıosmanoğlu, G.G.; Eroğlu, M.S.; Kasavi, C.; Genç, S.; Can, Z.S.; Toksoy Oner, E. *Halomonas smyrnensis* as a Cell Factory for Co-Production of PHB and Levan. *Int. J. Biol. Macromol.* **2018**, *118*, 1238–1246. [CrossRef]
40. Kulkarni, S.O.; Kanekar, P.P.; Jog, J.P.; Sarnaik, S.S.; Nilegaonkar, S.S. Production of Copolymer, Poly (Hydroxybutyrate-Co-Hydroxyvalerate) by *Halomonas campisalis* MCM B-1027 Using Agro-Wastes. *Int. J. Biol. Macromol.* **2015**, *72*, 784–789. [CrossRef]
41. Jiang, X.R.; Yao, Z.H.; Chen, G.Q. Controlling Cell Volume for Efficient PHB Production by *Halomonas*. *Metab. Eng.* **2017**, *44*, 30–37. [CrossRef]
42. Trakunjae, C.; Boondaeng, A., Apiwatanapiwat, W.; Kosugi, A.; Arai, T.; Sudesh, K.; Vaithanomsat, P. Enhanced Polyhydroxybutyrate (PHB) Production by Newly Isolated Rare Actinomycetes *Rhodococcus* sp. Strain BSRT1-1 Using Response Surface Methodology. *Sci. Rep.* **2021**, *11*, 1896. [CrossRef] [PubMed]
43. Radhika, D.; Murugesan, A.G. Bioproduction, Statistical Optimization and Characterization of Microbial Plastic (Poly 3-Hydroxy Butyrate) Employing Various Hydrolysates of Water Hyacinth (*Eichhornia crassipes*) as Sole Carbon Source. *Bioresour. Technol.* **2012**, *121*, 83–92. [CrossRef]
44. Wellen, R.M.R.; Rabello, M.S.; Fechine, G.J.M.; Canedo, E.L. The Melting Behaviour of Poly(3-Hydroxybutyrate) by DSC. Reproducibility Study. *Polym. Test* **2013**, *32*, 215–220. [CrossRef]
45. Wendlandt, K.D.; Geyer, W.; Mirschel, G.; Al-Haj Hemidi, F. Possibilities for Controlling a PHB Accumulation Process Using Various Analytical Methods. *J. Biotechnol.* **2005**, *117*, 119–129. [CrossRef]
46. Alsafadi, D.; Ibrahim, M.I.; Alamry, K.A.; Hussein, M.A.; Mansour, A. Utilizing the Crop Waste of Date Palm Fruit to Biosynthesize Polyhydroxyalkanoate Bioplastics with Favorable Properties. *Sci. Total Environ.* **2020**, *737*, 139716. [CrossRef]
47. Arminda, M.; Josúe, C.; Cristina, D.; Fabiana, S.; Yolanda, M. Use of Activated Carbons for Detoxification of a Lignocellulosic Hydrolysate: Statistical Optimisation. *J. Environ. Manag.* **2021**, *296*, 113320. [CrossRef] [PubMed]
48. Sahu, A.K.; Srivastava, V.C.; Mall, I.D.; Lataye, D.H. Adsorption of Furfural from Aqueous Solution onto Activated Carbon: Kinetic, Equilibrium and Thermodynamic Study. *Sep. Sci. Technol.* **2008**, *43*, 1239–1259. [CrossRef]
49. Tan, D.; Xue, Y.S.; Aibaidula, G.; Chen, G.Q. Unsterile and Continuous Production of Polyhydroxybutyrate by *Halomonas* TD01. *Bioresour. Technol.* **2011**, *102*, 8130–8136. [CrossRef]
50. Park, Y.L.; Song, H.S.; Choi, T.R.; Lee, S.M.; Park, S.L.; Lee, H.S.; Kim, H.J.; Bhatia, S.K.; Gurav, R.; Park, K.; et al. Revealing of Sugar Utilization Systems in *Halomonas* sp. YLGW01 and Application for Poly(3-Hydroxybutyrate) Production with Low-Cost Medium and Easy Recovery. *Int. J. Biol. Macromol.* **2021**, *167*, 151–159. [CrossRef]
51. Parawira, W.; Tekere, M. Biotechnological Strategies to Overcome Inhibitors in Lignocellulose Hydrolysates for Ethanol Production: Review. *Crit. Rev. Biotechnol.* **2011**, *31*, 20–31. [CrossRef]
52. Lee, J.M.; Venditti, R.A.; Jameel, H.; Kenealy, W.R. Detoxification of Woody Hydrolyzates with Activated Carbon for Bioconversion to Ethanol by the Thermophilic Anaerobic Bacterium *Thermoanaerobacterium saccharolyticum*. *Biomass Bioenergy* **2011**, *35*, 626–636. [CrossRef]
53. El-malek, F.A.; Farag, A.; Omar, S.; Khairy, H. Polyhydroxyalkanoates (PHA) from *Halomonas Pacifica* ASL10 and *Halomonas salifodiane* ASL11 Isolated from Mariout Salt Lakes. *Int. J. Biol. Macromol.* **2020**, *161*, 1318–1328. [CrossRef] [PubMed]
54. Coronado, M.-J.; Vargas, C.; Hofemeister, J.; Ventosa, A.; Nieto, J.J. Production and Biochemical Characterization of an Î±-Amylase from the Moderate Halophile *Halomonas meridiana*. *FEMS Microbiol. Lett.* **2000**, *183*, 67–71. [CrossRef] [PubMed]

polymers

MDPI

Article

The Hyperproduction of Polyhydroxybutyrate Using *Bacillus mycoides* ICRI89 through Enzymatic Hydrolysis of Affordable Cardboard

Fady Abdelmalek [1], Alexander Steinbüchel [1] and Marian Rofeal [1,2,*]

[1] International Center for Research on Innovative Biobased Materials (ICRI-BioM)-International Research Agenda, Lodz University of Technology, Zeromskiego 116, 90-924 Lodz, Poland; fady.abdelmalek@p.lodz.pl (F.A.); alexander.steinbuchel@p.lodz.pl (A.S.)

[2] Department of Botany and Microbiology, Faculty of Science, Alexandria University, Moharam Bek, Alexandria 21521, Egypt

* Correspondence: marian.rofeal@p.lodz.pl

Abstract: Bioplastics are contemplated as remarkable substitutes for conventional plastics to accommodate green technological advancements. However, their industrial production has not been fully implemented owing to the cost of carbon resources. From another perspective, valorizing different paper mill wastes has become a prominent research topic. These materials may serve as an affording sustainable feedstock for bioplastic production. Adjustment of cardboard waste hydrolysate as suitable fermentation media for production of bacterial polyhydroxyalkanoates (PHAs) has been investigated. Cardboard samples were defibered and dried before enzymatic hydrolysis. The enzymatic degradation of commercial cellulase was monitored over 15 days. Interestingly, 18.2 ± 0.2 g/L glucose yield was obtained from 50 g cardboard samples using a 1.5% (v/v) enzyme concentration. The samples exhibited maximum weight loss values of 69–73%. Meanwhile, five soil samples were collected from local sites in Lodz, Poland. A total of 31 bacterial isolates were screened and cultured on Nile blue plates. Analysis of the 16S rRNA gene sequence of the most potent producer revealed 100% similarity to *Bacillus mycoides*. Cardboard hydrolysates whole medium, modified MSM with cardboard hydrolysate and nitrogen depleted MSM with cardboard hydrolysate were utilized for PHA production, followed by PHA productivity and cell dry weight (CDW) estimation compared to glucose as a standard carbon source. An impressive PHA accumulation of 56% CDW was attained when the waste hydrolysate was used as a carbon source. FTIR and NMR analysis of the isolated PHA indicated that functional groups of the polymer were related to PHB (polyhydroxybutyrate). Thermal analysis demonstrates that PHB and PHB-CB (PHB produced from cardboard hydrolysate) have degradation temperatures of 380 and 369 °C, respectively, which reflect the high thermal stability and heat resistance compared to the same properties for a standard polymer. This is the first demonstration of full saccharification of corrugated cardboard paper waste for high-level production of PHA. In addition, the attained PHB productivity is one of the highest levels achieved from a real lignocellulosic waste.

Keywords: polyhydroxyalkanoates; poly(3-hydroxybutrate); enzymatic hydrolysis; lignocellulosic waste; cardboard waste; waste management; *Bacillus* sp.

Citation: Abdelmalek, F.; Steinbüchel, A.; Rofeal, M. The Hyperproduction of Polyhydroxybutyrate Using *Bacillus mycoides* ICRI89 through Enzymatic Hydrolysis of Affordable Cardboard. *Polymers* **2022**, *14*, 2810. https://doi.org/10.3390/polym14142810

Academic Editor: Shashi Kant Bhatia

Received: 6 June 2022
Accepted: 8 July 2022
Published: 10 July 2022

1. Introduction

It is acknowledged that there has been a growing need for bioplastics and biodegradable polymers in numerous fields owing to their diversified merits [1]. Polyesters are in high demanded in several sectors, such as pharmaceutical nanodelivery, food safety, and biomedical applications [2,3]. PHAs are efficient substitutes for petrochemical plastics from fossil fuels. They have been exploited in 3D printing inks, tissue engineering scaffolds, smart packaging materials, and biocompatible implants [4]. Moreover, modified PHA variants have become potent components as antimicrobial and antiviral agents, as well as targeting anticancer ligands [5,6]. Nevertheless, elevated production costs continue to inhibit scale-up. From an environmental perspective, various nations have urgent policies to turn to natural plastics manufacturing to mitigate the widespread usage of synthetic plastics [7]. Almost 4.8–12.7 million tons of plastic wastes are expected to enter the ocean by 2025, with no expected measures to manage this situation. PHAs are microbial polyesters accumulated inside bacterial cells under nitrogen limitations and excess carbon supplementation. The bioprocess technology for these polymers is still facing significant hindrance due to the negative effect of high-level production economics [8].

In this context, the usage of raw and waste materials has often been investigated in an attempt to identify economical strategies for bioplastic production. A strong example of this is a recent project which valorized marine algal biomass hydrolysate as promising feed stock for sustainable PHA synthesis. A simple acid pre-treatment for three marine algal biomass was performed in search of affordable sources of carbon to support microbial metabolism. A reasonable content of PHA was successfully produced using 3 different algal hydrolysates media from *Halomonas pacifica* ASL10 and *Halomonas salifodiane* ASL11 [9]. Since the issue of PHA commercialization has not been resolved, affordable alternative carbon sources, such as the use of agricultural waste [10], household waste [11], fats, oils, industrial by-products [12], glycerol, sugars, wastewater, and lignocellulosic materials [13], have been considered.

One of the most promising sustainable feedstocks for microbial polymer synthesis is lignocellulosic materials. Recently, a 70% yield of CDW (cell dry weight) poly(3-hydroxybutyrate) (P(3HB)) was obtained using *Burkholderia sacchari* DSM 17165 to promote enzymatically hydrolyzed wheat straw as a substitute medium [14]. Pretreatment of rice straw with dilute aqueous acid as an abundant sugar resource to provide a carbon source for producing high-value products, such as PHB using *B. megaterium* [15]. Mechanical conversion of biomass into smaller particles and fibers is commonly used to disrupt the lignocellulosic matrix and to gain access to the carbohydrates. Chemical pretreatment is frequently used to assist lignin decomposition and removal [16]. These procedures can be expensive, and they constitute a major impediment to broad carbohydrate production from lignocellulosic sources. These simple sugars can be used for biofuel production or other goods, such as biodegradable and compostable plastics, after the carbohydrate polymers are broken down to monomers [14].

Among these materials, cardboard is a yet undiscovered substance for affordable glucose supplementation. It is mainly composed of cellulose 56.1%, hemicellulose 10.4%, and lignin 12.8% [17]. They have been successfully used to obtain fermentable sugars, such as glucose [18,19]. Cardboard and other lignocellulosic stocks might undergo pretreatment, such as alkaline or paper hydrolysis, to ameliorate the efficiency of the enzymatic hydrolysis of cellulose. However, owing to the potential risks related to the formation of toxic compounds in alkaline or acidic hydrolysis, commercial cellulases digestion could be the optimum and most economical approach for the generation of glucose or fermentable sugars [15].

Bacillus sp. Has been widely used in industry and in academia due to the stability of its replication and plasmid maintenance, as well as its importance and supremacy in PHA production [20]. Furthermore, *Bacillus* sp. has a significant advantage over other bacterial species to produce PHAs due to the absence of a lipopolysaccharide layer, which makes extraction easier, as well as its ability to grow in low-cost raw materials and a high growth rate compared to other bacteria [21]. *Bacillus* sp. metabolism has proven to be rich, generating high-value products, such as lipopeptides, biosurfactants, antiviral proteins, and enzymes [22,23]. Thus, thorough screening of possible *Bacillus* strains from stress-prone environments, improved PHA synthesis methodologies, and the addition of low-cost carbon sources may all contribute to make the entire process more cost-effective [20].

This study investigates a successful bioconversion of cardboard to one of the most important polyesters, PHB. The saccharification of cardboard samples was conducted via enzymatic hydrolysis. A set of experiments was performed to optimize the fermentation process of cardboard hydrolysate as a feedstock for a locally isolated *Bacillus* strain. This represents the first attempt to valorize cardboard as a reliable source of glucose for microbial PHAs synthesis.

2. Materials and Methods

2.1. Materials and Chemicals

A commercial cellulase enzyme NewCell Conc L (NewEnzymes, Portugal) was kindly provided by Professor Stanislaw Bielecki, Lodz University of Technology, Poland. The enzyme preparation was composed of Cellulase and, 2-Benzisothiazolin-3(2H)-one (>1%). Dinitrosalicylic acid (DNS) was purchased from Sigma-Aldrich, Germany. The glucose detecting kit was obtained from Biomaxima, Poland. All other chemicals and media components were purchased from Pol-Aura, Poland.

2.2. Cardboard Sample Preparation

Corrugated cardboard specimens were obtained from local public solid waste containers. The contents of cellulose, hemicellulose, and lignin were detected according to the Van Soest and Robertson method [24]. The specimens were cut into pieces before being suspended in water at 60 °C for 24 h at a solid concentration of 50 g/L and defibered for 1 min using an IKA T50 Ultra Turrax Mixer. Samples that had been defibered were filtered and air dried for further examination [25].

2.3. Enzymatic Degradation of Cardboard Samples

Enzymatic hydrolysis tests were conducted using a commercial cellulase enzyme preparation. The specific activity of 65 ± 0.3 FPU/mL and enzyme protein concentration of 122 ± 0.3 mg/mL were determined using filter paper activity and Bradford methods [26,27]. The hydrolysis process took place in Erlenmeyer flasks at 55 °C, 150 rpm. The pH was preserved at 5.5 using 0.05 N citric acid–sodium citrate buffer.

(a) Determination of optimum incubation time

To determine the optimum incubation time for the enzyme activity, 50 g samples of defibered cardboard were incubated with 1.5% (*v*/*v*) cellulase enzyme in 1 L at different time intervals (1–15 days). The released glucose concentrations were determined by calorimetric kit at 540 nm.

(b) Weight loss

Different sample weights (20, 50, 80, and 100 g) were used for enzyme activity assessment using different enzyme concentrations at 55 °C, 150 rpm for 7 days incubation. The remaining cardboard was filtrated, dried at 70 °C for 24 h and weighed [25]. The following equation was used to calculate the proportion of cardboard consumption:

$$(\mathrm{W0} - \mathrm{Wt}) \times 100/\mathrm{W0}$$

where W0 is the sample's dry weight at zero time and Wt is the sample's dry weight after 7 days incubation.

(c) Determination of reducing sugars

The quantity of reducing sugars released in the enzyme-cardboard mixture was determined by DNS reagent technique [28,29]. A total of 1 g dinitro salicylic acid was mixed with 10 mL distilled water and agitated for 10 min. A volume of 500 µL supernatant was mixed with 500 µL dinitrosalicylic acid reagent, left for 15 min at 28 °C, boiled for 5 min, then rapidly cooled, and the optical density was measured at 540 nm.

(d) Determination of liberated glucose

The enzymatic hydrolysis processes were stopped after enzymatic hydrolysis by keeping the solution at 4 °C for 2 h. The supernatants were recovered by centrifugation at

4 °C for 10 min at 8000 g. The glucose colorimetric kit (Biomaxima, Poland) was used to determine the glucose content of the hydrolysis liquors [30].

2.4. Collection of Samples, Growth Conditions, and Screening of PHA Producing Isolates

Soil samples were collected from several places in Lodz, Poland. In brief, samples were collected in sterile containers and transported directly to the lab, where they were serially diluted and inoculated on PHA screening plates. The screening medium is a Linko medium containing 2% glucose and 0.02 mg/L Nile blue stain as an indicator (Sigma-Aldrich-Darmstadt, Germany). Under UV light, the PHA synthesizing colonies appear orange [31]. Pure cultures of PHA-producing bacteria were then grown on Nutrient agar plates (pH 7.0 at 37 °C for 24 h) for future usage. Sudan black-B (SBB) stain (Sigma-Aldrich-Darmstadt, Germany) was used to perform a confirmatory screening test for PHA generating isolates. The accessibility of the C/N ratio in the medium plays a key role in PHA production. The use of a high carbon-containing medium promotes PHA accumulation. Thus, the medium employed was previously reported by [31] with minor adjustments. The production medium contained (in g/L) glucose (20), $(NH_4)_2SO_4$ (0.5), KH_2PO_4 (2.0), Na_2HPO_4 (2.0), $MgSO_4{\cdot}7H_2O$ (0.5), Na_2HCO_3 (0.5), and $CaCl_2$ (0.01) and a 100 mL trace element solution containing $ZnSO_4{\cdot}7H_2O$ (0.01), $MnCl_2{\cdot}4H_2O$ (0.003), H_3BO_4 (0.003), $CuCl_2{\cdot}2H_2O$ (0.001), and $NICl_2{\cdot}6H_2O$ (0.002). The incubation was carried out at 37 °C, pH 7 and 160 rpm.

2.5. Identification and Characterization of PHA Producing Isolates

The Genomic Mini kit (A&A biotechnology, Pomeranian Voivodeship, Poland) was used to extract genomic DNA from the selected bacterial strain, with minor adjustments to the first step: bacterial cells were treated with lysozyme and incubated at 37 °C for 20 min. The PCR was carried out using a MJ Mini Gradient Thermal Cycler (Bio-Rad, Hercules, CA, USA). The 16S rRNA gene was amplified using universal primers 27F and 1492R (5′-AGAGTTTGATCCTGGCTCAG-3′ 5′-GGTTACCTTGTTACGACTT-3′). Each PCR reaction contained 40 pmol of each primer, 1.5 U of RedTaq ReadyMix DNA polymerase (Sigma-Aldrich, St. Louis, MO, USA) and 20 ng of template DNA, and it was built up to 50 µL with PCR grade water. A 1.0% (*w*/*v*) agarose gel electrophoresis in 0.5 TBE buffer was used to identify PCR products (Sigma-Aldrich). The Big Dye Terminator Ready Reaction Cycle Sequencing kit was used to purify PCR products and extract gene nucleotide sequences (Applied Biosystems, Foster City, CA, USA). The PCR products were examined using an Applied Biosystems model 3730 Genetic Analyzer. The nucleotide sequences of the 16S rRNA gene were proofread, assembled, and aligned in Vector NTI Express Software (Life Technologies, Thermo Fisher Scientific Inc., Waltham, MA, USA), and they were compared with sequences available in the National Center for Biotechnology Information (NCBI, Bethesda, MD, USA), using the BLASTN algorithm (Version 2.2.30+) [31].

2.6. PHA Production Extraction and Purification

(a) MSM supplemented with glucose as a carbon source

The isolated PHA producer was cultured in Nutrient broth overnight at 37 °C and 160 rpm before being transferred to Mineral Salt Medium (MSM) to produce PHA [9]. The medium was supplemented with 20 g (2% *w*/*v*) glucose and 0.5 g (0.05% (*w*/*v*) $(NH_4)_2SO_4$ to maintain a higher C/N ratio. The production cultures were incubated at 37 °C, 160 rpm for 7 days.

(b) Cardboard hydrolysate as a whole medium

The production media were prepared by filtrating 1 L of the cardboard hydrolysate medium to remove the cardboard residues. The filtrate was then kept at 4 °C for 2 h to stop the enzyme activity. The filtrate medium was sterilized by filtration using 200 nm bacterial filters (Alchem, Poland), then inoculated with the PHA producing strain and incubated at 37 °C, 160 rpm for 7 days.

(c) Modified MSM with cardboard hydrolysate

PHA production was studied in a synthetic medium in which 1 L filtrate of cardboard hydrolysate was supplied with all components of MSM except glucose, where the cardboard hydrolysate would be the carbon source. The concentration of carbon source was adjusted to be almost 2% (v/v).

(d) Nitrogen depleted MSM with cardboard hydrolysate

The filtrate of cardboard hydrolysate (1 L) was supplied with all components of MSM except glucose and $(NH_4)_2SO_4$, where the cardboard hydrolysate would be the source of carbon and nitrogen. The concentration of carbon source was adjusted to be almost 2% (v/v), as previously mentioned.

(e) Extraction of PHA

The whole volume of the fermentation medium was centrifuged at 4 °C in a cooling centrifuge (4500 rpm for 15 min), and the cell pellets were freeze dried. The dry pellet weight was measured, and the bacterial cell wall was disrupted by treating it with hot acetone (50 °C) for 20 min. The suspension was centrifuged at 4500 rpm for 15 min before being dried to eliminate excess acetone. PHA was dissolved with chloroform at 37 °C for 48 h, while shaking at 160 rpm. The precipitation of PHA was done using cold methanol and water (7:3) [31]. The percentage of PHA yield was determined by the following equation:

$$\text{PHA yield\%} = W_{PHA}/W_{Cells} * 100$$

where W_{PHA} resembles the weight of the polymer recovered from the freeze-dried cells weight (W_{Cells}).

2.7. *Characterization of the Produced Polymer*

(a) Fourier Transform Infrared (FTIR)

The refined polymers were subjected to FTIR analysis. The study was carried out in a range of 400–4000 cm^{-1} using an FTIR Nicolet 6700 spectrophotometer and OMNIC 3.2 software (Thermo Scientific Products: Riviera Beach, FL, USA) [9].

(b) 1H and 13C NMR

Polymer samples (25 mg) were diluted in 1 mL deuterated chloroform ($CDCl_3$) and analyzed using NMR. The investigation was carried out using a JEOL JNM ECA 500 MHz (JEOL, Japan) to determine the chemical structure of the produced polymers [31].

(c) TGA, DTG, and DTA

Thermal analysis was performed on the purified polymers using the DSC Q20 and TGA Q50 analyzers. The analysis was carried out in the presence of a 20 mL/min N_2 flow. Using a sequential heating system (heating rate 10 °C/min, temperature range 0 to 600 °C), 10 mg of moisture-free distilled PHAs were heated. Following that, the polymer's degradation temperature (*Td*), glass transition temperature (*Tg*), enthalpy of fusion (DH_m), and melting temperature (*Tm*) were measured [31].

(d) XRD

The diffractograms were used to investigate the crystalline structure of the polyesters. A copper tube with a wavelength of 1.5418 A° was used to record the spectrum, which was operated at 30 kV and 10 mA. A 2 mm diameter capillary tube was employed, each scan was recorded in step-by-step mode from 0 to 100° (2θ) with 5° intervals, and the intensities were recorded [5].

2.8. *Statistical Analysis*

The statistically significant differences across the studies were examined using a one-way ANOVA with the Tukey test ($p < 0.05$ confidence). The process monitoring assays were performed in triplicate, and the results were reported as the mean value and its standard deviation. The data was analyzed using Prism 7 (GraphPad, Inc. San Diego, CA, USA).

3. Results and Discussion

3.1. Enzymatic Degradation of Cardboard Samples

Owing to the high percentage of cellulose in cardboard, which can reach up to 59% [25], it could be a promising source of carbon to produce value-added products. The complexity of polysaccharides in such waste (mainly cellulose and hemicellulose) has made cellulase systems paramount in liberating fermentable sugars. From an environmental perspective, the industrial implementation of the current study could assist in tackling waste accumulation issues in Europe, especially Poland. Paper and cardboard waste comprised approximately 15% of municipal solid waste from 2012 to 2017, whereas plastics resamples were 18%, and this data is expected to witness a considerable rise in the next years [32]. The chemical composition of corrugated cardboard showed that our samples contained 58.2% cellulose, 9.6% hemicellulose, 10.0% lignin, and 22.2% other components, which were in good line with another investigation [33]. Enzyme-assisted cardboard hydrolysis seems to be an appropriate approach for supplying PHA producers with sufficient amounts of carbon source in an economic manner. To assess cellulase activity on cardboard digestion, representative experimental concentration profiles of glucose and reducing sugars obtained in this set of experiments are shown in Figure 1.

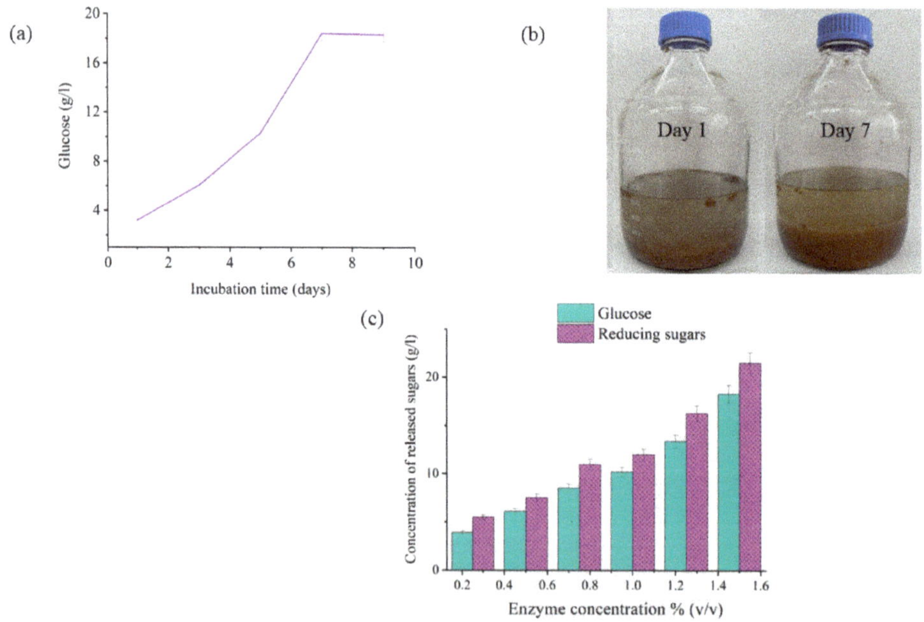

Figure 1. Enzymatic hydrolysis of cardboard, (**a**) Incubation time effect on the hydrolysis process, (**b**) presentation of the enzymatic hydrolysis process on cardboard fibers on the 1st and the 7th day and (**c**) colorimetric analysis of the enzymatic hydrolysis products, including reducing sugars and glucose.

Regarding the liberated glucose profile, the glucose concentration was directly proportional to incubation time till day 7 at pH 5.5 and 55 °C. Moreover, the highest released glucose concentration was reported to be 18.3 ± 0.2 g/L on day 7, which comprises approximately 62% of the total cellulose content in the investigated cardboard samples (Figure 1a), whereas it exhibited almost constant values from days 8 to 15 (data not shown). The reason for the positive correlation between the released glucose and incubation days could be attributed to cellulase activity on the cellulosic portion of cardboard [34]. Cellulases work synergistically to hydrolyze cellulose as it is degraded from crystalline or amorphous cellulose to small, soluble cellobiose fragments and finally to glucose [35]. The constant glucose records after day 7 demonstrated a complete saccharification of cellulose and hemicellulose contents, where the major morphological changes in the fiber structure took place. It might also be attributed to the depletion of cellulosic components in cardboard as well as the generation of metabolic by-products. (Figure 1b) [33]. Therefore, the incubation period of 7 days has been selected as the enzyme incubation time in the following experiments.

The effect of different enzyme concentrations was assessed on different samples' weights in terms of cardboard degradation. Maximum weight loss values for all samples were observed at an enzyme concentration of 1.5% (v/v), which was found to be in the range of 69 to 73% (Table 1). The maximum weight loss values that resulted from the highest enzyme concentration supports the fact that cellulase catalyzes the polysaccharide decomposition by simply cleaving β-1,4-glycosidic linkages [36]. The enzymatic action of cellulase could be seen in Figure 1b, where the firm fibers of cardboard were transformed to powder-like structures after 7 days incubation at 55 °C.

Table 1. Weight loss (%) of different cardboard samples after enzymatic hydrolysis. Mean and standard deviations were used to show the values. The presence of different superscript letters within the same column indicates significance ($p < 0.05$).

Cardboard Samples Weights (g)	**100**	**80**	**50**	**20**
Enzyme Concentration (%)	**Weight Loss (%)**			
0.25	20 ± 0.5^{a}	23 ± 0.3^{a}	26 ± 0.5^{a}	30 ± 0.1^{a}
0.5	32 ± 0.2^{b}	33 ± 0.9^{b}	36 ± 0.5^{b}	39 ± 0.3^{b}
0.75	51 ± 0.1^{c}	50 ± 0.5^{c}	50 ± 0.3^{c}	52 ± 0.9^{c}
1	60 ± 0.2^{d}	61 ± 0.4^{d}	64 ± 0.1^{d}	66 ± 0.1^{d}
1.25	65 ± 0.3^{e}	67 ± 0.1^{d}	68 ± 0.1^{d}	70 ± 0.6^{e}
1.5	69 ± 0.3^{e}	69 ± 0.2^{d}	70 ± 0.3^{d}	73 ± 0.3^{e}

The hydrolysis products, such as released glucose and reducing sugars, were determined in 50 g cardboard samples. The highest enzyme concentration (1.5% (v/v)) released 18.3 ± 0.3 g glucose, which was the maximum value compared with the lower cellulase concentrations. Moreover, the glucose percentage per 50 g cardboard sample was determined to be 36.5 ± 0.4%, which represents 18.3 ± 0.3 g glucose produced from a 50 g cardboard sample. This glucose content was estimated to be approximately 85.1 ± 0.1% of reducing sugars per 50 g sample (Figure 1c). Such data were consistent with the obtained glucose values measured at 36–63% (g/g of cellulose sample) using commercial enzymes cellulase CTec2 and hemicellulase HTec [33]. These results support the fact that the yielded carbon source would provide a sufficient supplement for PHA production.

3.2. Molecular Identification of the PHA Producer and Phylogenetic Analysis

The screened soil bacterial isolates resulted in 9 PHA producers, which were confirmed by orange color fluorescence on Nile blue plates and Sudan black staining. The bacterial isolate with the highest PHA productivity was selected for further molecular identification. The 16S rRNA study of the PHA producing strain exhibited a significant degree of similarity to the *Bacillus* sp. genera. A sequence comparison using BLAST showed strain ICRI89's close relationship to *Bacillus mycoides* (formerly known as *Bacillus weihenstephanensis* UT11) with 100% similarity. The 16S rRNA sequence of the newly isolated strain was submitted to GenBank, NCBI as *Bacillus mycoides* ICRI89. A neighbor-joining dendrogram with several *Bacillus* sp. as an outer group shows the phylogentic position of *B. mycoides* ICRI89 (Figure 2).

Figure 2. Neighbor-joining phylogenetic analysis based on 16S rRNA gene sequences showing the position of strain *B. mycoides* ICRI89. These phylogenetic relationships were identified by MEGA 11 sequence alignment editor (version 11.0.11).

3.3. Nucleotides Accession Numbers

Consensus sequences were generated with the MEGA 11 sequence alignment editor (version 11.0.11), and the sequences were then evaluated with the BLASTN software (NCBI) [37]. The 16S sequence was submitted to GenBank with the accession number ON231789.

3.4. MSM Supplemented with Glucose as a Carbon Source

B. mycoides ICRI89 was cultivated in 1 L MSM in 2 L flasks with 2% (*w*/*v*) glucose and 0.05% (*w*/*v*) $(NH_4)_2SO_4$ at 37 °C, pH = 7, 160 rpm for 7 days. The biomasses from PHA production were collected, lyophilized, and weighed [31]. The lyophilized mass of *B. mycoides* ICRI89 was 4.27 g/L, and the generated polyester was recovered from the cells. The pure polymers produced were 2.63 ± 0.1 g/L, corresponding to 61.7% CDW (Figure 3). These results are highly similar to those of a previous investigation [38] in which *B. megaterium* accumulated a maximum PHA weight of 2.74 g/L with glucose as the sole carbon source. When compared to other carbon sources, such as arabinose, starch, lactose, lactic acid, glycerol, or sodium acetate, *Bacillus* sp. can efficiently metabolize glucose for greater PHA synthesis. As a result, when glucose was added as a carbon source, it produced the highest PHA content when compared to other carbon sources [38]. The maximum PHA yield in the current study was detected upon using $(NH_4)_2SO_4$ as a nitrogen source. PHA synthesis was comparable when various nitrogen sources were used, including, with a minor difference, protease peptone, glycine, potassium nitrate, urea, and ammonium chloride. Previously, ammonium sulphate was shown to be the optimum nitrogen source for *B. mycoides* RLJ B017 [39] and *B. Megaterium* [40]. On the other hand, the current findings showed a higher PHA productivity than those of *Bacillus* sp. AZR-1, which used glucose for PHA production. *Bacillus* sp. AZR-1 produced CDW of 1.88 g/L with a PHA content up to 40%. It was shown to be extremely effective in exploiting soluble starch as a precursor for PHA accumulation, having a CDW of 0.87 g/L and a PHA content of 0.19 g/L, resembling 22% CDW. This amount is approximately half that when glucose was utilized as a carbon source, which is logical given that the starch hydrolysis process makes starch less useful when compared to simple monosaccharides, such as glucose [41]. This suggests that for growth and subsequent PHA synthesis, in general, these bacteria prefer simple monomers, such as glucose. Our isolate *B. mycoides* ICRI89 appears to have the most active metabolic machinery for synthesis of PHA. In many studies, there has been a positive correlation between the amount of glucose used as a carbon source and PHA production [42].

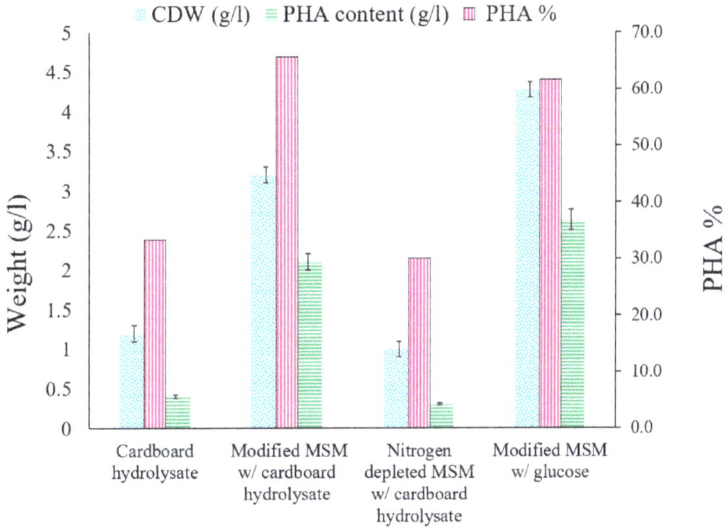

Figure 3. PHA content (g/L), cell dry weight (CDW) (g/L), and PHA productivity (%) in four different models.

3.5. Cardboard Hydrolysate as a Whole Medium

Enzymatically hydrolyzed cardboard was employed as a complete medium to evaluate our new strain's potential to accumulate PHA. When hydrolyzed cardboard was utilized as a PHA production medium for *B. mycoides* ICRI89, the highest PHA concentration was 0.4 ± 0.1 g/L, with 33.3% CDW PHA productivity (Figure 3). The PHA content witnessed an 84% reduction in comparison with that of glucose as a carbon source. In the same manner, there was an apparent decline in the yielded CDW to be 1.2 g/L compared to 4.27 g/L for the glucose standard medium. The major explanation for the low PHA concentration might be the lack of certain minerals and salts. Furthermore, the existence of KH_2PO_4 in the production medium at concentrations less than 0.1 g/L could reduce cellular proliferation, whilst concentrations higher than 0.1 g/L might improve PHA productivity. Thus, phosphate limitation (KH_2PO_4 and K_2HPO_4) was discovered to have an essential role in PHA build-up [43]. Enzymatically hydrolyzed waste paper was recently used as a sustainable feed stock for PHA synthesis by *B. Sacchari*. The bacterial strain accumulated 3.63 g/L PHA, which comprises approximately 44.2% CDW. The relatively higher yield compared to our study could be reasoned to performing enzymatic digestion for paper waste using an enzyme mixture of cellulase, β-glucosidase, and hemicellulose [44]. Such a mixture may have played a critical role in having a better saccharification of lignocellulosic biomass, including paper waste. That's why it is recommended to examine the effect of different enzyme cocktails on the saccharification of different kinds of cardboard in future studies.

3.6. Modified MSM with Cardboard Hydrolysate

The filtrate of cardboard hydrolysate (1 L) was supplied with all components of MSM except glucose, where the cardboard hydrolysate would be the carbon source. The concentration of carbon source was adjusted to be almost 2% (v/v), which represents the content of reducing sugar in cardboard hydrolysate. The PHA generating isolate *B. mycoides* ICRI89 had a maximum PHA concentration of 2.1 $\pm$ 0.1 g/L, which corresponds to 65.6% CDW (Figure 3). This experiment yielded almost the same PHA weights of that when glucose was used as the only carbon source. These findings are consistent with a previous study aiming at increasing PHA bioproduction using wheat straw lignocellulosic hydrolysates [14]. Since *B. sacchari* DSM 17165 can metabolize its primary carbohydrates, such as glucose, xylose, and arabinose, it was employed to generate PHA from wheat straw hydrolysates. When grown on a mixture of commercial C6 and C5 sugars as a control production medium, the *B. sacchari* cell weight of 6.0 g/L accumulated approximately 4.4 g/L PHA, resembling 70% CDW with a polymer on sugar yield of 0.18 g/g, whereas when wheat straw hydrolysates were used as the carbon source, these values reached approximately 4.4 g/L PHA, presenting 60% CDW [14]. Even though PHAs offer several environmental benefits, their high production costs limit their widespread application. One of the grand challenges facing these polymers scale-up is the high-cost feed stock, especially carbon sources, which comprise nearly the majority of the final cost [45,46]. Paper rejects, including cardboard, could be promising sources for essential and simple sugars for bacterial metabolism supporting polyester accumulation. Different cardboard rejects treatment procedures may provide more glucose and xylose per kilogram. For example, sodium hydroxide treatment of cardboard yielded approximately 373 g glucose and 61 g xylose per kilogram of rejects, which is regarded as an economical technique for PHA scale-up [47].

3.7. Nitrogen Depleted MSM with Cardboard Hydrolysate

Under nutrient-restricted environments, microbial polyesters accumulate with surplus carbon and polymerize as inclusion bodies [48]. To mimic such conditions, 1 L filtrate of cardboard hydrolysate was supplied with all components of MSM except glucose and $(NH_4)_2SO_4$, where the cardboard hydrolysate would be the carbon and nitrogen sources. The concentration of carbon source was adjusted to be almost 2% (v/v), which represents the content of reducing sugar in cardboard hydrolysate. After 7 days of incubation at 37 °C and 160 rpm, *B. mycoides* ICRI89 used this hydrolysate to generate a PHA content of 0.3 $\pm$ 0.1 g/L (30% CDW) (Figure 3). The hydrolysate contains not only carbon but also nitrogen. The amount of nitrogen in the hydrolysate is insufficient to support biomass growth. A recent study [49] reported a low 0.003% nitrogen content in corrugated carboard acidic hydrolysate. Since high nitrogen concentrations directly increase cell biomass rather than PHA formation [50], PHA generation and cell proliferation were severely hindered.

When all the preceding data are compared, it is clear that the PHA concentration achieved by employing modified MSM with cardboard hydrolysate was extremely near to that obtained when glucose was used as the only carbon source. Sugar consumption in the cardboard hydrolysate medium was the primary cause of PHA production and build-up. Thus, polymers extracted from cells grown on glucose and cardboard hydrolysate medium as a carbon source were used for further characterization.

3.8. Analysis and Characterisation of the Purified Polymer

3.8.1. FTIR

The chemical structure of the purified biopolymers was determined using FTIR. The IR spectra of polymer generated in a medium containing glucose as a carbon source revealed three prominent absorption bands at 1730, 765, and 710 cm^{-1} due to ester, CH group, and carbonyl group, respectively. These bands also appeared in the IR spectra of the polyester produced from cardboard hydrolysate medium as a carbon source at 1733, 769, and 708 cm^{-1}, respectively (Figure 4). The existence of -CH bonding is shown by the typical bands at 2955 and 2965 cm^{-1}, while the C=O and ester groups are represented by the bands at 1730 and 1733 cm^{-1}. The bands found at 1080 and 1090 cm^{-1} correspond to the C-O bonding for PHB derived from glucose and cardboard, respectively. The bands of the aforementioned polyester samples are quite similar to the bands of standard PHB, which affirms the high purity of the generated polymer. These findings are consistent with those reported by a recent study of PHB produced by *B. megaterium* MTCC 453 [51]. Because of the FTIR data, it is apparent both polymers produced from glucose (PHB) and cardboard hydrolysate (PHB-CB) as carbon sources are PHB, which is the common homopolymer of PHAs.

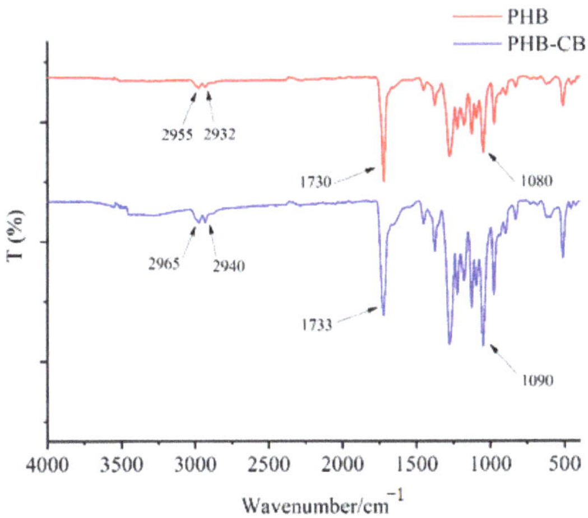

Figure 4. FTIR spectra of polymers isolated from *B. mycoides* ICRI89 grown in MSM containing either glucose (PHB) or cardboard hydrolysate (PHB-CB) as carbon sources.

3.8.2. NMR

PHB samples were characterized using ^1H NMR and ^{13}C NMR spectroscopic methods. The ^1H and ^{13}C NMR spectra of PHB produced from *B. mycoides* ICRI89 by microbial fermentation of glucose and cardboard hydrolysate are shown in Figure 5. PHB characteristic peaks were detected in both PHB and PHB-CB, such as δ = 5.21 and 5.23 ppm, which correspond to –CH doublet, δ = 2.50 and 2.51 ppm for $–CH_2$ multiplet, and δ = 1.21 and 1.22 of $–CH_3$ doublet for the PHB and PHB-CB, respectively. The large peaks at δ = 7.3 ppm (Figure 5a,c) indicate the solvent (CD_3Cl), while the small peaks at δ = 1.61 and 1.63 ppm are due to the minor H_2O contamination of the solvent. These findings were identical to those obtained using the PHB standard. Hence, we determined that the polyesters produced by *B. mycoides* ICRI89 strain cultivated in glucose and cardboard hydrolysate as the carbon source were PHBs [52]. ^{13}C NMR analysis also confirmed these findings. The functional groups C=O (170.5 and 170.3), CH (65.9 and 65.7 ppm), CH_2 (42.55 and 41.9 ppm), and CH_3 (19.5 and 19.6 ppm) peaks were assigned for PHB and PHB-CB, and they were similar to the PHB previously obtained from *Bacillus* sp. [53].

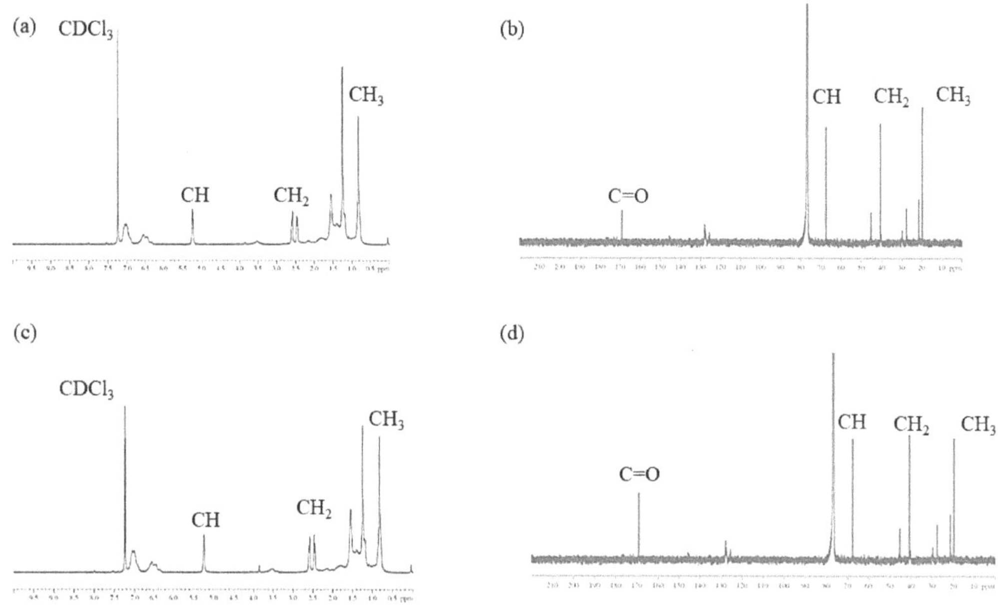

Figure 5. ^{1}HNMR and ^{13}CNMR analysis of PHB and PHB-CB, (**a**) ^{1}HNMR for PHB, (**b**) ^{13}CNMR for PHB, (**c**) ^{1}HNMR for PHB-CB and (**d**) ^{13}CNMR for PHB-CB.

3.8.3. TGA and DTG

TGA profiles of PHB and PHB-CB synthesized by *B. mycoides* ICRI89 are depicted in Figure 6a. The TGA curve represents the weight loss of the synthesized PHB in two phases for the two generated polyesters generated. The first step of mass loss occurred at temperatures ranging from 100 to 180 °C. For PHB and PHB-CB, the mass loss was approximately 1.5 and 1.3% of total mass, respectively. This loss is caused by the evaporation of physically adsorbed solvents, such as methanol, chloroform, and others that have formed on the polymer surface. Furthermore, the second or major step of polymer degradation started after 200 °C, which occurs after the melting point of PHB. The decomposition process involves a molecular weight decrease, which includes chain scission and hydrolysis. The random chain scission process, which involves the breakage of C=O and C-O bonds in ester moieties by β-scission, destruction of crystalline areas, and depolymerization, is responsible for the rapid heat breakdown of PHB at this stage [54]. The second stage of weight loss occurred when the temperature increased further, as hydrolysis, chain scission, and the synthesis of crotonic acid all contribute to the deterioration process. From the analysis of the initial and the maximum degradation temperatures of main step weight loss and residual mass percentage for PHB and PHB-CB, the maximum degradation temperatures for PHB and PHB-CB were found to be 380 and 369 °C, respectively. As a result, it can be inferred that both forms of PHB exhibited greater thermal stability when compared to standard PHB, which was found to have a decomposition temperature of 285 °C [55]. Furthermore, the residual mass of PHB and PHB-CB is less than 1.5%. The second stage of degradation for PHB produced from *Bacillus* sp. ranges between 237 and 320 °C which is lower than our records. This implies that the synthesized PHBs have higher degrees of thermal stability than the PHB produced by *Bacillus* sp. N-2 [56].

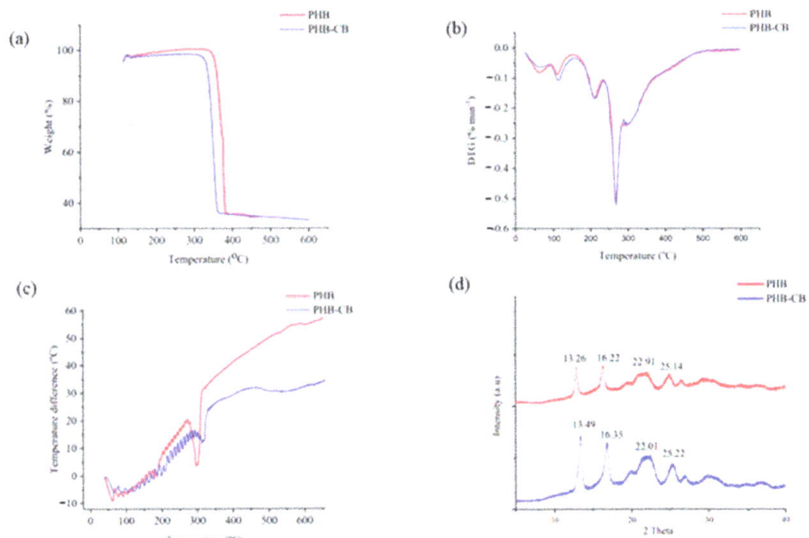

Figure 6. Characterization of PHB and PHB-CB demonstrating (**a**) TGA, (**b**) DTG, (**c**) DTA, and (**d**) XRD.

The rate of mass loss of a polymer sample with relation to temperature was investigated using differential thermogravimetric (DTG) analysis (Figure 6b). The DTG curve peaks reflect the thermal stability of PHB in relation to the temperature at which the highest breakdown rate of the polymer matrix occurs. The DTG characteristic curves, as the TGA curves, revealed three distinct phases. The mass loss rate in the first phase was approximately 0.16 to 0.20 mass%/min until 190 to 200 °C, and the amount of residue is quite high in both PHB and PHB-CB samples. The maximum degradation temperature for PHB in the second stage of degradation was approximately 266 °C, with a maximum mass loss rate of 34%/min. PHB-CB's maximum degradation temperature was approximately 268 °C, with a maximum mass loss rate of 32%/min. It has previously been reported that PHB standard has a degradation temperature of roughly 236 °C, with a maximum mass loss rate of 30%/min [51]. According to the results of the foregoing investigation, the PHB and PHB-CB produced by *B. mycoides* ICRI89 indicate strong thermal stability or resistance to heat deterioration.

3.8.4. DTA

Differential Thermal Analysis (DTA) aids in determining breakdown heat (Figure 6c). This experiment was carried out to assess the cross-linking capabilities and the heat stability of the generated polymer. Due to the existence of cross-linking events during PHB degradation, an exothermic peak is found in the DTA thermogram. The curing temperatures, which are 331 and 325 °C for both PHB and PHB-CB, are the temperatures at which cross-linkage occurs. It is the most essential attribute that appears to be a major impediment to the commercial application of PHB, generating thermal instability due to a lack of cross-link capacity [57].

3.8.5. XRD

The XRD spectra (Figure 6d) presents X-ray diffraction of PHB samples from both pure polyesters, PHB and PHB-CB. The observed peaks in XRD spectra for PHB are $2\theta = 3.69°$, 13.26°, 16.22°, 22.91°, and 25.14°, while the observed peaks in XRD spectra for PHB-CB are $2\theta = 3.61°$, 13.49°, 16.35°, 22.01°, and 25.22°. The peaks at $2\theta = 13.26°$,13.49°,16.22°, and 16.35°, which are the most intense and scattering peaks, each indicate an orthorhombic unit cell. The relatively weaker peaks observed at 2θ = 22.91° and 22.01° correspond to α-PHB crystal, while the minor spectra observed at 2θ = 25.14° and 25.22° denote PHB's partly crystalline nature. The polymer matrix adopts a regular helicoidal shape with two antiparallel chains in the rhombic unit cell inside the crystalline domain [51–58]. Pradhan et al. [44] showed that the diffractogram of the produced PHBs in the current study is nearly equivalent to that of PHB produced from *Bacillus* sp. [51] Collectively, more efforts should be made to recycle environmental waste [59].

4. Conclusions

The current study proposes corrugated cardboard waste hydrolyzed by commercial cellulase as a cheap and readily available substrate for microbial PHB synthesis. This would remove one of the major barriers facing PHB scaling-up. The enzyme concentration affected the reducing sugars, reaching a satisfactory yield of 21.3 ± 0.1 g/L at an enzyme concentration of 1.5% (v/v) at 55 °C. Moreover, the newly isolated *B. mycoides* ICRI89 accumulated 2.1 ± 0.2 g/L of PHAs when grown on modified MSM containing cardboard hydrolysate, which was highly close to that produced when grown on MSM supplemented with glucose. It is important to note that the polymers purified from B. mycoides ICRI89 cells were almost entirely composed of PHB, according to FTIR, NMR, and XRD findings.

5. Future Prospects

The present research has several future prospects, including the industrial fermentation of cardboard and paper waste in humongous bioreactors, using several enzyme mixtures for promoting polyester synthesis. In addition, the methods of PHB purification from bacterial cells have to be improved to obtain pure polymer batches at a large scale.

Author Contributions: F.A., A.S. and M.R. conceived and designed the research. F.A. and M.R. conducted the experiments. F.A. and M.R. contributed to analyzing the data. F.A. and M.R. wrote the manuscript. All authors have read and agreed to the published version of the manuscript.

Funding: This research was conducted as a part of the International Research Agendas PLUS program of the Foundation for Polish Science, co-financed by the European Union under the European Regional Development Found (MAB PLUS/2019/11).

Institutional Review Board Statement: Not applicable.

Informed Consent Statement: Not applicable.

Data Availability Statement: The accession number(s) can be found on the GenBank. Available online: https://www.ncbi.nlm.nih.gov/, ON231789 (Accessed on 17 May 2022).

Acknowledgments: The authors gratefully acknowledge Stanisław Bielecki, Institute of Molecular and Industrial Biotechnology, Lodz University of Technology for providing the cellulase enzyme.

Conflicts of Interest: The authors declare no conflict of interest.

References

1. Abd El-Malek, F.; Rofeal, M.; Zabed, H.M.; Nizami, A.-S.; Rehan, M.; Qi, X. Microorganism-mediated algal biomass processing for clean products manufacturing: Current status, challenges and future outlook. *Fuel* **2022**, *311*, 122612. [CrossRef]
2. Rofeal, M.; Abdelmalek, F.; Steinbüchel, A. Naturally-Sourced Antibacterial Polymeric Nanomaterials with Special Reference to Modified Polymer Variants. *Int. J. Mol. Sci.* **2022**, *23*, 4101. [CrossRef] [PubMed]
3. Koller, M.; Mukherjee, A. Polyhydroxyalkanoates (PHAs)—Production, Properties, and Biodegradation. In *Biodegradable Polymers in the Circular Plastics Economy*; Wiley: Hoboken, NJ, USA, 2022; pp. 145–204. [CrossRef]
4. El-malek, F.A.; Khairy, H.; Farag, A.; Omar, S. The sustainability of microbial bioplastics, production and applications. *Int. J. Biol. Macromol.* **2020**, *157*, 319–328. [CrossRef] [PubMed]
5. Rofeal, M.; Abd El-Malek, F.; Qi, X. In vitro assessment of green polyhydroxybutyrate/chitosan blend loaded with kaempferol nanocrystals as a potential dressing for infected wounds. *Nanotechnology* **2021**, *32*, 375102. [CrossRef] [PubMed]
6. Prakash, P.; Lee, W.-H.; Loo, C.-Y.; Wong, H.S.J.; Parumasivam, T. Advances in Polyhydroxyalkanoate Nanocarriers for Effective Drug Delivery: An Overview and Challenges. *Nanomaterials* **2022**, *12*, 175. [CrossRef] [PubMed]

7. Kumar, R.; Verma, A.; Shome, A.; Sinha, R.; Sinha, S.; Jha, P.K.; Kumar, R.; Kumar, P.; Shubham; Das, S.; et al. Impacts of Plastic Pollution on Ecosystem Services, Sustainable Development Goals, and Need to Focus on Circular Economy and Policy Interventions. *Sustainability* **2021**, *13*, 9963. [CrossRef]
8. Kalia, V.C.; Singh Patel, S.K.; Shanmugam, R.; Lee, J.-K. Polyhydroxyalkanoates: Trends and advances toward biotechnological applications. *Bioresour. Technol.* **2021**, *326*, 124737. [CrossRef]
9. Abd El-malek, F.; Rofeal, M.; Farag, A.; Omar, S.; Khairy, H. Polyhydroxyalkanoate nanoparticles produced by marine bacteria cultivated on cost effective Mediterranean algal hydrolysate media. *J. Biotechnol.* **2021**, *328*, 95–105. [CrossRef] [PubMed]
10. Wang, J.; Liu, S.; Huang, J.; Qu, Z. A review on polyhydroxyalkanoate production from agricultural waste Biomass: Development, Advances, circular Approach, and challenges. *Bioresour. Technol.* **2021**, *342*, 126008. [CrossRef] [PubMed]
11. Revelles, O.; Beneroso, D.; Menendez, J.A.; Arenillas, A.; García, J.L.; Prieto, M.A. Syngas obtained by microwave pyrolysis of household wastes as feedstock for polyhydroxyalkanoate production in Rhodospirillum rubrum. *Microb. Biotechnol.* **2017**, *10*, 1412–1417. [CrossRef]
12. Elain, A.; Le Grand, A.; Corre, Y.-M.; Le Fellic, M.; Hachet, N.; Le Tilly, V.; Loulergue, P.; Audic, J.-L.; Bruzaud, S. Valorisation of local agro-industrial processing waters as growth media for polyhydroxyalkanoates (PHA) production. *Ind. Crops Prod.* **2016**, *80*, 1–5. [CrossRef]
13. Vigneswari, S.; Noor, M.S.M.; Amelia, T.S.M.; Balakrishnan, K.; Adnan, A.; Bhubalan, K.; Amirul, A.-A.A.; Ramakrishna, S. Recent advances in the biosynthesis of polyhydroxyalkanoates from lignocellulosic feedstocks. *Life* **2021**, *11*, 807. [CrossRef]
14. Cesário, M.T.; Raposo, R.S.; De Almeida, M.C.M.D.; Van Keulen, F.; Ferreira, B.S.; Da Fonseca, M.M.R. Enhanced bioproduction of poly-3-hydroxybutyrate from wheat straw lignocellulosic hydrolysates. *New Biotechnol.* **2014**, *31*, 104–113. [CrossRef]
15. Li, J.; Yang, Z.; Zhang, K.; Liu, M.; Liu, D.; Yan, X.; Si, M.; Shi, Y. Valorizing waste liquor from dilute acid pretreatment of lignocellulosic biomass by *Bacillus megaterium* B-10. *Ind. Crops Prod.* **2021**, *161*, 113160. [CrossRef]
16. Zoghlami, A.; Paës, G. Lignocellulosic biomass: Understanding recalcitrance and predicting hydrolysis. *Front. Chem.* **2019**, *7*, 874. [CrossRef] [PubMed]
17. Suthar, S.; Kishore Singh, N. Fungal pretreatment facilitates the rapid and valuable composting of waste cardboard. *Bioresour. Technol.* **2022**, *344*, 126178. [CrossRef]
18. Wang, L.; Sharifzadeh, M.; Templer, R.; Murphy, R.J. Bioethanol production from various waste papers: Economic feasibility and sensitivity analysis. *Appl. Energy* **2013**, *111*, 1172–1182. [CrossRef]
19. Ramamoorthy, N.K.; Nagarajan, R.; Ravi, S.; Sahadevan, R. An innovative plasma pre-treatment process for lignocellulosic bio-ethanol production. *Energy Sources Part A Recovery Util. Environ. Eff.* **2020**, 1–15. [CrossRef]
20. Mohapatra, S.; Pattnaik, S.; Maity, S.; Mohapatra, S.; Sharma, S.; Akhtar, J.; Pati, S.; Samantaray, D.P.; Varma, A. Comparative analysis of PHAs production by Bacillus megaterium OUAT 016 under submerged and solid-state fermentation. *Saudi J. Biol. Sci.* **2020**, *27*, 1242–1250. [CrossRef] [PubMed]
21. Vu, D.H.; Wainaina, S.; Taherzadeh, M.J.; Åkesson, D.; Ferreira, J.A. Production of polyhydroxyalkanoates (PHAs) by *Bacillus megaterium* using food waste acidogenic fermentation-derived volatile fatty acids. *Bioengineered* **2021**, *12*, 2480–2498. [CrossRef] [PubMed]
22. Rofeal, M.; El-Malek, F.A. Valorization of Lipopeptides Biosurfactants as Anticancer Agents. *Int. J. Pept. Res. Ther.* **2021**, *27*, 447–455. [CrossRef]
23. Rofeal, M.; El-Malek, F.A. Ribosomal proteins as a possible tool for blocking SARS-COV 2 virus replication for a potential prospective treatment. *Med. Hypotheses* **2020**, *143*, 109904. [CrossRef]
24. Van Soest, P.; Robertson, J. Systems of analysis for evaluating fibrous feeds. In *Standardization of Analytical Methodology for Feeds: Proceedings*; IDRC: Ottawa, ON, Canada, 1979.
25. Yáñez, R.; Alonso, J.L.; Parajó, J.C. Production of hemicellulosic sugars and glucose from residual corrugated cardboard. *Process Biochem.* **2004**, *39*, 1543–1551. [CrossRef]
26. Adney, B.; Baker, J. *Measurement of Cellulase Activities*; Laboratory Analytical Procedure (LAP): Location Golden, CO, USA, 2008.
27. Bradford, M.M. A rapid and sensitive method for the quantitation of microgram quantities of protein utilizing the principle of protein-dye binding. *Anal. Biochem.* **1976**, *72*, 248–254. [CrossRef]
28. Yousef, N.; Mawad, A.; Abeed, A. Enhancement the cellulase activity induced by endophytic bacteria using calcium nanoparticles. *Curr. Microbiol.* **2019**, *76*, 346–354. [CrossRef] [PubMed]
29. Ghose, T. Measurement of cellulase activities. *Pure Appl. Chem.* **1987**, *59*, 257–268. [CrossRef]
30. Martinez-Trujillo, M.A.; Bautista-Rangel, K.; García Rivero, M.; Martínez-Estrada, A.; Cruz-Díaz, M.R. Enzymatic saccharification of banana peel and sequential fermentation of the reducing sugars to produce lactic acid. *Bioprocess Biosyst. Eng.* **2020**, *43*, 413–427. [CrossRef] [PubMed]
31. El-malek, F.A.; Farag, A.; Omar, S.; Khairy, H. Polyhydroxyalkanoates (PHA) from Halomonas pacifica ASL10 and Halomonas salifodiane ASL11 isolated from Mariout salt lakes. *Int. J. Biol. Macromol.* **2020**, *161*, 1318–1328. [CrossRef] [PubMed]
32. Przydatek, G. Assessment of changes in the municipal waste accumulation in Poland. *Environ. Sci. Pollut. Res.* **2020**, *27*, 25766–25773. [CrossRef] [PubMed]
33. Kinnarinen, T.; Häkkinen, A. Influence of enzyme loading on enzymatic hydrolysis of cardboard waste and size distribution of the resulting fiber residue. *Bioresour. Technol.* **2014**, *159*, 136–142. [CrossRef] [PubMed]

34. Zhao, S.; Chen, W.; Luo, W.; Fang, H.; Lv, H.; Liu, R.; Niu, Q. Anaerobic co-digestion of chicken manure and cardboard waste: Focusing on methane production, microbial community analysis and energy evaluation. *Bioresour. Technol.* **2021**, *321*, 124429. [CrossRef] [PubMed]
35. Sambasivarao, S.V.; Granum, D.M.; Wang, H.; Maupin, C.M. Identifying the enzymatic mode of action for cellulase enzymes by means of docking calculations and a machine learning algorithm. *AIMS Mol. Sci.* **2014**, *1*, 59–80. [CrossRef]
36. Jayasekara, S.; Ratnayake, R. Microbial cellulases: An overview and applications. *Cellulose* **2019**, *22*, 92. [CrossRef]
37. Altschul, S.F.; Gish, W.; Miller, W.; Myers, E.W.; Lipman, D.J. Basic local alignment search tool. *J. Mol. Biol.* **1990**, *215*, 403–410. [CrossRef]
38. Mohanrasu, K.; Rao, R.G.R.; Dinesh, G.H.; Zhang, K.; Prakash, G.S.; Song, D.-P.; Muniyasamy, S.; Pugazhendhi, A.; Jeyakanthan, J.; Arun, A. Optimization of media components and culture conditions for polyhydroxyalkanoates production by *Bacillus megaterium*. *Fuel* **2020**, *271*, 117522. [CrossRef]
39. Borah, B.; Thakur, P.; Nigam, J. The influence of nutritional and environmental conditions on the accumulation of poly-β-hydroxybutyrate in *Bacillus mycoides* RLJ B-017. *J. Appl. Microbiol.* **2002**, *92*, 776–783. [CrossRef]
40. Rodríguez-Contreras, A.; Koller, M.; Miranda-de Sousa Dias, M.; Calafell-Monfort, M.; Braunegg, G.; Marqués-Calvo, M.S. High production of poly (3-hydroxybutyrate) from a wild *Bacillus megaterium* Bolivian strain. *J. Appl. Microbiol.* **2013**, *114*, 1378–1387. [CrossRef] [PubMed]
41. Ali, I.; Jamil, N. Biosynthesis and characterization of poly3-hydroxyalkanote (PHA) from newly isolated bacterium *Bacillus* sp. AZR-1. *Iran. J. Sci. Technol. Trans. A Sci.* **2018**, *42*, 371–378. [CrossRef]
42. Khosravi-Darani, K.; Mokhtari, Z.-B.; Amai, T.; Tanaka, K. Microbial production of poly (hydroxybutyrate) from C1 carbon sources. *Appl. Microbiol. Biotechnol.* **2013**, *97*, 1407–1424. [CrossRef] [PubMed]
43. Sangkharak, K.; Prasertsan, P. Nutrient optimization for production of polyhydroxybutyrate from halotolerant photosynthetic bacteria cultivated under aerobic-dark condition. *Electron. J. Biotechnol.* **2008**, *11*, 83–94. [CrossRef]
44. Al-Battashi, H.; Annamalai, N.; Al-Kindi, S.; Nair, A.S.; Al-Bahry, S.; Verma, J.P.; Sivakumar, N. Production of bioplastic (poly-3-hydroxybutyrate) using waste paper as a feedstock: Optimization of enzymatic hydrolysis and fermentation employing Burkholderia sacchari. *J. Clean. Prod.* **2019**, *214*, 236–247. [CrossRef]
45. Tan, D.; Wang, Y.; Tong, Y.; Chen, G.Q. Grand challenges for industrializing polyhydroxyalkanoates (PHAs). *Trends Biotechnol.* **2021**, *39*, 953–963. [CrossRef]
46. Chen, G.Q.; Chen, X.Y.; Wu, F.Q.; Chen, J.C. Polyhydroxyalkanoates (PHA) toward cost competitiveness and functionality. *Adv. Ind. Eng. Polym. Res.* **2020**, *3*, 1–7. [CrossRef]
47. Rauzi, J.; Tschirner, U. Enzymatic Glucose and Xylose Production from Paper Mill Rejects. *Recycling* **2022**, *7*, 24. [CrossRef]
48. Abd El-malek, F.; Steinbüchel, A. Post-Synthetic Enzymatic and Chemical Modifications for Novel Sustainable Polyesters. *Front. Bioeng. Biotechnol.* **2021**, *9*, 817023. [CrossRef]
49. Ibrahim, H.G.; Ouiminga, S.K.; Yonli, A.; Sanogo, O.; Daho, T.; Koulidiati, J. Study of temperature fields and heavy metal content in the ash and flue gas produced by the combustion of briquettes coming from paper and cardboard waste. *Recycling* **2018**, *3*, 32. [CrossRef]
50. Albuquerque, M.G.E.; Torres, C.A.V.; Reis, M.A.M. Polyhydroxyalkanoate (PHA) production by a mixed microbial culture using sugar molasses: Effect of the influent substrate concentration on culture selection. *Water Res.* **2010**, *44*, 3419–3433. [CrossRef] [PubMed]
51. Pradhan, S.; Dikshit, P.K.; Moholkar, V.S. Production, ultrasonic extraction, and characterization of poly (3-hydroxybutyrate) (PHB) using Bacillus megaterium and Cupriavidus necator. *Polym. Adv. Technol.* **2018**, *29*, 2392–2400. [CrossRef]
52. Li, R.; Jiang, Y.; Wang, X.; Yang, J.; Gao, Y.; Zi, X.; Zhang, X.; Gao, H.; Hu, N. Psychrotrophic *Pseudomonas mandelii* CBS-1 produces high levels of poly-β-hydroxybutyrate. *Springerplus* **2013**, *2*, 1–7. [CrossRef] [PubMed]
53. Ravuri, J.; Galla, M. Assessment of marine microbial polyhydroxybutyrate by employing high-performance liquid chromatography and nuclear magnetic resonance techniques. *Int. J. Environ. Sci. Technol.* **2022**, *19*, 85–94. [CrossRef]
54. Hablot, E.; Bordes, P.; Pollet, E.; Avérous, L. Thermal and thermo-mechanical degradation of poly (3-hydroxybutyrate)-based multiphase systems. *Polym. Degrad. Stab.* **2008**, *93*, 413–421. [CrossRef]
55. Mahansaria, R.; Dhara, A.; Saha, A.; Haldar, S.; Mukherjee, J. Production enhancement and characterization of the polyhydroxyalkanoate produced by *Natrinema ajinwuensis* (as synonym) ≡ *Natrinema altunense* strain RM-G10. *Int. J. Biol. Macromol.* **2018**, *107*, 1480–1490. [CrossRef] [PubMed]
56. Hassan, M.A.; Bakhiet, E.K.; Ali, S.G.; Hussien, H.R. Production and characterization of polyhydroxybutyrate (PHB) produced by *Bacillus* sp. isolated from Egypt. *J. Appl. Pharm. Sci.* **2016**, *6*, 046–051. [CrossRef]
57. Saranya, V.; Shenbagarathai, R. Production and characterization of PHA from recombinant *E. coli* harbouring phaC1 gene of indigenous *Pseudomonas* sp. LDC-5 using molasses. *Braz. J. Microbiol.* **2011**, *42*, 1109–1118. [CrossRef]
58. Škrbić, Z.; Divjaković, V. Temperature influence on changes of parameters of the unit cell of biopolymer PHB. *Polymer* **1996**, *37*, 505–507. [CrossRef]
59. Sabbagh, F.; Muhamad, I.I. Production of poly-hydroxyalkanoate as secondary metabolite with main focus on sustainable energy. *Renew. Sust. Energ. Rev.* **2017**, *72*, 95–104. [CrossRef]

polymers

Article

Bioconversion of Mixed Alkanes to Polyhydroxyalkanoate by *Pseudomonas resinovornas*: Upcycling of Pyrolysis Oil from Waste-Plastic

Jong-Min Jeon [1], So-Jin Park [1,2], Ye-Seung Son [1], Yung-Hun Yang [3] and Jeong-Jun Yoon [1,*]

1 Green & Sustainable Materials Research and Development Department, Korea Institute of Industrial Technology (KITECH), Cheonan 31056, Korea; j2pco@kitech.re.kr (J.-M.J.); p0215sj@kitech.re.kr (S.-J.P.); yaeseung@kitech.re.kr (Y.-S.S.)
2 School of Industrial technology, University of Science and Technology (UST), Daejeon 34113, Korea
3 Department of Biological Engineering, College of Engineering, Konkuk University, Seoul 05029, Korea; seokor@konkuk.ac.kr
* Correspondence: jjyoon@kitech.re.kr; Tel.: +82-41-589-8266

Abstract: Polyhydroxyalkanoate (PHA) is a biodegradable plastic that can be used to replace petroleum-based plastic. In addition, as a medium-chain-length PHA (mcl-PHA), it can be used to provide elastomeric properties in specific applications. Because of these characteristics, recently, there has been much research on mcl-PHA production using inexpensive biomass materials as substrates. In this study, mcl-PHA producers were screened using alkanes (n-octane, n-decane, and n-dodecane) as sources of carbon. The amount of PHA produced by *Pseudomonas resinovorans* using sole n-octane, n-decane, or n-dodecane was 0.48 g/L, 0.27 g/L, or 0.07 g/L, respectively, while that produced using mixed alkane was 0.74 g/L. As a larger amount of PHA was produced using mixed alkane compared with sole alkane, a statistical mixture analysis was used to determine the optimal ratio of alkanes in the mixture. The optimal ratio predicted by the analysis was a medium with 9.15% n-octane, 6.44% n-decane, and 4.29% n-dodecane. In addition, through several concentration-specific experiments, the optimum concentrations of nitrogen and phosphorus for cell growth and maximum PHA production were determined as 0.05% and 1.0%, respectively. Finally, under the determined optimal conditions, 2.1 g/L of mcl-PHA and 60% PHA content were obtained using *P. resinovorans* in a 7 L fermenter.

Keywords: polyhydroxyalkanoate; mcl-PHA; *Pseudomonas resinovorans*; mixed alkane

Citation: Jeon, J.-M.; Park, S.-J.; Son, Y.-S.; Yang, Y.-H.; Yoon, J.-J. Bioconversion of Mixed Alkanes to Polyhydroxyalkanoate by *Pseudomonas resinovornas*: Upcycling of Pyrolysis Oil from Waste-Plastic. *Polymers* **2022**, *14*, 2624. https://doi.org/10.3390/polym14132624

Academic Editor: Adriana Kovalcik

Received: 8 June 2022
Accepted: 22 June 2022
Published: 28 June 2022

1. Introduction

World plastic production has been increasing every year and reached 368 million tons in 2019 [1]. Among this plastic, 79% is dumped in landfills or the environment, where it takes from around 20 to 600 years to degrade [2]. In addition, the incineration of plastic waste causes serious emissions of greenhouse gases, which accelerates global warming and abnormal climate change [3]. According to the sixth assessment report of the Intergovernmental Panel on Climate Change (IPCC), the usage of petroleum-based plastic materials is still increasing, and the report warned that this will lead to irreversible climate change within around 10 years [4]. As a result, much attention is being paid to the development and use of alternative eco-friendly plastic materials, such as biodegradable plastic from renewable resources.

Polyhydroxyalkanoate (PHA) is a biodegradable plastic that is regarded as a source of alternative materials because it has similar physical properties to petroleum-based plastic [5]. PHA is biosynthesized and accumulated by many bacteria in their cytoplasm as carbon storage materials when they encounter harsh growth conditions in the presence of excess carbon sources, which are classified into two groups, based on the numbers of carbon atoms composed of monomers, with different material properties: short-chain-length PHA (scl-PHA) composed of monomers with 3 to 5 carbon atoms and medium-chain-length PHA (mcl-PHA) composed of 6 to 14 carbon atoms [6]. Among such materials, mcl-PHA has numerous industrial applications, such as coating materials, pressure-delicate glues, and polymer-binding agents in organic solvent-free paints, and also it can be used in a series of biomedical applications [7]. It is produced by PHA-accumulating bacteria such as *Pseudomonas* species, and it has various physicochemical properties depending on the monomer composition and ratio [8,9]. The overall cost of the PHA production process is still expensive; therefore, there have been many studies that have attempted to use various carbon materials such as organic waste, seaweed biomass, animal fat, chitin, and oil waste [5,10]. Among them, pyrolysis oil based on plastic waste also has the potential for use as a carbon source for the production of PHA by micro-organisms. Pyrolysis oil contains aliphatic and aromatic hydrocarbon compounds depending on the catalyst type, plastic waste type, and cracking conditions [11]. The aliphatic hydrocarbon compounds from pyrolysis of plastic waste are predominantly composed of olefins (C20+), and these can be converted to low-carbon alkane or alkene compounds via hydrocracking [12].

PHA production based on alkanes has been studied since the 1980s, and it was discovered that many hydrocarbon-degrading bacteria can degrade and utilize various alkanes as a carbon source to grow and accumulate PHA in vivo [13]. Most alkanes can be utilized through the β-oxidation pathway with conversion to the carboxylic acid formed by alkane monooxygenase and then utilized to acyl-CoA, which can be used as a monomer for PHA accumulation [14]. Therefore, numerous studies have demonstrated the production of PHA from single alkanes by various *Pseudomonas* species, and n-octane was shown to be an economical carbon source for the production of mcl-PHA by *P. oleovorans* [15]. Since then, the maturation of plastic-waste-based pyrolysis oil conversion technology has drawn attention to the use of organic resources containing large amounts of alkane compounds. However, its uses have been limited to applications such as heating oil due to problems such as the fact that it must be additionally purified with a single compound for use in the chemical process. In the case of conversion to biodegradable plastic materials, polyethylene (PE) pyrolysis wax contains a low level of n-octane, and it was demonstrated that *P. oleovorans* is not suitable for producing mcl-PHA from PE pyrolysis wax as a sole carbon source [16]. There are differences in the efficiency of conversion of various alkanes to PHA monomers depending on the affinity with polymerase, and the portion of each alkane in the monomer composition of the produced PHA may be different. Consequently, each alkane affects the monomer composition of the produced PHA, which is an important factor for determining its physical properties [16]. Therefore, if waste oil or plastic pyrolysis oil, which contain mixtures of alkanes, are used for PHA production, it is necessary to find and evaluate a suitable PHA-producing strain for application to the alkane mixture resources.

In this study, *Pseudomonas* species were evaluated for production of mcl-PHA using alkane mixtures containing n-octane, n-decane, and n-dodecane as a sole carbon source, and the optimal condition for increasing PHA production was determined. In addition, the relationship between the ratio of various alkanes and the composition of the produced PHA monomer was determined.

2. Materials and Methods

2.1. Micro-Organism and Culture Conditions

P. fluorescens (ATCC 42821), *P. putida* (ATCC 1751), *P. resinovorans* (ATCC 12498), and *P. stutzeri* (ATCC 1066) were used to screen for producers of mcl-PHA from alkanes as a carbon source. All the strains and cultures were incubated with a working volume of 50 mL in a 250 mL flask at 30 °C for 48 h, and the initial pH was set to 7. Cell growth was monitored by measuring optical density at 600 nm (OD_{600}). All the strains used in this study were precultured in LB medium at 30 °C for 24 h; then, 1% (v/v) of cultured cells were used for inoculation for further study. All components were sterilized via autoclaving for at least 20 min at 121 °C. As a preculture, 1% (v/v) frozen stock was incubated overnight at 30 °C in a shaking incubator in 14 mL round bottom tubes with 5 mL of Luria–Bertani medium broth (LB) (Difco, Detroit, MI, USA). To screen for the optimal strain, each strain was cultured in glucose-free M9 media containing either 10% n-octane, 10% n-decane, or 10% n-dodecane as a sole carbon source at 30 °C for 48 h. To compare cell growth and mcl-PHA production, *P. resinovorans* was cultured in LB medium, cultured in glucose free M9 media with various carbon sources, including 2% (*w*/*w*) of glucose, 10% (*v*/*v*) of mixed alkane (n-octane, n-decane, and n-dodecane of the same volume), and 10% of each alkane.

The amount of nitrogen and phosphorus needed to maximize PHA production was determined using various concentrations in the medium and the determined mixed alkane. The effect of nitrogen and phosphorus concentration (0 to 1.0% and 0 to 5%, respectively) on *P. resinovorans* growth, PHA accumulation, and its monomer composition was studied. Finally, *P. resinovorans* was cultured in a 7 L bioreactor (GF Fermentech, Cheongju, Korea) with a working volume of 3 L at the optimized culture condition. It was operated at 30 °C for 72 h, with a stirring speed of 300 rpm and 3 V/min of gas flow for aeration.

2.2. mcl-PHA Recovery from Biomass

Methyl ethyl ketone (MEK), methanol, and chloroform were used as extraction and purification solvents for mcl-PHA recovery. MEK was added to the lyophilized cell and the mixture maintained at 60 °C for 6 h in sealed screw-top test tubes. Then, the tubes were briefly vortexed and incubated at room temperature. After mixing, the tubes were centrifuged for 10 min at 2500× *g*. The supernatant was transferred to a sealed screw-top tube, and then 3 volumes of methanol were added to remove the lipids remaining in the solvent. The PHA was dissolved with chloroform by heating at 100 °C in a heat block for 4 h. The tubes were incubated at room temperature until the solvent evaporated.

2.3. Design of Experiments and Mixture Analysis

To develop a strategy for optimizing cell growth and PHA production, a mixture analysis model of three alkanes (n-octane, n-decane, and n-dodecane) was developed and populated using a standard mixture-analysis methodology and the Minitab V19 program. For the design of mixture-analysis experiments to populate the model, we used a simplex lattice method. The degree of the lattice for this mixture analysis was 2; therefore, the experimental design contained the set of all 10 combinations. All experiments were performed using 50 mL cultures with 20% total alkane content, and the cultures were cultured at 30 °C for 48 h in duplicate. To plot mixture contours, a mixture regression using the model-fitting method was applied with full quadratic component terms initially included. In the data analysis, the coefficients with p value below 0.1 were used as parameters.

2.4. Characterization of Obtained mcl-PHA

The quantity and composition of PHA were determined by gas chromatography (GC) and gas chromatography-mass spectrometry (GC-MS), using a slight modification of a method described previously [17]. For analysis, the microbial culture after the completion of growth was centrifuged at 10,000× *g* for 30 min, washed with deionized water two times, and suspended in 1 mL of water. The suspended samples were subjected to lyophilization, and the freeze-dried cells from each experiment were subjected to methanolysis. A weighed sample was placed in a Teflon-stoppered glass vial, and 1 mL chloroform and 1 mL methanol/sulfuric acid (85:15 *v*/*v*) were added to the vial. The samples were incubated at 100 °C for 2 h, cooled to room temperature, and incubated on ice for 10 min. After adding 1 mL of ice-cold water, the samples were mixed thoroughly using a vortex for 1 min and then centrifuged at 2000× *g*. The organic phase (bottom) was carefully extracted using a pipette and was moved to clean borosilicate glass tubes. A 2 μL portion of the organic phase of these samples was then injected into a gas chromatograph (6090N, Agilent Technologies, Santa Clara, CA, USA) using a flame ionization detector (FID) and a 30 m × 250 μm DB-FFAP capillary column with hydrogen as the carrier gas. The inlet of the gas chromatograph was maintained at 250 °C, and the oven was held at 80 °C for 5 min, heated to 220 °C at 20 °C min^{-1}, and then held at 220 °C for 5 min. Peak detection was performed using a flame ionization detector, which was maintained at 300 °C. The fatty acid content was analyzed via GC–MS chromatography (Perkin Elmer Clarus 500, Waltham, MA, USA) according to the modified method previously reported in [17]. About 1 uL of methanolized sample was injected into the Clarus 680 GC-MS equipped with triple axis detector carrying Elite 5 ms column (30 mm length × 0.25 mm internal diameter × 0.25 mm film) at a split ratio of 10:1 with column flow 1.0 mL min^{-1}. The injector temperature was set at 280 °C while the oven and column temperatures were programmed as 10 °C for 1 min, then increased to 130 °C at 11 °C min^{-1}, held for 2 min, and increased to 310 °C at 10 °C min^{-1}, and held for 10 min. Helium was used as carrier gas at 47.3 mL min^{-1} and 0.40 bar pressure. Mass spectra were acquired at 1250 scan speed using electron-impact energy of 70 eV at 200Uc ion source and 280 °C interface temperatures, respectively. Complete instrument control was available through TurboMass™ driver. NIST/EPA/NIH library was used to predict the methylated PHAs and their corresponding mass ion. Statistical analysis was carried out through one-way ANOVA, where $p < 0.05$ was considered to be statistically significant.

To study the melting behavior of synthesized polymer, differential scanning calorimetry (DSC) analysis were performed by Discovery DSC (TA Instruments, Bellefonte, PA, USA) in the temperature range from −80 to +100 °C. The glass transition temperature (Tg) was determined at a heating rate of 20 °C/min. In this study, Tg was taken as the midpoint of the step-transition. The weight-average molecular weight (Mw), number-average molecular weight (Mn), and polydispersity (Mw/Mn) were determined by gel permeation chromatography (GPC) conducted in THF solution at 35 °C and a flow rate of 1 mL/min. A 10 µL sample in THF at a concentration of 1% *w*/*v* was injected. Polystyrene standards with narrow molecular-mass distribution were used to generate a calibration curve.

3. Results and Discussion

3.1. Screening of Alkane-Based PHA-Producing Strains

After the discovery in 1983 that *P. oleovorans* could produce mcl-PHA using n-octane, various *Pseudomonas* species were found to be capable of PHA production using alkane compounds [18–20]. However, few studies have reported on the use of various alkane compounds as a carbon source because most studies were focused on the use of n-octane. The composition of pyrolysis oil from polyethylene-based waste plastics is determined by its process conditions, and C8- to C22-saturated hydrocarbons are predominantly included in the case of pyrolysis oil received from the Korea Institute of Industrial Technology (Supplementary Figure S1). To find a suitable strain that is capable of mcl-PHA production using alkanes, *P. fluorescens*, *P. resinovorans*, *P. stutzeri*, and *P. putida* were selected and cultured in minimal media with 10 (*v*/*v*)% of either n-octane, n-decane, or n-dodecane as the sole carbon source. Each species showed different growth activity, and *P. resinovorans* showed better growth than the others (Figure 1). In addition, *P. resinovorans* was cultured in range of 1 to 50% of n-octane to investigate the relationship between the concentration of alkane and growth. The highest cell growth (1.34 g/L), PHA production (0.31 g/L), and PHA content (approximately 20 (*w*/*w*)%) were produced with 20 (*v*/*v*)% of n-octane (Figure 2). Meanwhile, with more than 30% n-octane in the medium, the cell dry weight (CDW), PHA amount, and PHA content decreased sharply to 0.68 g/L, 0.09 g/L, and 13.45%, respectively, which is attributed to the oxygen rate being rapidly reduced as the oil and liquid ratio increased [21].

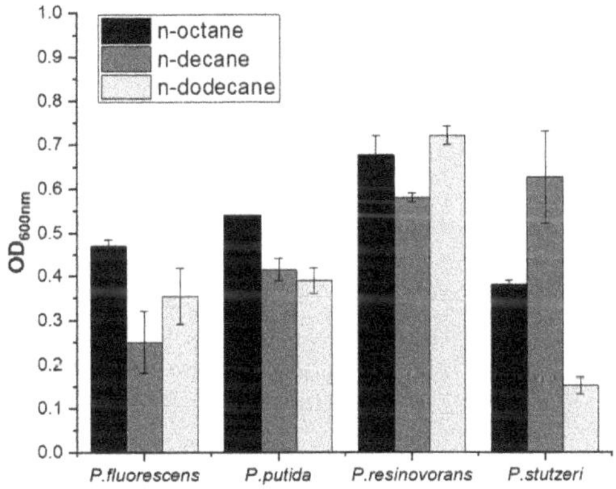

Figure 1. Cell growth of *Pseudomonas* species using n-octane, n-decane, and n-dodecane as a sole carbon source. Each alkane is present as 10% (*v*/*v*) in minimal medium.

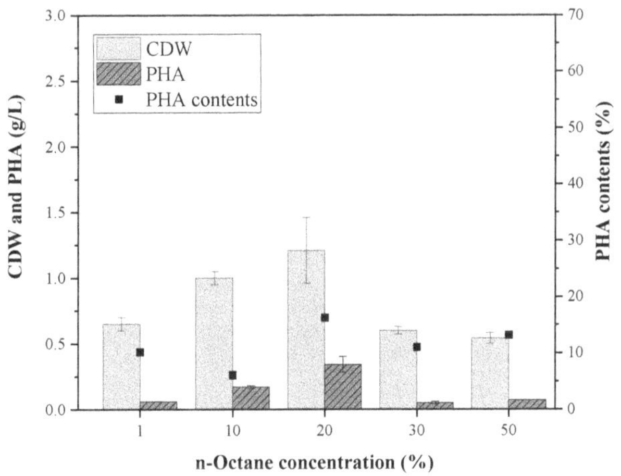

Figure 2. Cell growth and PHA production by *P. resinovorans* in the range of n-octane.

P. resinovorans was selected for the recycling of pyrolysis oil, which contains various olefin compounds, because it is able to utilize alkanes and produces more mcl-PHA than the other *Pseudomonas* species, including *P. oleovorans*. In addition, to maximize the production of mcl-PHA, the optimum concentration of alkanes in the medium was determined to be 20%.

3.2. PHA Production by P. resinovorans Using Various Types of Carbon

Microbial metabolism for the accumulation of PHA occurs in two different ways: (1) short-chain carbon sources are usually utilized by biosynthesis for accumulation of PHA by the phaABC pathway, through acetyl-CoA to acetoacetyl-CoA by phaA (thiolase), while phaB (oxidoreductase) and phaC (polymerase) are mainly involved in intracellular processes; and (2) long-chain carbon sources, such as fatty acids or alkanes, are utilized through the β-oxidation pathway, and various acyl-CoA transferases convert to inter metabolites, including the mcl-PHA precursor [17]. *P. resinovorans* has both metabolisms for PHA accumulation; therefore, there is a need to evaluate the PHA production of different carbon sources. *P. resinovorans* was cultured using 2% of glucose, 20 (*v*/*v*)% of each alkane (n-octane, n-decane, or n-dodecane), and 20 (*v*/*v*)% of mixed alkane to compare PHA production and the monomer composition in accordance with the carbon source (Table 1). With glucose as the carbon source, 1.56 g/L of PHA was produced that contained 1.47% 3-hydroxyhexanoate (3HHx), 10.60% 3-hydroxyoctanoate (3HO), and 87% 3-hydroxyoctanoate (3HD). Among the carbon sources with 20% of a single alkane (n-octane/n-decane/n-docane), the best cell growth and PHA production were found using n-octane as the sole carbon source. This produced a CDW of 1.74 g/L, PHA of 0.37 g/L, and PHA content of 22.4% that contained 4.59% of 3HHx, 86.05% of 3HO, and 9.36% of 3HD. When the mixed alkane was used, 0.69 g/L of PHA was produced that contained 11.77% of 3HHx, 74.39% of 3HO, and 13.84% of 3HD. Although cell growth was not the highest, the mixed alkane showed an approximately 50% increase in PHA production compared with using n-octane.

Table 1. Cell growth, PHA production, and monomer composition by *P. resinovorans* in various carbon sources.

Substrates	mol% of 3HHx	mol% of 3HO	mol% of 3HD	CDW (g/L)	PHA (g/L)	PHA Content (%)
2% of glucose	1.47 ± 0.03	10.60 ± 1.76	87.93 ± 3.22	2.68 ± 0.09	1.56 ± 0.04	57.76 ± 2.35
20% of mixed alkane	11.77 ± 0.26	74.39 ± 3.15	13.84 ± 2.12	1.46 ± 0.03	0.69 ± 0.09	47.46 ± 1.12
20% of n-octane	4.59 ± 0.12	86.05 ± 4.33	9.36 ± 0.06	1.74 ± 0.07	0.37 ± 0.01	22.35 ± 0.08
20% of n-decane	-	40.42 ± 1.43	61.18 ± 6.34	1.23 ± 0.11	0.24 ± 0.01	12.99 ± 0.06
20% of n-dodecane	-	-	97.50 ± 0.05	0.39 ± 0.09	0.08 ± 0.04	22.66 ± 0.02

3HHx: 3-hydroxyhexanoate, 3HO: 3-hydroxyoctanoate, 3HD: 3-hydroxydecanoate.

3.3. Mixture Analysis of Alkanes (n-octane, n-decane, and n-dodecane) as a Carbon Source for mcl-PHA Production

To investigate the effect of the three alkanes on cell growth, PHA production, and PHA content, *P. resinovorans* was grown as 10 culture compositions based on the mixture analysis model, as described in "Materials and methods" (Table 2). For each culture, we measured the cell growth, amount of PHA, and PHA content. The results of the statistical analysis were shown by contour plots and the predicted highest cell growth, PHA titer, and PHA content using *P. resinovorans* were determined (Figure 3).

Table 2. Monomer composition of 10 conditions by mixture analysis.

ID #	n-octane (*v*/*v*%)	n-decane (*v*/*v*%)	n-dodecane (*v*/*v*%)	mol% of 3HHx	mol% of 3HO	mol% of 3HD	PHA (g/L)	PHA Content (%)
1	20	0	0	3.55	69.44	27.01	0.42	21.69
2	10	10	0	5.43	59.41	35.16	1.25	34.72
3	10	0	10	4.14	67.09	28.81	0.63	26.53
4	0	20	0	-	-	100	0.03	2.21
5	0	10	10	-	44.88	55.12	0.21	14.26
6	0	0	20	-	-	-	-	-
7	6.66	6.66	6.66	6.02	84.32	9.67	1.68	41.78
8	13.33	3.33	3.33	10.28	80.18	9.54	2.86	50.46
9	3.33	13.33	3.33	41.03	32.21	47.28	1.64	40.69
10	3.33	3.33	13.33	4.43	80.99	14.58	1.17	29.33

3HHx: 3-hydroxyhexanoate, 3HO: 3-hydroxyoctanoate, 3HD: 3-hydroxydecanoate.

Contour plots for cell growth predicted the best ratio to be when n-octane, n-decane, and n-dodecane are 10.36%, 5.48%, and 4.5%, respectively, which produced CDW of 5.28 g/L (Figure 3A). In the case of 9.15% of n-octane, 6.44% of n-decane, and 4.29% of n-dodecane, the PHA titer was 2.39 g/L of PHA (Figure 3B). In addition, the PHA content calculated by the amount of PHA contained in the cell was predicted to be a maximum of 52.33% content (*w*/*v*) when n-octane, n-decane, and n-dodecane are 8.63%, 6.86%, and 4.50%, respectively (Figure 3C). The optimal alkane composition for total PHA production predicted by the model is a combination of n-octane, n-decane, and n-dodecane, rather than pure alkane or a combination of two different alkanes [16,22,23]. The reason why a combination of alkane compounds in the medium is beneficial is not currently clarified, but is estimated to be due to the better activities of alkane hydroxylase of *P. resinovorans* when the mixed alkane exists [24,25].

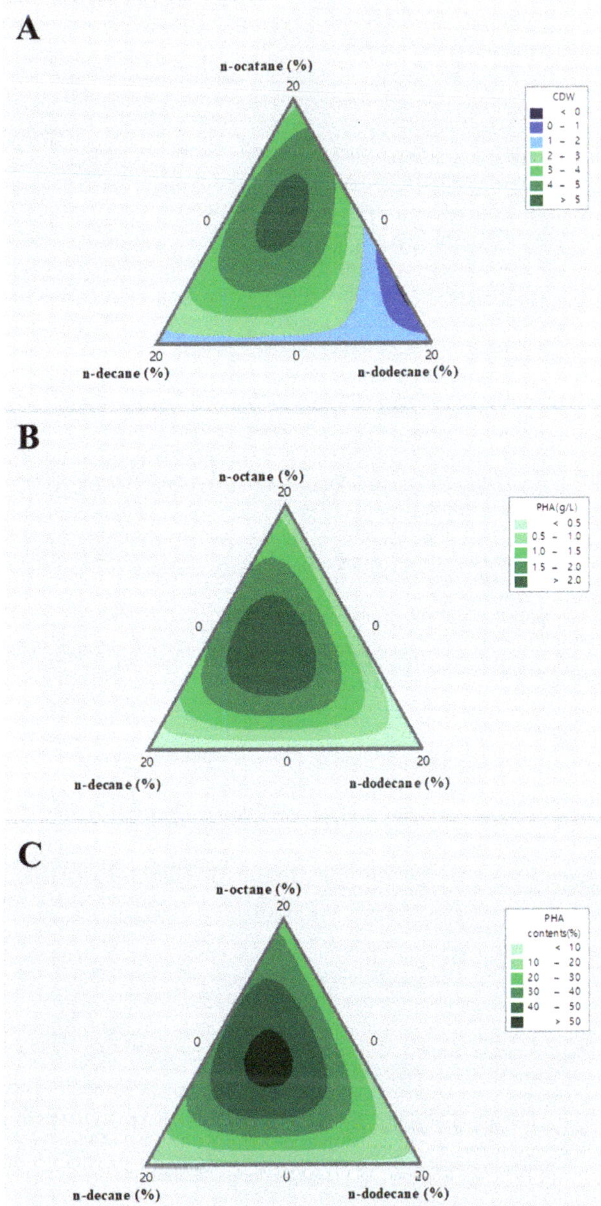

Figure 3. Mixture contour plots of mixed alkane composition for PHA production. The optimal composition ratio between n-octane, n-decane, and n-dodecane was determined as 20% (v/v) as a total concentration. (**A**): CDW (g/L), (**B**): PHA (g/L), and (**C**): PHA content (%) of different mixed alkane compositions.

3.4. Jar-Scale Fermentation for mcl-PHA Production in Media-Optimized Conditions

Carbon, nitrogen, and phosphorus are essential nutrients for growth, and their limitation can trigger mcl-PHA production in micro-organisms. However, the specific molecular mechanisms that drive this synthesis in *Pseudomonas* species under unfavorable growth conditions remain poorly understood. Therefore, it is necessary to determine the optimal concentration of nitrogen and phosphorus, because mcl-PHA production is related to their concentration when alkanes are used as carbon sources. To determine the optimal conditions, nitrogen concentrations were observed from 0% to 1.0% and phosphorus concentrations from 0% to 3.0% (Figure 4). When the concentration of nitrogen in the culture medium increased from 0 to 0.05%, PHA production tended to increase accordingly. The maximized PHA produced in 0.05% nitrogen was 1.31 g/L, but in the medium where the nitrogen concentration was more than 0.1%, PHA production decreased (Figure 4A). When the concentration of phosphorus in the medium increased from 0 to 1%, the production of PHA increased as well (Figure 4B). The maximum PHA production was 1.14 g/L at a concentration of 1% phosphorus. However, PHA production tended to decrease when the concentration of phosphorus was more than 2%. These results suggest that the optimal phosphorus concentration in the medium is 1.0%.

Using the optimized alkane combination and nitrogen and phosphorus concentration, *P. resinovorans* was cultured in a 7 L jar fermenter (3 L of working volume). The CDW, amount of PHA, and PHA content reached 3.5 g/L, 2.1 g/L, and 60%, respectively (Figure 5). Previous reports focused on *P. oleovorans* for the production of mcl-PHA with n-octane, and these show production of up to 16.8 g/L of PHA by fed-batch culture, while n-decane and n-dodecane were not considered because of low PHA production (Table 3). Compared with other *Pseudomonas* species, it was proven that use of *P. resinovorans* is more effective for production of mcl-PHA when mixed alkane is used as a carbon source.

Table 3. mcl-PHA production using various alkanes by Pseudomonas species.

Organism	Carbon Source	CDW	Amt of PHA	Monomer Composition	Cultivation Mode	Reference
P. oleovorans	n-octane	-	13.4 g/100 g of CDW	3HHx, 3HO	Batch	[22]
P. oleovorans	n-decane	-	5.1 g/100 g of CDW	3HO, 3HD	Batch	[22]
P. oleovorans	n-dodecane	-	-	-	Batch	[22]
P. aeruginosa GL-1	Pyrolysis oil (C8 to C27 alkane)	0.39 (g/L)	0.07 (g/L)	3HHx, 3HO, 3HN, 3HD, 3HUD, 3HDD, 3HTD	Batch	[16]
P. oleovorans	Pyrolysis oil (C8 to C27 alkane)	-	-	-		[16]
P. resinovorans	octanoic acid	4.6 (g/L)	0.4 (g/L)	3HB, 3HHx, 3HO, 3HD	Continuous	[23]
P. resinovorans	n-octane, n-decane and n-dodecane	3.5 (g/L)	2.1 (g/L)	3HHx, 3HO, 3HD	Batch	In this study

CDW: final cell dry weight, PHA formed: final concentration of PHA, 3HB: 3-hydroxybutyrate, 3HHx: 3-hydroxyhexanoate, 3HO: 3-hydroxyoctanoate, 3HN: 3-hydroxynonanoate, 3HD: 3-hydroxydecanoate, 3HUD: 3-hydroxyundecanoate, 3HDD: 3-hydroxydodecanoate and 3HTD: 3-hydroxytetradecanoate.

3.5. Physical Properties of Produced mcl-PHA by P. resinovorans

The thermal properties of the produced mcl-PHA were determined by means of DSC (Table 4 and Supplementary Figure S2). The Tg of the samples ranged from −43.44 to −34.29, which can be regarded as values typical for this type of PHA. The sample also showed a Tm in the range of 39.62 with a ΔHm of 13.2 J/g. In addition, these values are typical for mcl-PHAs and indicate classical rubber- to latex-like characteristics [26]. The Mw, Mn, and polydispersity (Mw/Mn) were similar for the copolyesters, despite the differences in their monomer compositions. The Mn and Mw were 267,649 and 630,526 Da, respectively.

Table 4. Physical properties of mcl-PHA from *P. resinovorans* using mixed alkane.

T_g (°C)	T_m (°C)	ΔHm (J/g)	Mw (Da)	Mn (Da)	PDI
−38.9	48.2	13.2	267,649	630,526	2.36

In addition, Z average (Mz) and Z + 1 average (Mz + 1) molar mass were determined as 1,110,977 and 1,729,603 Da, respectively. The polydispersity index (PDI) was calculated (PDI = Mw/Mn) as 2.36, more or less similar to that of the PHA value of mcl-PHAs biosynthesized by other known *Pseudomonas* species.

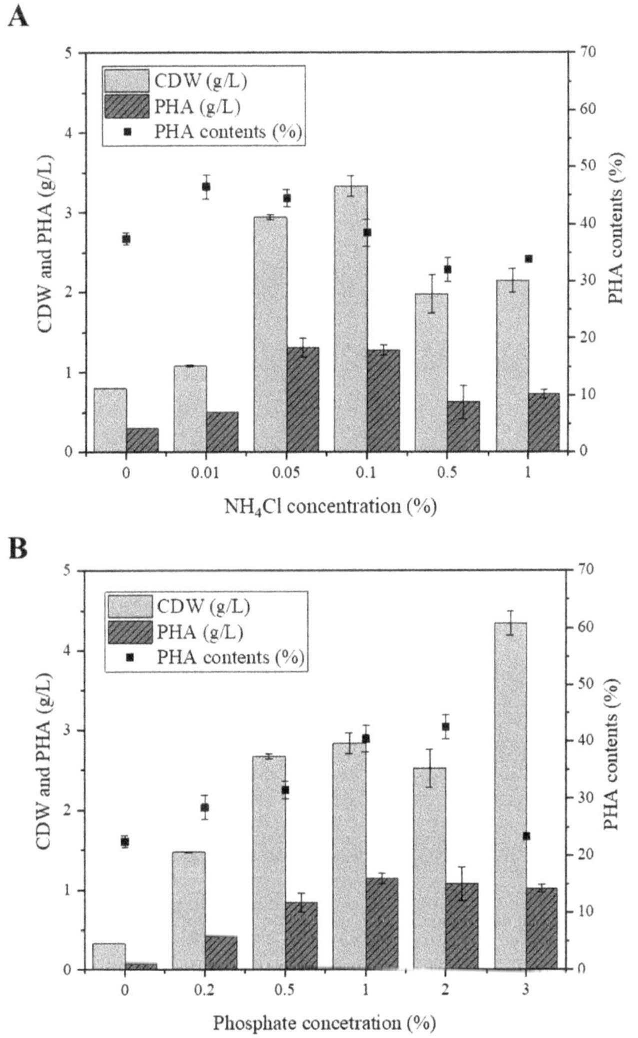

Figure 4. Effects of nitrogen and phosphorus limitation on PHA production. (**A**): CDW, PHA titer, and PHA content (%) in range of 0 to 1% nitrogen concentration. (**B**): CDW, PHA titer, and PHA content (%) in range of 0 to 3% phosphate concentration.

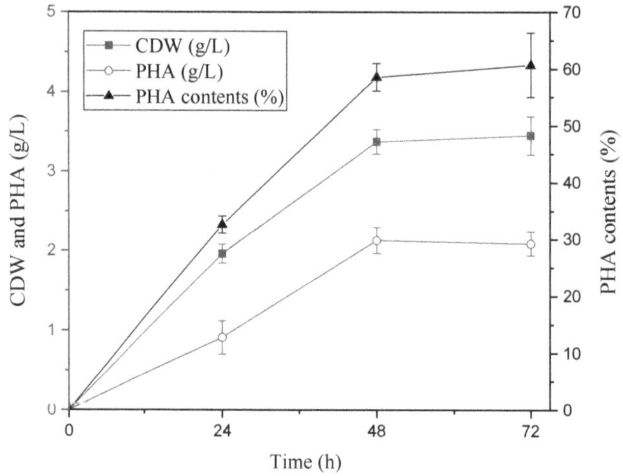

Figure 5. PHA production by *P. resinovorans* in optimized condition. Batch culture was performed in 7 L scale jar fermenter (3 L as working volume) with optimized culture medium.

4. Conclusions

Recent studies have shown that the use of renewable carbon sources can produce a significant reduction in actual production costs. A very interesting research topic would be for industrially produced waste oil or oil produced by pyrolysis of waste plastics to be used as a fermentation substrate for production of PHA, which is a promising bioplastic. Therefore, it is important to find PHA-producing micro-organisms and establish culture conditions in accordance with the carbon composition of waste oil or oil from waste plastics.

In this study, the ability of *P. resinovorans* to produce PHA from alkanes (n-octane, n-decane, and n-dodecane), especially mixed alkane, was identified, and a model was built to improve mcl-PHA production. Compared with other *Pseudomonas* species, *P. resinovorans* was much more effective in utilizing mixed alkane, and it can be an attractive microbial host for production of PHA. Although the regulation of PHA production based on alkane utilization has not been discovered, the results of this study are meaningful for use as basic data for the application of various renewable carbon resources containing alkanes.

Supplementary Materials: The following supporting information can be downloaded at: https://www.mdpi.com/article/10.3390/polym14132624/s1, Figure S1: Pyrolysis oil of waste-plastic and its composition, Figure S2: DSC analysis of purified PHA produced by *P. resinovorans* using mixed alkane.

Author Contributions: J.-M.J.: Writing—Original Draft, Investigation, Data Curation. S.-J.P.: Methodology, Investigation, Data Curation. Y.-S.S.: Data Curation. Y.-H.Y.: Writing—Review and Editing, Methodology. J.-J.Y.: Conceptualization, Supervision, Project administration, Funding acquisition. All authors have read and agreed to the published version of the manuscript.

Funding: This work was supported by the National Research Foundation of Korea (NRF) grant funded by the Korean government (Ministry of Science and ICT or MSIT; NRF-2020R1A2C2102381) and the Korea Institute of Industrial Technology (KITECH) through Development of eco-friendly production system technology for a total periodic resource cycle (21-0140).

Institutional Review Board Statement: Not applicable.

Informed Consent Statement: Not applicable.

Conflicts of Interest: The authors declare no conflict of interest.

References

1. Garside, M. *Global Plastic Production 1950–2019*; Statista: New York, NY, USA, 2020. Available online: https://www.statista.com/statistics/282732/global-production-of-plastics-since-281950/ (accessed on 11 January 2022).
2. Mazhandu, Z.S.; Muzenda, E.; Mamvura, T.A.; Belaid, M.; Nhubu, T. Integrated and Consolidated Review of Plastic Waste Management and Bio-Based Biodegradable Plastics: Challenges and Opportunities. *Sustainability* **2020**, *12*, 8360. [CrossRef]
3. Shen, M.C.; Huang, W.; Chen, M.; Song, B.; Zeng, G.M.; Zhang, Y.X. (Micro)plastic crisis: Unignorable contribution to global greenhouse gas emissions and climate change. *J. Clean Prod.* **2020**, *254*, 120138. [CrossRef]
4. Michelozzi, P.; De' Donato, F. IPCC Sixth Assessment Report: Stopping climate change to save our planet. *Epidemiol. Prev.* **2021**, *45*, 227–229. [CrossRef] [PubMed]
5. Bhatia, S.K.; Otari, S.V.; Jeon, J.M.; Gurav, R.; Choi, Y.K.; Bhatia, R.K.; Pugazhendhi, A.; Kumar, V.; Banu, J.R.; Yoon, J.J.; et al. Biowaste-to-bioplastic (polyhydroxyalkanoates): Conversion technologies, strategies, challenges, and perspective. *Bioresour. Technol.* **2021**, *326*, 124733. [CrossRef] [PubMed]
6. Park, S.J.; Kim, T.W.; Kim, M.K.; Lee, S.Y.; Lim, S.C. Advanced bacterial polyhydroxyalkanoates: Towards a versatile and sustainable platform for unnatural tailor-made polyesters. *Biotechnol. Adv.* **2012**, *30*, 1196–1206. [CrossRef]
7. Koller, M. Advances in Polyhydroxyalkanoate (PHA) Production, Volume 2. *Bioengineering* **2020**, *7*, 24. [CrossRef]
8. Chen, G.Q. A microbial polyhydroxyalkanoates (PHA) based bio- and materials industry. *Chem. Soc. Rev.* **2009**, *38*, 2434–2446. [CrossRef]
9. Sudesh, K.; Abe, H.; Doi, Y. Synthesis, structure and properties of polyhydroxyalkanoates: Biological polyesters. *Prog. Polym. Sci.* **2000**, *25*, 1503–1555. [CrossRef]
10. Suriyamongkol, P.; Weselake, R.; Narine, S.; Moloney, M.; Shah, S. Biotechnological approaches for the production of polyhydroxyalkanoates in microorganisms and plants—A review. *Biotechnol. Adv.* **2007**, *25*, 148–175. [CrossRef]
11. Papari, S.; Bamdad, H.; Berruti, F. Pyrolytic Conversion of Plastic Waste to Value-Added Products and Fuels: A Review. *Materials* **2021**, *14*, 2586. [CrossRef]
12. Kumar, H.; Gilbert, F.F. A Generalized Mechanistic Kinetic Model for the Hydroisomerization and Hydrocracking of Long-Chain Paraffins. *Ind. Eng. Chem. Res.* **2007**, *46*, 4075–4090. [CrossRef]
13. De Smet, M.J.; Eggink, G.; Witholt, B.; Kingma, J.; Wynberg, H. Characterization of intracellular inclusions formed by *Pseudomonas oleovorans* during growth on octane. *J. Bacteriol.* **1983**, *154*, 870–878. [CrossRef]
14. Jeon, J.M.; Kim, H.J.; Bhatia, S.K.; Sung, C.; Seo, H.M.; Kim, J.H.; Park, H.Y.; Lee, D.; Brigham, C.J.; Yang, Y.H. Application of acetyl-CoA acetyltransferase (AtoAD) in *Escherichia coli* to increase 3-hydroxyvalerate fraction in poly(3-hydroxybutyrate-co-3-hydroxyvalerate). *Bioproc. Biosyst. Eng.* **2017**, *40*, 781–789. [CrossRef] [PubMed]
15. Durner, R.; Witholt, B.; Egli, T. Accumulation of Poly[(R)-3-hydroxyalkanoates] in *Pseudomonas oleovorans* during growth with octanoate in continuous culture at different dilution rates. *Appl. Environ. Microbiol.* **2000**, *66*, 3408–3414. [CrossRef] [PubMed]
16. Guzik, M.W.; Kenny, S.T.; Duane, G.F.; Casey, E.; Woods, T.; Babu, R.P.; Nikodinovic-Runic, J.; Murray, M.; O'Connor, K.E. Conversion of post consumer polyethylene to the biodegradable polymer polyhydroxyalkanoate. *Appl. Microbiol. Biot.* **2014**, *98*, 4223–4232. [CrossRef]
17. Jeon, J.M.; Brigham, C.J.; Kim, Y.H.; Kim, H.J.; Yi, D.H.; Kim, H.; Rha, C.; Sinskey, A.J.; Yang, Y.H. Biosynthesis of poly(3-hydroxybutyrate-co-3-hydroxyhexanoate) (P(HB-co-HHx)) from butyrate using engineered *Ralstonia eutropha*. *Appl. Microbiol. Biotechnol.* **2014**, *98*, 5461–5469. [CrossRef]
18. Diard, S.; Carlier, J.P.; Ageron, E.; Grimont, P.A.D.; Langlois, V.; Guerin, P.; Bouvet, O.M.M. Accumulation of poly(3-hydroxybutyrate) from octanoate, in different *Pseudomonas* belonging to the rRNA homology group I. *Syst. Appl. Microbiol.* **2002**, *25*, 183–188. [CrossRef]
19. Hazenberg, W.; Witholt, B. Efficient production of medium-chain-length poly(3-hydroxyalkanoates) from octane by *Pseudomonas oleovorans*: Economic considerations. *Appl. Microbiol. Biot.* **1997**, *48*, 588–596. [CrossRef]
20. Silva, J.B.; Pereira, J.R.; Marreiros, B.C.; Reis, M.A.; Freitas, R.F. Microbial production of medium-chain length polyhydroxyalkanoates. *Process. Biochem.* **2021**, *102*, 393–407. [CrossRef]
21. Preusting, H.; van Houten, R.; Hoefs, A.; van Langenberghe, E.K.; Favre-Bulle, O.; Witholt, B. High cell density cultivation of *Pseudomonas oleovorans*: Growth and production of poly (3-hydroxyalkanoates) in two-liquid phase batch and fed-batch systems. *Biotechnol. Bioeng.* **1993**, *41*, 550–556. [CrossRef]
22. Lageveen, R.G.; Huisman, G.W.; Preusting, H.; Ketelaar, P.; Eggink, G.; Witholt, B. Formation of Polyesters by *Pseudomonas oleovorans*: Effect of Substrates on Formation and Composition of Poly-(R)-3-Hydroxyalkanoates and Poly-(R)-3-Hydroxyalkenoates. *Appl. Environ. Microbiol.* **1988**, *54*, 2924–2932. [CrossRef] [PubMed]
23. Ramsay, B.A.; Saracovan, I.; Ramsay, J.A.; Marchessault, R.H. Continuous Production of Long-Side-Chain Poly-beta-Hydroxyalkanoates by *Pseudomonas oleovorans*. *Appl. Environ. Microbiol.* **1991**, *57*, 625–629. [CrossRef] [PubMed]
24. Smits, T.H.; Balada, S.B.; Witholt, B.; van Beilen, J.B. Functional analysis of alkane hydroxylases from gram-negative and gram-positive bacteria. *J. Bacteriol.* **2002**, *184*, 1733–1742. [CrossRef] [PubMed]
25. Grund, A.; Shapiro, J.; Fennewald, M.; Bacha, P.; Leahy, J.; Markbreiter, K.; Nieder, M.; Toepfer, M. Regulation of alkane oxidation in *Pseudomonas putida*. *J. Bacteriol.* **1975**, *123*, 546–556. [CrossRef]
26. Kathiraser, Y.; Aroua, M.K.; Ramachandran, K.B.; Tan, I.K.P. Chemical characterization of medium-chain-length polyhydroxyalkanoates (PHAs) recovered by enzymatic treatment and ultrafiltration. *J. Chem. Technol. Biot.* **2007**, *82*, 847–855. [CrossRef]

polymers

Article

Combination of Hypotonic Lysis and Application of Detergent for Isolation of Polyhydroxyalkanoates from Extremophiles

Ivana Novackova [1], Xenie Kourilova [1], Katerina Mrazova [2], Petr Sedlacek [1], Michal Kalina [1], Vladislav Krzyzanek [2], Martin Koller [3,4] and Stanislav Obruca [1,*]

1 Faculty of Chemistry, Brno University of Technology, Purkynova 118, 612 00 Brno, Czech Republic; xcnovackova@fch.vut.cz (I.N.); xckourilovax@fch.vut.cz (X.K.); sedlacek-p@fch.vut.cz (P.S.); kalina-m@fch.vut.cz (M.K.)
2 Institute of Scientific Instruments of the Czech Academy of Sciences, v.v.i., Kralovopolska 147, 612 64 Brno, Czech Republic; mrazova@isibrno.cz (K.M.); krzyzanek@isibrno.cz (V.K.)
3 Research Management and Service, c/o Institute of Chemistry, NAWI Graz, University of Graz, Heinrichstrasse 28/IV, 8010 Graz, Austria; martin.koller@uni-graz.at
4 ARENA—Arbeitsgemeinschaft für Ressourcenschonende & Nachhaltige Technologien, Inffeldgasse 21b, 8010 Graz, Austria
* Correspondence: obruca@fch.vut.cz; Tel.: +420-541-149-354

Abstract: Production of polyhydroxyalkanoates (PHA), microbial biopolyesters, employing extremophilic microorganisms is a very promising concept relying on robustness of such organisms against microbial contamination, which provides numerous economic and technological benefits. In this work, we took advantage of the natural susceptibility of halophilic and thermophilic PHA producers to hypotonic lysis and we developed a simple and robust approach enabling effective isolation of PHA materials from microbial cells. The method is based on the exposition of microbial cells to hypotonic conditions induced by the diluted solution of sodium dodecyl sulfate (SDS) at elevated temperatures. Such conditions lead to disruption of the cells and release of PHA granules. Moreover, SDS, apart from its cell-disruptive function, also solubilizes hydrophobic components, which would otherwise contaminate PHA materials. The purity of obtained materials, as well as the yields of recovery, reach high values (values of purity higher than 99 wt.%, yields close to 1). Furthermore, we also focused on the removal of SDS from wastewater. The simple, inexpensive, and safe technique is based on the precipitation of SDS in the presence of KCl. The precipitate can be simply removed by decantation or centrifugation. Moreover, there is also the possibility to regenerate the SDS, which would substantially improve the economic feasibility of the process.

Keywords: polyhydroxyalkanoate (PHA); sodium dodecyl sulfate (SDS); PHA isolation; extremophiles; halophiles; thermophiles; *Halomonas halophila*; *Schlegelella thermodepolymerans*

Citation: Novackova, I.; Kourilova, X.; Mrazova, K.; Sedlacek, P.; Kalina, M.; Krzyzanek, V.; Koller, M.; Obruca, S. Combination of Hypotonic Lysis and Application of Detergent for Isolation of Polyhydroxyalkanoates from Extremophiles. *Polymers* **2022**, *14*, 1761. https://doi.org/10.3390/polym14091761

Academic Editor: Shashi Kant Bhatia

Received: 21 March 2022
Accepted: 23 April 2022
Published: 26 April 2022

1. Introduction

Polyhydroxyalkanoates (PHAs) are microbial storage polyesters, which are accumulated by various prokaryotic microorganisms. Apart from their primary storage function for carbon, energy, and reduction equivalents, PHAs also significantly enhance the stress robustness of microorganisms and accumulation of them is probably a part of adaptation to some extreme conditions such as high salinity [1]. PHAs are, therefore, widespread metabolites among extremophilic prokaryotes [2]. Moreover, they represent a very promising ecologically friendly alternative to traditional petrochemical polymers–PHAs are completely biodegradable and compostable materials, which can be produced from totally renewable resources including also waste products stemming from food production and other industries. Their properties and application fields are very similar to petrochemical plastics such as polypropylene [3].

Nevertheless, the main problem prohibitive for large-scale production of PHAs is their high production cost. There are several strategies how to overcome this limitation. One of them is the employment of extremophiles as PHA-producing chassis. The main advantage of extremophiles is their capability of growth and production of various metabolites including but not limited to PHA under conditions, which reduce or even eliminate the risk of contamination of the process by common mesophilic microflora, e.g., high or low temperature, high salinity, extreme pH-values, etc. Due to their robustness against unfavorable microbial contamination, which can destroy whole fermentation batches, biotechnological processes using extremophilic microorganisms can be operated under reduced requirements for sterility, or eventually even completely without sterilization of the cultivation media or equipment. Consequently, it has a significantly positive impact on the energy and economic balance of these processes. Moreover, high tolerance of the process against contamination allows running the cultivation process in highly productive continuous or semi-continuous mode. Therefore, employing extremophilic microorganisms in biotechnology is nowadays a modern trend, which is called the concept of industrial biotechnology of the next-generation ("Next-Generation Industrial Biotechnology", NGIB) [4].

In this context, the employment of extremophilic microorganisms can significantly reduce the price of PHA. It can take advantage of the ability of numerous extremophiles to produce PHA natively. Predominantly many halophilic microorganisms (microorganisms adapted to the high salinity of environments) belong among promising PHA producers–PHA production ability was described for several representatives of genus *Halomonas* such as *Halomonas boliviensis* [5], *Halomonas neptunia* and *Halomonas hydrothermalis* [6], *Halomonas bluephagenesis* [7], *Halomonas halophila* [8], or extremely halophilic representatives of the Archaea domain, such as *Haloferax mediterranei* [9]. As mentioned above, the accumulation of PHA is a part of the adaptation strategy of halophiles to hyperosmotic environment; however, the main adaptation strategy is usually the accumulation of either organic (osmolytes) or inorganic (K^+ ions) substances, which compensate for the osmotic pressure of the surrounding, as observed especially for extreme haloarchea [10].

Instead of halophiles, the production of PHAs was also observed in a restricted number of thermophilic bacteria, which are adapted to high temperature condions [2]. The most studied strains belong to the genera *Caldimonas* [11], *Tepidimonas* [12], in addition to the very promising PHA producer *Schlegelella thermodepolymerans*, which showed expedient potential for PHA production using lignocellulose-based substrates with high xylose content [13]. For thermophiles, PHA most likely does not directly take the role in the adaptation to high temperature, since the adaptation is usually connected with changes of protein structure, enhancement of the activity of chaperones and other heat-shock proteins, and also accumulation of organic compatible molecules similarly to halophilic prokaryotes [14].

Apart from the microbial biosynthesis of PHA, also the expenses associated with the isolation of PHA, an intracellular product, from microbial biomass contribute to the high cost of PHA. Extraction of the polymer from the microbial cells is often performed by organic solvents; however, considering limited polymer solubility, the extraction is limited to selected, predominantly chlorinated, solvents such as chloroform or dichloromethane. Alternatively, compounds typically described as "PHA anti-solvents", such as acetone, can be employed for PHA extraction under conditions of highly elevated temperature (120 °C) and pressure (7 bar) [15]. Nevertheless, the approaches based on organic solvents are technologically difficult, and the use of toxic and hazardous solvents eliminates the positive ecological impact of final materials [16]. The other possibility is the utilization of supercritical CO_2 for PHA extraction. Despite a very positive ecological connotation, due to the low solubility of PHA in supercritical CO_2 and the need for additional compounds acting as solubility mediators ("modifiers"), such as methanol, the process shows low efficiency [17,18]. Also, an alternative PHA isolation approach based on the removal of other components of biomass can be used, when PHA granules remain in the intact, solid state and can be further separated by centrifugation or filtration. Form an economical viewpoint, this approach makes definitely sense: Considering the fact that the PHA fraction in biomass can exceed 90 wt.%, it appears cumbersome to recover these 90 wt.% instead of simply removing the minor (10%) fraction of biomass. Indeed, many reagents and strategies can be used to remove other cell components instead of PHA. For instance, in literature the use of hydrolytic enzymes (predominantly commercially available proteases–Protease L330, Esparase, Alcalase, Neutrase, Allprotease, etc.) is described [19], but in practice, these approaches are limited by the high cost of enzymes. An alternative way is the digestion of biomass using various chemical reagents such as sodium hydroxide [20], sodium hypochlorite [21], EDTA, and detergents such as SDS [22] or ammonium laurate [23]. Individual approaches can be combined as was mentioned in Kathiraser et al., who used enzymes together with detergents and EDTA for PHA isolation [24].

Hypotonic lysis is an additional tool that can be used for the simple disruption of microbial cells. This approach is based on the exposition of cells to surrounding with significantly lower osmotic strength than osmolarity of cultivation medium. During sudden exposition to a hypotonic environment, water begins to penetrate into cells, which leads to disruption of their integrity, rupture of cells, and release of the intracellular cell content. This hypoosmotic cell lysis approach was described predominantly for extreme halophilic prokaryotes such as *Hfx. mediterrenei*, which are cultivated in media with salt contents of more than 200 g/L and are, therefore, extremely sensitive to hypotonic treatment. During exposure to distilled water, cells disruption and release of PHA granules into the solution occur immediately. After centrifugation, there are two phases generated—the bottom pinkish phase represents cell mass residues (colorized by carotenoid pigments), while the upper white phase consists of PHA granules [25]. However, the purity of PHA isolated in the described way is not too high due to the PHA granules being covered by a proteinaceous membrane; therefore, the following purification using for instance hypochlorite, H_2O_2, or other reagents is usually necessary [8,26].

In this work, we decided to investigate the possibility of utilization of hypotonic lysis to isolate PHA from moderately extremophilic microorganisms—halophilic *Halomonas halophila* and also thermophilic *Schlegelella thermodepolymerans*. We hypothesized that both microorganisms could be partially sensitive to hypotonic lysis due to the accumulation of compatible solutes as a strategy of their adaptation to extreme conditions. To further support the process of cell disruption and increase the purity of the isolated materials, we have also introduced a low concentration of 10 g/L of sodium dodecyl sulfate (SDS). Since SDS might reduce the positive environmental impact of the process by serious contamination of wastewater, we introduced a step in which SDS leftovers after PHA isolation are precipitated with KCl and can be easily removed from wastewater and potentially even regenerated for repeated use.

2. Materials and Methods

2.1. Cultivation of Microorganisms

For experimental work, two extremophilic strains were used, the moderately halophilic strain *Halomonas halophila* (CCM 3662) obtained from Czech Collection of Microorganisms, Brno, Czech Republic, and the moderately thermophilic strain *Schlegelella thermodepolymerans* (DSM 15344) purchased from Leibnitz Institute DSMZ—German Collection of Microorganisms and Cell Cultures, Braunschweig, Germany.

The first cultivation step, preparation of inoculum, was performed in 100 mL Erlenmeyer flasks with 50 mL of media. Inoculum culture of *H. halophila* was prepared in a medium consisting of yeast extract (3 g/L), peptone (15 g/L), glucose (1 g/L), and sodium chloride (66 g/L) during 24 h in 30 °C, permanent shaking 180 rpm. Inoculum of *S. thermodepolymerans* was prepared in 25 g/L Nutrient Broth medium (10 g/L peptone, 10 g/L beef extract and 5 g/L NaCl) during 24 h in 50 °C, permanent shaking 180 rpm.

After 24 h lasting cultivation, mineral salts media were inoculated by 10% (*v*/*v*) of inoculum, cultivations were performed in 250 mL Erlenmeyer flasks with 100 mL of medium. The basic mineral salt medium for *H. halophila* cultivation consisted of $(NH_4)_2SO_4$ (3.0 g/L), KH_2PO_4 (1.02 g/L), $Na_2HPO_4 \cdot 12H_2O$ (11.1 g/L), $MgSO_4 \cdot 7H_2O$ (0.2 g/L), NaCl

(66 g/L), glucose (20 g/L) and 1mL/L of microelements solution ($FeCl_3$ (9.7 g/L), $CaCl_2$ (7.8 g/L), $CuSO_4$ (0.156 g/L),$CoCl_2$ (0.119 g/L), $NiCl_2$ (0.118 g/L) and $CrCl_2$ (0.062 g/L) dissolved in 0.1 M HCl) and 20 g/L of glucose. The mineral salt medium for cultivations of *S. thermodepolymerans* consisted of $Na_2HPO_4 \cdot 12\ H_2O$ (9.0 g/L), KH_2PO_4 (1.5 g/L), NH_4Cl (1.0 g/L), $MgSO_4 \cdot 7\ H_2O$ (0.2 g/L), $CaCl_2 \cdot 2\ H_2O$ (0.02 g/L), Fe(III)NH_4citrate (0.0012 g/L), yeast extract (0.5 g/L), 20 g/L xylose and 1 mL/L of microelements solution (EDTA (50.0 g/L), $FeCl_3 \cdot 6\ H_2O$ (13.8 g/L), $ZnCl_2$ (0.84 g/L), $CuCl_2 \cdot 2\ H_2O$ (0.13 g/L), $CoCl_2 \cdot 6\ H_2O$ (0.1 g/L), $MnCl_2 \cdot 6\ H_2O$ (0.016 g/L), H_3BO_3 (0.1 g/L), dissolved in distilled water) (all used chemicals including hydrochloric acid, salts, glucose and xylose were purchased from LachNer, Neratovice, CZE; yeast extract, peptone and Nutrient Broth were purchased from HiMedia, Brno, CZE). Inoculated media were cultivated during 72 h under permanent shaking at 180 rpm at the optimal growth temperature, 30 °C for *H. halophila* and 50 °C for *S. thermodepolymerans*, respectively.

2.2. Basic Characterization of Cultures

After 72 h, cultivations were finished. For determination of growth of culture, optical density of cultures was measured using a spectrophotometer (P300, Implen, Munich, Germany) at 630 nm against distilled water as a blank. Gravimetrical determination of cell dry matter (CDM) was performed by centrifugation of 10 mL of culture at 3460× *g*, 5 min (EBA 200, Hettich, Spenge, Germany). The generated supernatant was discarded and the pellet was washed with distilled water and centrifuged again. The pellet was dried to constant weight at 70 °C in drying chamber (IP60, LTE Scientific, Oldham, UK) and then the amount of biomass (g/L) was obtained by weighing on an analytical scale (Pioneer PA224C, Ohaus, Parsippany, NJ, USA). Moreover, the content of PHA in dried biomass was measured by gas chromatography with flame ionization detection (GC-FID; gas chromatograph Trace 1300, Thermo Scientific, column: DB-WAX 30 m by 0.25 mm, Thermo Scientific, Waltham, MA, USA) as reported previously in Brandtl et al. (1988) [27]. All determinations were performed in dublicate.

2.3. Isolation of PHA from Bacterial Biomass

Right after cultivation, the isolation of PHA from wet bacterial biomass is performed. Predominantly for halophilic strain *H. halophila*, but also for *S. thermodepolymerans*, a very important strategy is the hypotonic environment, which induces disruption of cells leading to release of the intracellular cell content including PHA granules. The main idea of isolation is based on the use of hypotonic conditions together with SDS (TCI, Zwijndrecht, Belgium) applied in low but still sufficient concentration along with the elevated temperature. These factors should lead to cell disruption and the release of water-insoluble PHA granules. For both microorganisms, we focused at the beginning on the optimization of SDS concentration (in range 1 to 10 g/L), temperature treatment (higher and lower than cultivation temperature), and the ratio of microbial biomass vs. volume of SDS solution.

The first optimized parameter was the most efficient concentration of SDS. The experiment was performed in this way: 10 mL of the culture after cultivation was centrifuged in a plastic test tube with a screw cap (3460× *g*, 10 min), the supernatant was discarded and replaced by prepared aqueous solution of SDS (1; 2.5; 5 and 10 g/L), in replicates for each SDS concentration. Pellets were resuspended and suspensions were incubated in a water bath at 70 °C for 120 min. After incubation, test tubes with suspensions were centrifuged again (3460× *g*, 5 min), supernatants were discarded and preserved for further use, the pellets containing predominantly isolated PHA granules were washed with distilled water and then dried to constant weight at 70 °C. The amount of the product was determined gravimetrically and the purity of obtained PHA was determined by GC-FID as the same as described above.

The second, optimized parameter was the temperature of isolation or combination of different temperatures, i.e., temperature treatment. For both strains, the same treatments were tested: 50 °C (120 min), 70 °C (120 min), 90 °C (120 min) and a combination of cooling-down and heating-up; 4 °C (120 min) followed by 70 °C (120 min). SDS was applied at a concentration of 5 g/L, otherwise, the experiment was performed in the same way as described above for optimization of SDS concentration.

To reduce the amount of detergent (SDS) using a higher biomass amount, the third series of optimization experiments was based on a comparison of the effectivity of isolation with different ratios of initial biomass concentration and 10 mL of aqueous SDS (5 g/L) solution in the most to verify the purity of obtained material and compare product yield. Also, the purpose of this experiment was to verify the robustness of the isolation process concerning different initial biomass concentrations—using a higher ratio of biomass to SDS solution can significantly reduce the cost of the isolation process (amount of SDS per biomass) and also significantly reduce environmental burden caused by SDS in wastewater. In practice, different amounts of biomass were obtained by centrifugation of a larger volume of cell culture when the volume of SDS solution was still 10 mL. However, the approach was the same as previous ones, using optimal SDS solution concentration of 5 g/L for both strains and optimal "heat treatment": 70 °C for 2 h for *H. halophila* and keeping at temperature 90 °C for 2 h for *S. thermodepolymerans*. All isolation experiments based on long-term exposition to elevated temperatures were performed in a water bath (BL 4/150, WSL, Czestochowa, Poland).

2.4. Characterization of Isolated PHA Polymer

In this work, we produced a homopolymer of 3-hydroxybutyrate-poly(3-hydroxybutyrate) (PHB), the most common representative of PHA. Quantification of the polymer in bacterial biomass and determination of purity of the isolated samples were performed by GC-FID as described above.

Further, we used Fourier transform infrared spectroscopy (FT-IR, Nicolet iS50, Thermo Scientific, Waltham, MA, USA) with the built-in single-reflection diamond attenuated total reflectance (ATR) crystal to collect infrared spectra of the original biomass, polymer isolates, and the reference material, commercially available P(3HB) (Biomer, Schwalbach am Taunus, Germany). Each spectrum was collected in the range of 4000–400 cm^{-1} as an average of 16 scans with a resolution of 4 cm^{-1} (data spacing 0.5 cm^{-1}).

For determination of molecular weight and its distribution (polydispersity index PDI) of the polymer, we used size-exclusion chromatography with multi-angle light scattering detector (SEC-MALS) (SEC chromatography, column PLgel mixed-C 5 μm, 300 × 7.5 mm, Agilent Technologies, Wilmington, DE, USA; detectors: MALS–DAWN HELEOS II, differential refractometer OPTILAB T-REX, Wyatt Technology, Dernbach, Germany). Using this method, we compared polymers isolated by different approaches, which were the most promising based on results from GC connected with the highest material purity. Approximately 200 mg of sample (biomass, isolated material) was weighed into glass pyrex test tubes, 10 mL of chloroform was added, then test tubes were tightly screwed with caps and during 12 h incubated at 70 °C in a thermoblock. After 12 h, the mixture was filtered through a paper filter on Petri dishes, and chloroform was freely evaporated. Then 1.5 mg of foil was weighed to the vial (volume 4 mL) and 1.5 mL of chloroform (LachNer, Neratovice, CZE) was added. The mixture was incubated at 50 °C in the thermoblock till all foil was dissolved, then the solution was filtered using nylon filters with 0.45 μm pore size into small vials, and the samples were analyzed.

Selected samples were also analyzed by electron microscopy. Bacterial cultures and isolated PHA samples were fixed using the high-pressure freezing method (EM ICE, Leica Microsystems, Wetzlar, Germany). Samples were centrifuged for 3 min at 4000 rpm and the generated pellet was pipetted on the 0.2 mm side of the 6 mm Al carrier type A and closed with the flat side of carrier type B (without using any cryo-protectant) for further processing for observation in the cryogenic scanning electron microscope (cryo-SEM). For freeze-substitution procedure followed by observation in the transmission electron microscope (TEM), samples were frozen in 0.2 mm side of the 3 mm Au carrier type A and closed with the flat side of carrier B. Both of the carriers were pre-treated with 1%

aqueous solution of lecithin (Sigma-Aldrich, Darmstadt, Germany) in chloroform. For cryo-SEM, high-pressure-frozen samples were transferred into a cryo vacuum preparation chamber (ACE 600, Leica Microsystems, Wetzlar, Germany), where they underwent freeze-fracture followed by sublimation for 7 min at −95 °C. Samples were then transferred into a scanning electron microscope equipped with a cryo stage (Magellan 400/L, FEI, Hillsboro, OR, USA) and observed using 1 keV electron beam at −120 °C. For TEM, high-pressure-frozen samples were transferred into a freeze substitution unit (AFS2, Leica Microsystems, Wetzlar, Germany) containing a freeze-substitution solution of 1.5% OsO_4 in acetone. The freeze-substitution protocol was set as previously described in Kourilova et al. [28]. Thus prepared freeze-substitution samples were washed in pure acetone (3 × 15 min) and gradually infiltrated with epoxy resin (Epoxy Embedding Medium, Sigma-Aldrich, Darmstadt, Germany), mixtures with acetone 1:2, 1:1, 2:1 and pure resin for 1 h each. After the final exchange of pure resin, samples were left overnight under vacuum, then embedded in fresh resin and cured at 62 °C heat for 48 h. Cured blocks of samples were cut to ultrathin sections using a diamond knife with cutting angle 45° (Diatome, Nidau, Switzerland) and ultramicrotome (UTC 7, Leica Microsystems, Wetzlar, Germany), and stained using solutions of uranyl acetate and lead citrate. Contrasted ultrathin sections were imaged using a transmission electron microscope (Talos F200C, Thermo Fisher Scientific, Waltham, MA, USA) using a 200 keV beam.

2.5. Precipitation of Potassium Dodecyl Sulfate (KDS) and Determination of Residual SDS Concentration

SDS is a water-soluble molecule, whereas the salt potassium dodecyl sulfate (KDS) is water-insoluble. This is the main idea of how to get rid of SDS from wastewater after isolation of PHA from bacterial biomass. A simple addition of KCl to supernatant containing SDS results in the formation of an insoluble precipitate of KDS, which can conveniently be separated by sedimentation or centrifugation. For this purpose, we prepared 4 M aqueous KCl solution and after isolation and following centrifugation, we added it in a 1:1 ratio to the supernatant to obtain a precipitate.

Residual concentration of SDS after PHA isolation procedure and also the efficiency of KDS precipitation from SDS solutions with different initial amounts of the detergent were determined employing a spectroscopic assay based on the use of Stains-All dye (Sigma-Aldrich, Darmstadt, Germany) as described by Rupprech et al. [29]. The analysis was performed using a microplate reader (ELISA reader, ELx808, BioTek, Winooski, VT, USA).

3. Results and Discussion

Every isolation experiment was preceded by the cultivation of microorganisms *H. halophila* and *S. thermodepolymerans* in mineral media with the most suitable carbon sources (glucose for *H. halophila* and xylose for *S. thermodepolymerans*) to obtain biomass with high PHA content. Within individual cultivations, we tried to keep the cultures under the same conditions for gaining results of biomass and PHA content as similar as possible, despite we measured both characteristics (biomass content gravimetrically and PHA content by GD-FID) within each experiment for independent comparison among results of cultivations. For each experiment, amounts of PHA in biomass are stated under almost all tables for the comparison with the purity of isolated polymer using different treatments, because for individual experiments all the yields were related to individual cultivation results. Based on the data from experiments, for *H. halophila* average concentration of biomass was (5.5 ± 0.3) g/L with a PHA fraction in biomass of about (78 ± 7) wt.%. For *S. thermodepolymerans* average biomass concentration was (6.1 ± 0.1) g/L with (70 ± 4) wt.% PHA.

3.1. Optimization of SDS Concentration

Within the whole PHA isolation development process, the first step was focused on the optimal concentration of detergent SDS. The idea was to expose the extremophilic bacterial cells naturally containing compatible solutes as a part of their adaptation strategy to hypotonic conditions with low amounts of SDS to improve cell disruption and enable solubilization of hydrophobic cell components which would otherwise get attached to PHA granules reducing the purity of the materials. For both used strains, tested SDS concentrations were 1; 2.5; 5 and 10 g/L; results for individual strains are listed in Table 1. For *H. halophila* the process was carried out in a water bath during a 2 h lasting process at 70 °C; for *S. thermodepolymerans*, the process lasting the same time was carried out at 90 °C.

Table 1. Effect of different SDS concentrations on PHA purity and yield during 2 h lasting isolation at 70 °C for *H. halophila* and 90 °C for *S. thermodepolymerans*.

Strain	SDS (g/L)	PHA Purity (wt.%)	Yield (−) [1]
H. halophila	1	88.0 ± 0.4	0.99
	2.5	91.4 ± 2.7	0.99
	5	97.1 ± 3.7	1.11
	10	89.6 ± 1.4	1.09
S. thermodepolymerans	1	65.7 ± 0.8	0.95
	2.5	87.7 ± 3.6	0.97
	5	86.4 ± 1.4	1.00
	10	85.5 ± 0.4	0.92

[1] Yield is defined as PHA (g/L) for individual isolation way divided by PHA (g/L) obtained from biomass without previous isolation step. PHA content in the biomass: *H. halophila* (84.02 ± 0.61) wt.%, *S. thermodepolymerans* (64.43 ± 5.96) wt.%.

In the case of *H. halophila*, the most effective SDS concentration for isolation of PHA was 5 g/L, when the purity of the polymer was more than 97 wt.% together with the highest product yield considering purity and also amount of the isolated material. For this moderately halophilic strain with optimal salt concentration for growth 66 g/L, utilization of 5 g/L of SDS induces effective hypotonic cell lysis and results in high purity of the isolated material. Therefore, the most suitable concentration of SDS was used for all the following experiments employing the halophilic strain.

For *S. thermodepolymerans*, the highest purity was determined for a SDS concentration of 2.5 g/L. In comparison with the halophilic culture, the concentration is 2 times less, which might be caused by the lower content of the compatible solutes in the cells of thermophilic culture. It can be expected that thermophilic bacteria are less prone to hypotonic lysis as compared to halophiles. This expectation is also supported by the fact that purities of PHA materials obtained for thermophilic *S. thermodepolymerans* are substantially lower than those obtained in its halophilic counterpart *H. halophila*. Further, aside from the purity of the material, we have also considered the yield of the isolation process which was highest for 5 g/L of SDS leading to the establishment of the most suitable concentration of SDS of 5 g/L—the same as for *H. halophila*.

The team of Arikawa et al. (2017) focused on using SDS in combination with sonication for the isolation of PHA from biomass of the mesophilic bacterium *Cupriavidus necator*, when the optimal concentration of detergent was set on 33 g/L [30]. In our case, we used a more than 6-times lower amount of detergent due to the advantageous effect of hypotonic shock, which is applicable on strains producing compatible solutes (for instance halophiles and thermophiles). Further, our approach relying on hypotonic shock does not require any special equipment, on the contrary, sonification of the bacterial cells can be challenging especially in scales of industrial production of PHA.

3.2. Optimization of "Heat Treatment"

Subsequently, the second step in the isolation process development was to optimize temperature and time of isolation; together, this process can be called "heat treatment". Within the first experiment focused on SDS concentration, "heat treatment" was represented by the application of a temperature of 70 °C for 2 h, but despite quite high yields, using different temperature can be more effective, or lowering the temperature could be more economically suitable, eventually. Based on previous results, time of exposition was set at

2 h, strain *H. halophila* was exposed to temperature 50 °C, 70 °C (as control—for comparison see Table 1), 90 °C and combination of cooling-down at 4 °C and following heating-up at 70 °C. The optimization process for *S. thermodepolymerans* included a little different strategy, when, in contrast to the halophilic strain, a lower temperature of 30 °C was also tested (we hypothesized that even low temperature could damage cell envelope of thermophilic bacterium). Other tested temperatures were 50 °C, 70 °C, 90 °C (as control—for comparison see Table 1) and also a combination of cooling-down at 4 °C and following heating-up at 70 °C as same as for *H. halophila*. The results of the experiments are listed in Table 2.

Table 2. Effect of different "heat treatments" on PHA purity and yield during isolation in 5 g/L SDS solution.

Strain	"Heat Treatment"	PHA Purity (wt.%)	Yield (−)
H. halophila	50 °C (120 min)	97.5 ± 1.2	0.92
	70 °C (120 min)	96.7 ± 0.3	0.97
	90 °C (120 min)	99.7 ± 0.1	0.90
	4 °C (120 min) + 70 °C (120 min)	94.3 ± 1.5	0.92
S. thermodepolymerans	30 °C (120 min)	67.6 ± 0.8	0.73
	50 °C (120 min)	81.4 ± 4.3	0.78
	70 °C (120 min)	96.7 ± 2.4	0.79
	90 °C (120 min)	99.8 ± 2.8	0.91
	4 °C (120 min) + 90 °C (120 min)	86.2 ± 2.1	0.87

PHA content in the biomass: *H. halophila* (82.27 ± 0.09) wt.%, *S. thermodepolymerans* (69.57 ± 0.79) wt.%.

Based on the results presented in Table 2, the most successful "heat treatment" for isolation of PHA from the halophilic strain considering the purity of isolated material was a temperature of 90 °C for 2 h, when the purity was higher than 99 wt.%. Despite a little lower purity after isolation at 70 °C, the yield of the material was higher than for the highest tested temperature 90 °C and also higher than for the combination of cooling-down and following heating-up. For subsequent isolation experiments, a treatment representing 2 h of heating at 70 °C was applied for strain *Halomonas halophila*.

The optimization process for *S. thermodepolymerans* included a little different strategy; in contrast to the halophilic strain, we tested a lower temperature (30 °C), but the effectivity of isolation was not too high (the purity was around 67 wt.% and the yield was 0.73). Based on these data, the most successful "heat treatment" was keeping the sample at 90 °C for 2 h; here, the purity exceeded 99 wt.%, hence, it was very similar to using a temperature of 70 °C for *H. halophila*. The combination of cooling followed by heating was not too successful for the thermophilic strain, whereas exposition to low temperature presumably led to the development of enhanced resistance to thermal treatment as was already reported in the literature [31]. This effect was described for mesophilic bacterial strains *E. coli* [32] and *C. necator* [33] after treatment at low temperature (0–5 °C). It was proved that enhanced resistance against elevated temperature was caused by the changes in the fatty acids composition in cell membranes [31]. Despite the fact that the effect was not described for thermophilic strains in detail, it is very likely that also in this case the low-temperature treatment resulted in active modulation of cell membrane lipids, which negatively impacted the susceptibility of the bacterial cells to hypotonic lysis. For our experiment, lower efficiency of combined "heat treatment" can be observed compared with keeping the cells at 90 °C for 2 h. To conclude, the most suitable "heat treatment" procedure for the thermophilic strain *S. thermodepolymerans* was keeping the isolation mixture at 90 °C for 2 h.

3.3. Testing of Different Ratios SDS Solution: Biomass

After optimization of SDS concentration and "heat treatment", the following step was focused on the influence of initial biomass amount on the purity of isolated material and the yield of the process. This is very important since PHA production is usually performed as a high cell density cultivation [34], therefore the robustness of the isolation process with respect to the amount of biomass employed per isolation batch is a very important factor determining the viability and feasibility of the process. Therefore, we tested different initial concentrations of biomass, while keeping the volume of SDS solution constant (5 g/L). Moreover, an increasing portion of biomass per isolation batch can also reduce the cost of the isolation process (lower SDS amount) and significantly decrease the environmental burden caused by the presence of SDS in wastewater.

Based on the results listed in Table 3, it is possible to state that the process is very robust regarding the initial biomass concentration; purities of material isolated from different biomass amounts were greatly similar. The same goes for almost all yields for both strains. Therefore, the developed isolation procedure can be advantageously applied on biomass obtained from high-cell-density cultivation (e.g., fed-batch cultivation).

Table 3. Effect of different biomass content on PHA purity and yield applied during 2 h lasting isolation in 5 g/L SDS solution at 70 °C for *H. halophila* and 90 °C for *S. thermodepolymerans*.

Strain	Biomass Amount in SDS Solution (g/L)	PHA Purity (wt.%)	Yield (−)
H. halophila	1.83	96.2 ± 1.9	0.86
	4.21 (standard process)	96.6 ± 0.3	0.99
	8.40	95.9 ± 7.1	1.04
	11.86	91.5	0.96
	17.72	94.9	1.03
	22.65	95.9 ± 1.0	1.06
S. thermodepolymerans	1.91	96.3 ± 1.3	0.89
	3.76 (standard process)	101.1 ± 4.7	0.92
	8.88	93.4 ± 0.5	0.77
	11.72	84.1	0.80
	16.01	93.3 ± 2.4	1.01
	20.99	94.1	0.94

PHA content in the biomass: *H. halophila* (75.5 ± 0.1) wt.%, *S. thermodepolymerans* (68.5 ± 0.0) wt.%.

3.4. Effect of Different Isolation Approaches on the Molecular Weight of the Polymer

The following experiment was focused on the determination of the molecular weight of obtained PHA materials; the main goal was to evaluate the influence of the isolation procedure on the molecular weight of the polymer.

Isolation of polymer using SDS solution (5 g/L) did not manifest in the decrease of molecular weight of PHA as can be seen in Table 4. For comparison, PHA was isolated from dried biomass using a procedure involving 12 h lasting incubation in chloroform at 70 °C in a thermoblock. To eliminate differences from sample preparation, all analyzed PHA samples were subjected to the same treatment. Despite the slight differences, values of molecular weight were in principle very similar at least for polymers isolated from *S. thermodepolymerans*. Only a little decrease can be observed which could be caused by enhanced temperature 90 °C used for isolation with SDS. In the case of *H. halophila*, isolation with SDS led to enhancement of molecular weight up to 1.5-times. It should be pointed out that the M_w of the polymer produced by *H. halophila* is very high as compared to other PHA producers [8] including *S. thermodepolymerans*. It can be assumed that the higher molecular weight of polymer after SDS isolation treatment might be caused by enhanced solubility of longer chains in chloroform after elimination of other cells components. 12 h lasting isolation at 70 °C in chloroform is possibly more efficient for shorter polymer chains with lower molecular weight. PDI values were very low for all studied isolation setups, indicating the high uniformity of generated biopolyesters in terms of molecular mass distribution, which is a beneficial feature for further polymer processing.

Table 4. The molecular weight of isolated PHA polymer.

Strain	"Treatment"	M_w (kDa)	PDI [2] (–)
H. halophila	PHA isolated from dried biomass	1050.4 ± 1.8	1.20 ± 0.04
	70 °C (120 min)—standard process	1581.9 ± 5.3	1.46 ± 0.19
S. thermodepolymerans	PHA isolated from dried biomass	711.5 ± 1.6	1.17 ± 0.01
	90 °C (120 min)—standard process	686.4 ± 16.1	1.23 ± 0.03

[2] PDI (polydispersity index) is defined as M_w/M_n. M_w: weight average molecular mass; M_n: number average molecular mass.

3.5. Determination of Material Purity by Infrared Spectroscopy

FTIR technique was applied on samples of dried biomass, PHA polymer isolated by optimized process for each strain (2 h lasting treatment at 70 °C for *H. halophila* (HH) or 90 °C for *S. thermodepolymerans* (ST) in solution with 5 g/L SDS), and also for commercial P(3HB) (Biomer) as a reference material. Result are shown in Figures 1 and 2.

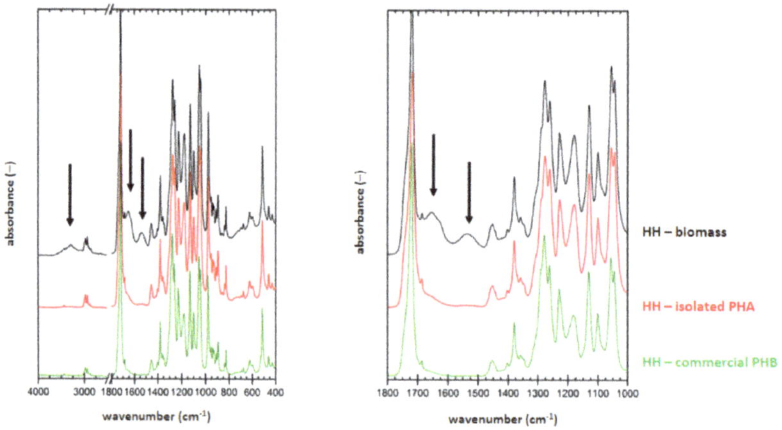

Figure 1. FTIR spectra of original biomass of *H. halophila* (black), polymer isolated via optimized approach (red) and commercial P(3HB) sample (Biomer) (green); arrows show absorption bands for components of biomass (3300; 1640 and 1540 cm^{-1}—proteins).

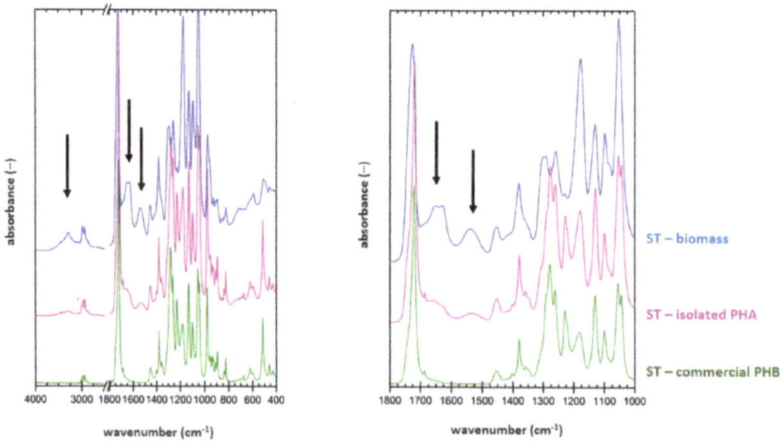

Figure 2. FTIR spectra of original biomass of *S. thermodepolymerans* (blue), polymer isolated via optimized approach (magenta) and commercial P(3HB) sample (Biomer) (green); arrows show absorption bands for components of biomass (3300; 1640 and 1540 cm^{-1}—proteins).

To evaluate the purity of the polymer isolates from FTIR spectra, characteristic spectral signatures of proteins can be used. These can be found at about 3300 (referred to as amide A band), 1640 cm^{-1} (referred to as amide I band) and 1540 cm^{-1} (referred to as amide II band). All these characteristic bands are marked with arrows in the biomass spectra in Figures 1 and 2. As expected, the protein signal is completely absent in spectrum of the reference P(3HB) material. In the FTIR spectra of polymer isolates obtained from both tested strains, significant decrease of the protein signal, compared to the original biomass, indicates high effectivity of the polymer purification during the extraction process. As can be seen from the comparison of Figures 1 and 2, lower content of the residual biomass was found for the polymer isolated from *H. Halophila*. This is in good agreement with the results of GC-FID. Based on the gravimetric yield of the chromatographic analysis, we determined polymer purity (86.4 ± 1.4) wt.% for *S. thermodepolymerans* and (91.3 ± 1.5) wt.% for *H. halophila*.

3.6. Electron Microscopy Imaging

Selected samples were analyzed using electron microscopy techniques, specifically, samples of microbial cells of *H. halophila* and *S. thermodepolymerans* and isolates of PHA from both strains.

Cryo-SEM image of *H. halophila* (Figure 3A) shows rod-shaped cells containing several granules of PHA. Most of the cells were fractured during the freeze-fracturing procedure and revealed the intracellular content of PHA. As previously described [35], PHA remain elastic even at temperatures of liquid nitrogen and can be observed being pulled out of the cells, showing "needle type" deformation (arrowhead in Figure 3) or as holes in cells (concave deformation) when the granule was pulled out of the cell completely. Cells of *S. thermodepolymerans* (Figure 3C), containing 1 or 2 granules of PHA also deformed by freeze-fracturing. In the image, it is possible to observe also fracture of the cell wall revealing the surface of the plasma membrane, marked with O. The image of isolated granules from *H. halophila* (Figure 3B) shows that the granules merged, but still remained elastic. PHA isolated from *S. thermodepolymerans* (Figure 3D), however, stayed separate after the extraction as individual oval granules, which were also affected by freeze-fracturing and show needle deformation.

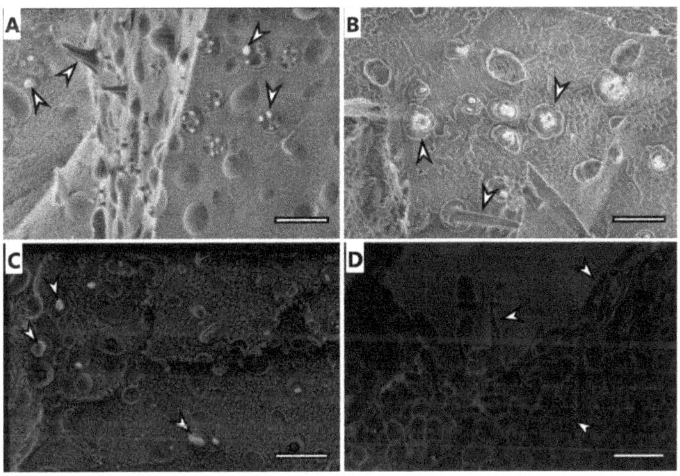

Figure 3. Cryo-SEM image of: (**A**) *H. halophila* biomass before extraction of PHA granules, (**B**) PHA granules isolated from *H. halophila*, (**C**) *S. thermodepolymerans* biomass before extraction of PHA granules, (**D**) PHA granules isolated from *S. thermodepolymerans*, needle deformation of PHA granules marked with arrowhead, cells with fractured cell wall revealing the surface of plasma membrane marked with O, scale bar 2 μm.

TEM observation confirmed previous findings of cryo-SEM. *H. halophila* contained several smaller granules in their cells (Figure 4A), while *S. thermodepolymerans* contained 1–3 larger granules. Generally, it is described that larger PHA granules are beneficial for the recovery process [36], which makes *S. thermodepolymerans* of even higher interest for an eventual industrial-scale application. Images of isolated granules also confirm the shape observed in cryo-SEM: isolates from *H. halophile* merged into irregularly shaped granules, while granules from *S. thermodepolymerans* of regular oval shape stayed separate. In Figure 4B,D, it is possible to see shades surrounding the granules. Based on the results of purification, the shades could be caused by contamination originating from disrupted cellular fragments, or by PHA granule membrane remnants. It is also possible to observe a few intact cells (Figure 4B marked with arrow), thus confirming the results of the determination of purity of the isolates—for instance (91.3 ± 1.5) wt.% of PHA for *H. halophila* using standard optimized process.

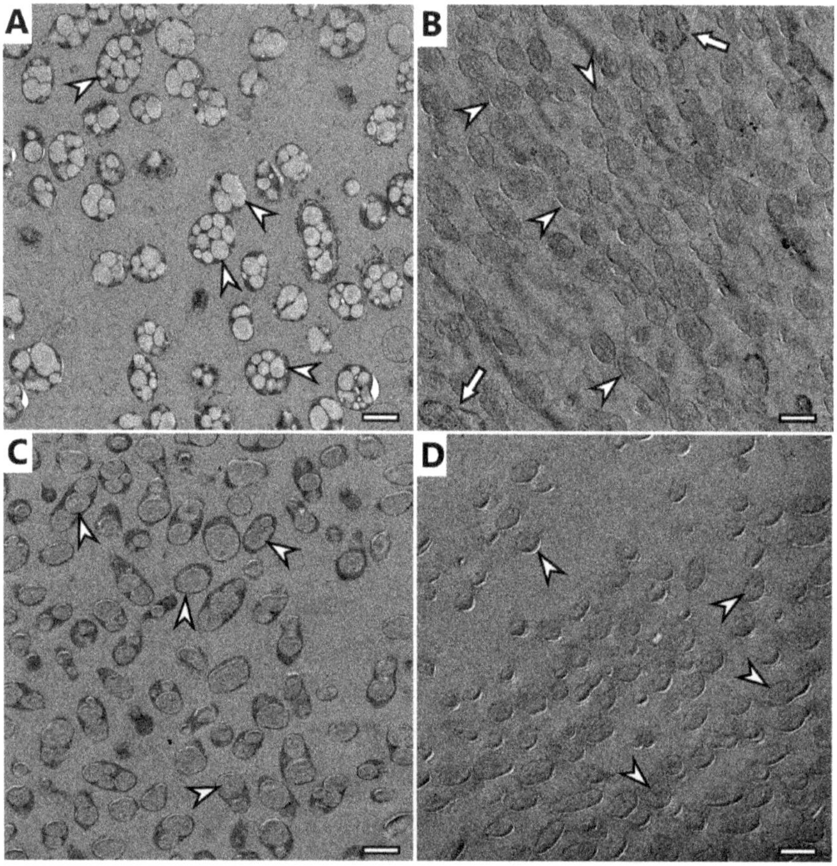

Figure 4. TEM image of (**A**) *H. halophila* biomass before extraction of PHA granules, (**B**) PHA granules isolated from *H. halophila*, (**C**) *S. thermodepolymerans* biomass before extraction of PHA granules, (**D**) PHA granules isolated from *S. thermodepolymerans*, PHA granules marked with arrowhead, intact cells after extraction marked with arrow, scale bar 1 μm.

3.7. *Wastewater Management—Precipitation of KDS from SDS after Isolation*

As same as other detergents, also SDS serves as a contaminant of wastewater with possible harmful effects on wastewater management. Therefore, we focused on the removal of the irritant compound SDS from the supernatant after the isolation process. We have tested several approaches (data not shown), and the most successful procedure was simple and feasible precipitation of SDS in the form of poorly water-soluble potassium dodecylsulfate (KDS) in the presence of 2 M KCl. The efficiency of the SDS removal process was determined via measurement of SDS concentration using the protocol of Rupprech et al. (2015) [29]. Data are demonstrated in Table 5.

Table 5. Results of SDS determination before and after KDS precipitation.

Parameter	SDS (g/L)			
Concentration used for isolation	1	2.5	5	10
In supernatant after isolation	0.6	1.8	4.9	10.7
Residual SDS concentration after KDS precipitation	0.02	n.d. [3]	n.d.	n.d.
KDS yield (mg)	26.8	64.9	169.9	316.0

[3] n.d.—no detectable.

Based on presented data, it seems that removal of SDS from a wastewater stream by simple addition of non-toxic and cheap KCl is a very effective process since after this treatment the residual concentration of SDS in the supernatant after isolation was not detectable for almost all the samples. The formation of KDS precipitate is a very quick and straightforward method to remove SDS used for isolation of PHA granules, since the formation of the precipitate can be observed immediately after the addition of KCl into supernatants after isolation as shown in Figure 5.

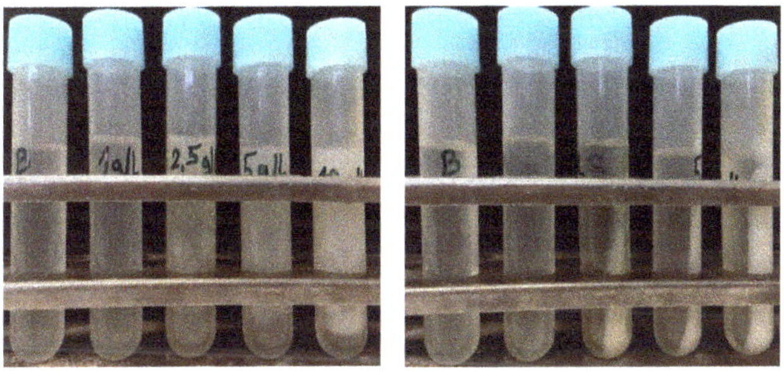

(**A**) (**B**)

Figure 5. KDS precipitate using 2 M KCl after isolation with different SDS concentrations—from the left: blank, 1 g/L, 2.5 g/L, 5 g/L, 10 g/L: (**A**) before centrifugation; (**B**) after centrifugation.

A different strategy to get rid of the detergent after isolation was used by the team of Samorì et al. (2015), who isolated PHA polymer from bacterial biomass. For this purpose, they used ammonium laurate (most efficient at a concentration of 200 wt.%), which acts similarly to SDS in terms of microbial cell disruption and solubilization of hydrophobic contaminants. After separation of the polymer by centrifugation, the addition of CO_2 into supernatant leads to a decrease of the pH-value, and switches the detergent into poorly water soluble neutral (protonated) lauric acid with the possibility of separation by centrifugation; resulting water phase containing ammonium hydrogen carbonate can serve as a nitrogen source for microorganisms [37]. Our approach relies on the application of the lower amount of inexpensive and abundant detergent SDS, which can be simply removed from wastewater by precipitation. It should be pointed out that unlike for ammonium laurate, the removal of the SDS from the supernatant in form of KCl occurs in a non-destructive way, and it is very likely that SDS could be regenerated for instance by ultradialysis in excess of Na^+ ions. Hence, our concept enables not only removal of detergent from wastewater, but also holds a promise of SDS recovery, which would be very beneficial with respect to economic aspects of the PHA isolation process. Nevertheless, the possibility of SDS recovery from KDS precipitate deserves further investigation, which is out of the scope of this preliminary work.

4. Conclusions

In this work, we developed a procedure for the isolation of PHA materials from extremophilic microbial cells. The method is based on the exposition of the bacterial cells naturally containing high intracellular concentrations of compatible solutes as part of their adaptation strategy to extreme hypotonic conditions induced by the diluted solution of SDS (optimal concentration is about 5 g/L of SDS) at elevated temperature. Our results indicate that such conditions lead to disruption of the cells and release of PHA granules. Moreover, SDS apart from its cell-disruptive function also solubilizes hydrophobic cell components which would otherwise get attached as contaminants to PHA materials. The procedure seems to be simple, robust, and feasible in industrail condtions. The purity of obtained materials (usually above 95%), as well as yields of the procedure (usually about 0.9), reach high values. If even higher purity of the material is needed, additional steps such as the washing of the materials with proper organic solvents or other agents can be involved. Furthermore, since leftovers of the detergent SDS in process wastewater might dramatically decrease the positive ecological features of the process, we also focused on the removal of SDS. We have developed a simple, cheap, and safe technique that is based on the precipitation of SDS in the presence of KCl. The resulting precipitate can be simply removed by decantation or centrifugation, hence, SDS does not contaminate the wastewater of the process. Moreover, there is also the possibility to regenerate SDS which would substantially improve the economic feasibility and overal sustainability of the process. To sum up, the developed strategy for PHA isolation is compatible with Next-Generation Industrial Biotechnology concepts, it is simple, cheap, robust, provides PHA materials in high purity and at high yields and is also ecologically friendly.

Author Contributions: Conceptualization, S.O.; investigation, I.N., X.K., K.M., P.S., M.K. (Michal Kalina) and V.K.; writing—original draft preparation, S.O., I.N. and M.K. (Martin Koller); writing—review and editing, X.K., P.S. and M.K. (Martin Koller); visualization, S.O., I.N. and K.M.; supervision, S.O.; project administration, I.N.; funding acquisition, I.N. All authors have read and agreed to the published version of the manuscript.

Funding: This work was supported by Grant FCH-K-21-6952 implemented within the project Quality Internal Grants of BUT (KInG BUT), Reg. No. CZ.02.2.69/0.0/0.0/19_073/0016948, which is financed from the OP RDE. Also CIISB research infrastructure project LM2018127 funded by MEYS CR is gratefully acknowledged for the financial support of the measurements at the CF CryoEM.

Conflicts of Interest: The authors declare no conflict of interest. The funders had no role in the design of the study; in the collection, analyses, or interpretation of data; in the writing of the manuscript, or in the decision to publish the results.

References

1. Obruca, S.; Sedlacek, P.; Koller, M.; Kucera, D.; Pernicova, I. Involvement of polyhydroxyalkanoates in stress resistance of microbial cells: Biotechnological consequences and applications. *Biotechnol. Adv.* **2018**, *36*, 856–870. [CrossRef]
2. Obruca, S.; Dvorak, P.; Sedlacek, P.; Koller, M.; Sedlar, K.; Pernicova, I.; Safranek, D. Polyhydroxyalkanoates synthesis by halophiles and thermophiles: Towards sustainable production of microbial bioplastics. *Biotechnol. Adv.* **2022**, *59*, 107906. [CrossRef]
3. Kourmentza, C.; Plácido, J.; Venetsaneas, N.; Burniol-Figols, A.; Varrone, C.; Gavala, H.N.; Reis, M.A.M. Recent Advances and Challenges towards Sustainable Polyhydroxyalkanoate (PHA) Production. *Bioengineering* **2017**, *4*, 55. [CrossRef]
4. Chen, G.-Q.; Jiang, X.-R. Next generation industrial biotechnology based on extremophilic bacteria. *Curr. Opin. Biotechnol.* **2018**, *50*, 94–100. [CrossRef]
5. Quillaguamán, J.; Hashim, S.; Bento, F.; Mattiasson, B.; Hatti-Kaul, R. Poly(β-hydroxybutyrate) production by a moderate halophile, *Halomonas boliviensis* LC1 using starch hydrolysate as substrate. *J. Appl. Microbiol.* **2005**, *99*, 151–157. [CrossRef]
6. Pernicova, I.; Kucera, D.; Nebesarova, J.; Kalina, M.; Novackova, I.; Koller, M.; Obruca, S. Production of polyhydroxyalkanoates on waste frying oil employing selected *Halomonas* strains. *Bioresour. Technol.* **2019**, *292*, 122028. [CrossRef]

7. Ren, Y.; Ling, C.; Hajnal, I.; Wu, Q.; Chen, G.-Q. Construction of *Halomonas bluephagenesis* capable of high cell density growth for efficient PHA production. *Appl. Microbiol. Biotechnol.* **2018**, *102*, 4499–4510. [CrossRef]
8. Kucera, D.; Pernicová, I.; Kovalcik, A.; Koller, M.; Mullerova, L.; Sedlacek, P.; Mravec, F.; Nebesarova, J.; Kalina, M.; Marova, I.; et al. Characterization of the promising poly(3-hydroxybutyrate) producing halophilic bacterium *Halomonas halophila*. *Bioresour. Technol.* **2018**, *256*, 552–556. [CrossRef]
9. Alsafadi, D.; Al-Mashaqbeh, O. A one-stage cultivation process for the production of poly-3-(hydroxybutyrate-*co*-hydroxyvalerate) from olive mill wastewater by *Haloferax mediterranei*. *New Biotechnol.* **2017**, *34*, 47–53. [CrossRef]
10. Shivanand, P.; Mugeraya, G. Halophilic Bacteria and Their Compatible Solutes—Osmoregulation and Potential Applications. *Curr. Sci.* **2011**, *100*, 1516–1521.
11. Hsiao, L.-J.; Lee, M.-C.; Chuang, P.-J.; Kuo, Y.-Y.; Lin, J.-H.; Wu, T.-M.; Li, S.-Y. The production of poly(3-hydroxybutyrate) by thermophilic *Caldimonas manganoxidans* from glycerol. *J. Polym. Res.* **2018**, *25*, 85. [CrossRef]
12. Kourilova, X.; Pernicova, I.; Vidlakova, M.; Krejcirik, R.; Mrazova, K.; Hrubanova, K.; Krzyzanek, V.; Nebesarova, J.; Obruca, S. Biotechnological conversion of grape pomace to poly(3-hydroxybutyrate) by moderately thermophilic bacterium *Tepidimonas taiwanensis*. *Bioengineering* **2021**, *8*, 141. [CrossRef]
13. Kourilova, X.; Pernicova, I.; Sedlar, K.; Musilova, J.; Sedlacek, P.; Kalina, M.; Koller, M.; Obruca, S. Production of polyhydroxyalkanoates (PHA) by a thermophilic strain of *Schlegelella thermodepolymerans* from xylose rich substrates. *Bioresour. Technol.* **2020**, *315*, 123885. [CrossRef]
14. Urbieta, M.S.; Donati, E.R.; Chan, K.-G.; Shahar, S.; Sin, L.L.; Goh, M.M. Thermophiles in the genomic era: Biodiversity, science, and applications. *Biotechnol. Adv.* **2015**, *33*, 633–647. [CrossRef]
15. Koller, M.; Bona, R.; Chiellini, E.; Braunegg, G. Extraction of short-chain-length poly-[(*R*)-hydroxyalkanoates] (scl-PHA) by the "anti-solvent" acetone under elevated temperature and pressure. *Biotechnol. Lett.* **2013**, *35*, 1023–1028. [CrossRef]
16. Ramsay, J.A.; Berger, E.; Voyer, R.; Chavarie, C.; Ramsay, B.A. Extraction of poly-3-hydroxybutyrate using chlorinated solvents. *Biotechnol. Tech.* **1994**, *8*, 589–594. [CrossRef]
17. Khosravi-Darani, K.; Vasheghani-Farahani, E.; Shojaosadati, S.A.; Yamini, Y. Effect of process variables on supercritical fluid disruption of *Ralstonia eutropha* cells for poly(*R*-hydroxybutyrate) recovery. *Biotechnol. Prog.* **2004**, *20*, 1757–1765. [CrossRef]
18. Hejazi, P.; Vasheghani-Farahani, E.; Yamini, Y. Supercritical fluid disruption of *Ralstonia eutropha* for poly(β-hydroxybutyrate) recovery. *Biotechnol. Prog.* **2003**, *19*, 1519–1523. [CrossRef]
19. Holmes, P.A.; Lim, G.B. Separation Process. U.S. Patent 4910145, 20 March 1990.
20. Choi, J.I.; Lee, S.Y. Efficient and economical recovery of poly(3-hydroxybutyrate) from recombinant *Escherichia coli* by simple digestion with chemicals. *Biotechnol. Bioeng.* **2000**, *62*, 546–553. [CrossRef]
21. Berger, E.; Ramsay, B.A.; Ramsay, J.A.; Chavarie, C.; Braunegg, G. PHB recovery by hypochlorite digestion of non-PHB biomass. *Biotechnol. Tech.* **1989**, *3*, 227–232. [CrossRef]
22. Marudkla, J.; Patjawit, A.; Chuensangjun, C.; Sirisansaneeyakul, S. Optimization of poly(3-hydroxybutyrate) extraction from *Cupriavidus necator* DSM 545 using sodium dodecyl sulfate and sodium hypochlorite. *Agric. Nat. Resour.* **2018**, *52*, 266–273. [CrossRef]
23. Mannina, G.; Presti, D.; Montiel-Jarillo, G.; Suárez-Ojeda, M.E. Bioplastic recovery from wastewater: A new protocol for polyhydroxyalkanoates (PHA) extraction from mixed microbial cultures. *Bioresour. Technol.* **2019**, *282*, 361–369. [CrossRef]
24. Kathiraser, Y.; Aroua, M.K.; Ramachandran, K.B.; Tan, I.K.P. Chemical characterization of medium-chain-length polyhydroxyalkanoates (PHAs) recovered by enzymatic treatment and ultrafiltration. *J. Chem. Technol. Biotechnol.* **2007**, *82*, 847–855. [CrossRef]
25. Rodriguez-Valera, F. Halobacteria as producers of polyhydroxyalkanoates. *FEMS Microbiol. Lett.* **1992**, *103*, 181–186. [CrossRef]
26. Bhattacharyya, A.; Pramanik, A.; Maji, S.K.; Haldar, S.; Mukhopadhyay, U.K.; Mukherjee, J. Utilization of vinasse for production of poly-3-(hydroxybutyrate-*co*-hydroxyvalerate) by *Haloferax mediterranei*. *AMB Express* **2012**, *2*, 34. [CrossRef]
27. Brandl, H.; Gross, R.A.; Lenz, R.W.; Clinton Fuller, A. *Pseudomonas oleovorans* as a Source of Poly(β-Hydroxyalkanoates) for Potential Applications as Biodegradable Polyesters. *Appl. Environ. Microbiol.* **1988**, *54*, 1977–1982. [CrossRef]
28. Kourilova, X.; Schwarzerova, J.; Pernicova, I.; Sedlar, K.; Mrazova, K.; Krzyzanek, V.; Nebesarova, J.; Obruca, S. The first insight into polyhydroxyalkanoates accumulation in multi-extremophilic *Rubrobacter xylanophilus* and *Rubrobacter spartanus*. *Microorganisms* **2021**, *9*, 909. [CrossRef]
29. Rupprecht, K.R.; Lang, E.Z.; Gregory, S.D.; Bergsma, J.M.; Rae, T.D.; Fishpaugh, J.R. A precise spectrophotometric method for measuring sodium dodecyl sulfate concentration. *Anal. Biochem.* **2015**, *486*, 78–80. [CrossRef]
30. Arikawa, H.; Sato, S.; Fujiki, T.; Matsumoto, K. Simple and rapid method for isolation and quantitation of polyhydroxyalkanoate by SDS-sonication treatment. *J. Biosci. Bioeng.* **2017**, *124*, 250–254. [CrossRef]
31. Middelberg, A.P.J. Process-scale disruption of microorganisms. *Biotechnol. Adv.* **1995**, *13*, 491–551. [CrossRef]
32. Katsui, N.; Tsuchido, T.; Takano, M.; Shibasaki, I. Effect of preincubation temperature on the heat resistance of *Escherichia coli* having different fatty acid compositions. *J. Gen. Microbiol.* **1981**, *122*, 357–361. [CrossRef]
33. Rees, P.; Cumming, R.H.; Watson, J.S. Rheology of heated bacterial DNA. In Proceedings of the 1994 ICheme Research Event, London, UK, 5–6 January 1994; Volume 1, pp. 183–185.
34. Koller, M.; Marsalek, L.; Miranda de Sousa Dias, M.; Braunegg, G. Producing microbial polyhydroxyalkanoate (PHA) biopolyesters in a sustainable manner. *New Biotechnol.* **2017**, *37*, 24–38. [CrossRef]

35. Sedlacek, P.; Slaninova, E.; Enev, V.; Koller, M.; Nebesarova, J.; Marova, I.; Hrubanova, K.; Krzyzanek, V.; Samek, O.; Obruca, S. What keeps polyhydroxyalkanoates in bacterial cells amorphous? A derivation from stress exposure experiments. *Appl. Microbiol. Biotechnol.* **2019**, *103*, 1905–1917. [CrossRef] [PubMed]
36. Shen, R.; Ning, Z.Y.; Lan, Y.X.; Chem, J.C.; Chen, C.G. Manipulation of polyhydroxyalkanoate granular sizes in Halomonas bluephagenesis. *Metab. Eng.* **2019**, *54*, 117–126. [CrossRef] [PubMed]
37. Samorì, C.; Basaglie, M.; Casella, S.; Favaro, L.; Galletti, P.; Giorgini, L.; Marchi, D.; Mazzocchetti, L.; Torri, C.; Tagliavini, E. Dimethyl carbonate and switchable anionic surfactants: Two effective tools for the extraction of polyhydroxyalkanoates from microbial biomass. *Green Chem.* **2015**, *17*, 1047–1056. [CrossRef]

polymers

Article

Two-Stage Bio-Hydrogen and Polyhydroxyalkanoate Production: Upcycling of Spent Coffee Grounds

Beom-Jung Kang [1,†], Jong-Min Jeon [1,†], Shashi Kant Bhatia [2], Do-Hyung Kim [3], Yung-Hun Yang [2], Sangwon Jung [4] and Jeong-Jun Yoon [1,*]

1 Green & Sustainable Materials R&D Department, Korea Institute of Industrial Technology (KITECH), Chunan-si 31056, Republic of Korea
2 Department of Biological Engineering, Konkuk University, Seoul 27478, Republic of Korea
3 Sustainable Technology and Wellness R&D Group, Korea Institute of Industrial Technology (KITECH), Jeju-si 63243, Republic of Korea
4 Department of Bio and Fermentation Convergence Technology, Kookmin University, Seoul 02707, Republic of Korea
* Correspondence: jjyoon@kitech.re.kr; Tel.: +82-41-589-8266
† These authors contributed equally to this work.

Abstract: Coffee waste is an abundant biomass that can be converted into high value chemical products, and is used in various renewable biological processes. In this study, oil was extracted from spent coffee grounds (SCGs) and used for polyhydroxyalkanoate (PHA) production through *Pseudomonas resinovorans*. The oil–extracted SCGs (OESCGs) were hydrolyzed and used for biohydrogen production through *Clostridium butyricum* DSM10702. The oil extraction yield through *n*–hexane was 14.4%, which accounted for 97% of the oil present in the SCGs. OESCG hydrolysate (OESCGH) had a sugar concentration of 32.26 g/L, which was 15.4% higher than that of the SCG hydrolysate (SCGH) (27.96 g/L). Hydrogen production using these substrates was 181.19 mL and 136.58 mL in OESCGH and SCGH media, respectively. The consumed sugar concentration was 6.77 g/L in OESCGH and 5.09 g/L in SCGH media. VFA production with OESCGH (3.58 g/L) increased by 40.9% compared with SCGH (2.54 g/L). In addition, in a fed–batch culture using the extracted oil, cell dry weight was 5.4 g/L, PHA was 1.6 g/L, and PHA contents were 29.5% at 24 h.

Keywords: bio-hydrogen; polyhydroxyalkanoate; mcl-PHA; *Pseudomonas resinovorans*

Citation: Kang, B.-J.; Jeon, J.-M.; Bhatia, S.K.; Kim, D.-H.; Yang, Y.-H.; Jung, S.; Yoon, J.-J. Two-Stage Bio-Hydrogen and Polyhydroxyalkanoate Production: Upcycling of Spent Coffee Grounds. *Polymers* **2023**, *15*, 681. https://doi.org/10.3390/polym15030681

Academic Editor: Hu Li

Received: 13 December 2022
Revised: 20 January 2023
Accepted: 20 January 2023
Published: 29 January 2023

1. Introduction

Coffee is the world's most-consumed drink, and about 167.17 million metric tons of coffee were produced in 2021 [1]. This is three times the world's average coffee consumption, and about 190,000 metric tons of green beans were imported in 2021 [2]. The problem with coffee consumption is that spent coffee grounds (SCGs) are produced in large quantities. After extracting the drink from the coffee bean, a large amount of SCGs remain that are known to have a dry weight of about 65% of the initial cherries [3]. Based on import figures, it is estimated that about 150,000 metric tons of SCGs were produced in Korea in 2021. Most of the SCGs are dumped as garbage and buried or incinerated [4]. Since coffee is not produced in Korea, importing coffee is tantamount to importing garbage and means that coffee imports lead to 150,000 metric tons of waste every year. From an environmental engineering perspective, SCGs are organic waste resources, of which the main components are carbohydrates, proteins, and fats. Therefore, it is necessary to devise a plan to utilize SCGs in various ways.

Organic waste and biomass resources (rice straw, marine algae, food waste, etc.) usually have a large amount of carbohydrates that can be used by microorganisms, so they are used in the form of monosaccharides or oligomers through hydrolysis [5–7]. This can be applied equally to SCGs, and there are several studies on using SCGs as a resource.

Studies have been conducted to utilize SCGs that have undergone hydrolysis to produce butanol, ethanol, methane, and polyhydroxyalkanoate (PHA) [8–10]. However, the SCGs were not fully utilized by hydrolysis only. As SCGs are composed of carbohydrates (about 60%), proteins (about 15%), and fats (about 15%) [4,11], it is possible to devise a plan to separate and utilize the oil among the ingredients of SCGs. In previous studies, SCG oil was extracted and utilization measures such as PHA production were devised [9]. However, the studies only used SCG oil, and they did not fully use SCGs. Therefore, the extracted oil and the hydrolysates of the oil–extracted SCGs may be utilized to increase the value of the SCGs.

Recently, interest in hydrogen energy and bioplastics is increasing because of global issues such as environmental pollution and fossil fuel depletion [12,13]. Hydrogen is one of the promising eco-friendly resources, and has high energy yield of 122 kJ/g which is 2.75 times greater than that of hydrocarbon fuel [14]. It can be generated by fossil fuel, electrolysis of water or biological processes—among them, bio hydrogen, which can be produced through utilization of biomass hydrolysates by microbes [15]. Therefore, various biomass hydrolysates such as corn stover, marine algae, oil palm empty fruit bunches, rice straw and sugarcane bagasse were used to produce bio-hydrogen via dark fermentation, resulting in 0.76 to 2.24 hydrogen yield (mol/mol) [14–17]. As in the above-mentioned studies of bio-hydrogen production from various biomass hydrolysates, oil-extracted SCGs could be considered an alternative carbon source from which to produce bio-hydrogen.

Bioplastics are eco–friendly plastics that can be produced or decomposed by microorganisms [18]. Similar to fossil-fuel based plastics, they offer a variety of physical properties. PHA, a representative bioplastic, is a polymer in which 3–hydroxyalkanoate produced by microorganisms for storing carbon sources is ester–bonded under unfavorable conditions for growth [19]. PHA is classified depending on the number of carbon atoms in the monomers as short chain length (scl, C3–5), medium chain length (mcl, C6–14), and long chain length (lcl, C15–18) PHA; additionally, the type and physical properties of PHA vary depending on the metabolism or substrate provided by the bacteria [20]. Many studies have been conducted on scl–PHA and mcl–PHA because of their physical properties and production yield advantages; scl–PHA also has solid properties in the form of polymers such as 3-hydroxybutyrate (3HB) and 3-hydroxyvalerate (3HV) [21], while mcl–PHA has elastic and adhesive properties [18]. In order for bioplastics to have various physical properties and forms as with fossil fuel–based plastics, it is necessary to promote research on mcl–PHA, which shows relatively low yield. Mcl–PHA can be produced through various bacteria and with different carbon sources. Among such bacterial strains, *Pseudomonas* sp. is a representative strain known to produce mcl–PHA [22]. Several studies have shown that *Pseudomonas* sp. is able to use fatty acids such as olive oil and waste cooking oil, thereby enhancing the value of various organic waste resources [23].

Therefore, this study aims to (i) investigate SCG oil extraction and yield, (ii) hydrolyze oil–extracted SCGs and use them as hydrogen production substrates, (iii) investigate PHA production possibilities using extracted SCG oil, and (iv) investigate the possibility of mass culture using a 10 L fermenter.

2. Materials and Methods

2.1. Materials

All reagents and chemicals were procured at the highest purity. M9 minimal salts (5×), 3-Hydroxyoctanoic acid, 3-Hydroxydecanoic acid, and 3-Hydroxydodecanoic acid were purchased from Sigma Aldrich (St. Louis, MO, USA). Chloroform, H_2SO_4, *n*-hexane, HCl, NH_4Cl, NaOH, methanol, NaCl, sodium acetate, sodium bicarbonate, palmitic acid, oleic acid, stearic acid, linoleic acid, L–cysteine hydrochloride and antifoam were purchased from Daejung Chemical & Metals (Siheung-si, Republic of Korea). LB media, yeast extract, peptone, beef extract, soybean extract and urea were purchased from BD-DIFCO (Detroit, MI, USA).

2.2. SCGs Oil Extraction

SCGs were obtained from a local cafe in Cheonan-si, South Korea. To dry the SCGs, they were spread widely and sun–dried for 8 h and then dried in a drying oven at 105 °C for 24 h. Oil was extracted from the dried SCGs using a Soxhlet extractor with 200 mL of *n*–hexane as a solvent, and 25 g of SCGs was put into a weighed cellulose filter for oil extraction for 1.5 h. The heating mantle was kept at 70 °C. After oil extraction was completed, the filter containing SCGs was dried in a fume hood for 24 h to evaporate the *n*–hexane and then weighed. The oil mixed with the *n*–hexane was purified by evaporation at 50 °C and 50 rpm for 30 min using a rotary evaporator. The concentrated SCGs oil was weighed to confirm the amount of the *n*–hexane residue and compared with the SCGs weight change. Mass extraction was carried out for future experiments with the products sterilized at 121 °C for 15 min in an autoclave and then refrigerated.

$$\text{Weight of SCGs extract} = (\text{filter weight} + \text{SCGs weight}) - (\text{dried filter weight} + \text{SCGs weight after oil extraction}) \quad (1)$$

$$\text{The amount of } n\text{-hexane remaining in SCGs oil } (w/w\%) = (\text{SCGs oil weight} - \text{SCGs weight variation})/\text{SCGs oil weight} \times 100 \quad (2)$$

$$\text{SCGs oil extraction yield } (\%) = \text{SCGs weight variation}/\text{SCGs initial weight} \times 100 \quad (3)$$

Gas chromatography–mass spectrometry (GC/MS; Claus 500, Perkin Elmer, Waltham, MA, USA) analysis was performed to analyze the main components of the extracted SCGs oil, and a gas chromatograph (GC; 6890N, Agilent Technologies, Santa Clara, CA, USA) was used for quantitative analysis.

2.3. Oil–Extracted SCGs Hydrolysis and Hydrogen Production

The oil–extracted SCGs (OESCGs) were used as a substrate for hydrogen production. OESCGs were dried in the fume hood for 12 h to remove residual *n*–hexane. Then, hydrolysis was performed with a 10% solid/liquid (S/L) ratio and 1% H_2SO_4 at 130 °C for 1 h. OESCG hydrolysate (OESCGH) was centrifuged at 3500 rpm for 30 min, and the supernatant was recovered. The composition of the OESCGH was analyzed through high–performance liquid chromatography (HPLC; 1200 series, Agilent Technologies, Santa Clara, CA, USA). The OESCGH was mass–produced and then sterilized in an autoclave for 15 min at 121 °C before being refrigerated for use in subsequent experiments.

An experiment was conducted to compare the hydrogen productivity of general SCG hydrolysate (SCGH) and OESCGH. The culture medium was defined–reinforced clostridium medium (RCM). Here, SCGH or OESCGH was added and diluted to a total sugar concentration of 10 g/L. *Clostridium butyricum* DSM10702 was used for hydrogen production and was inoculated in the anaerobic chamber under 99.9% of N_2. The culture conditions were pH 5.5 (adjusted by HCl), 37 °C, and 150 rpm. Gas production, hydrogen production, and metabolite change were measured for 32 h. Defined RCM was composed as follows: 3 g/L yeast extract, 10 g/L peptone, 10 g/L beef extract, 0.5 g/L L–cysteine hydrochloride, 3 g/L sodium acetate, 5 g/L sodium chloride, 5 g/L sodium bicarbonate, and 100 μL/L antifoam.

2.4. PHA Production Using SCGs Oil

A PHA production test was conducted to confirm the suitability of the SCGs oil. *Pseudomonas resinovorans*, which is known to produce mcl–PHA using oil as a substrate, was used as the inoculum. A stock stored in a deep freezer was pre–cultured in 5 mL of lysogeny broth (LB) media. Pre–cultured bacteria were inoculated into the main culture medium after 24 h. The medium was M9 salt medium, and 2% oil was added as a substrate. After that, for C/N ratio optimization, various N sources were added at 0.1% and 0.5%, and productivity was confirmed. A fed–batch culture was performed using ammonium chloride as an N source and SCGs oil as a substrate. The reactor was a 10 L fermenter

(BIOCNS Co., Ltd., Deajeon, Republic of Korea), with a working volume of 3 L, in duplicate, and inoculation volume of 10%. The medium was M9 salt medium, initial SCGs oil concentration was 2%, and SCGs oil was fed at 5% at 4 h and at 18 h. An amount of 1 M HCl and 3 M NaOH were used to adjust the pH to 6.8 ± 0.2, and aeration was maintained at 10 L/3 L/1 min to adjust to 30% dissolved oxygen (DO). The reactor was operated at 600 rpm and 30 °C for 96 h, and samples were taken to analyze cell dry weight (CDW) and PHA contents. The culture medium was centrifuged at 4 °C and 3500 rpm, for 1 h. The supernatant was discarded, and the cell pellets were freeze–dried after washing. Acetone was added to the freeze–dried cells to dissolve PHA, and methanol was treated at 1:10 to recover PHA precipitated at the bottom.

2.5. Analysis

GC/MS was used for qualitative analysis using the method described in previous studies [24]. After obtaining fatty acid methyl ester (FAME) from the SCGs oil, the analysis was conducted. Peaks were identified by the mass spectrometric fragmentation data and confirmed by comparison to spectral data that was available from the online libraries of Wiley (http://www.palisade.com, accessed on 12 December 2022) and NIST (http://www.nist.gov, accessed on 12 December 2022).

For quantitative analysis of the SCGs oil, fatty acid (for standard curve), and PHA, each sample was analyzed using the GC. Samples of 10 mg of SCGs oil or fatty acid were contained in a 15 mL glass round bottom tube for pretreatment for FAME analysis.

For quantitative analysis of PHA, 1 mL of culture medium was centrifuged at 13,000 rpm for 5 min. Afterwards, the cell pellet was washed twice with deionized water (DW), and the resuspended cells were placed in a 15 mL glass round bottom tube that was sealed with teflon, frozen in a freezer, and freeze–dried. Afterwards, the CDW was measured, and the dried cells were treated for FAME analysis. For FAME, 1 mL of chloroform and 1 mL of sulfuric acid–methanol solution (15:85) were added and the sample was reacted at 100 °C for 2 h on a heating block. The sample was then cooled at room temperature for 10 min and then chilled on ice for 10 min. After adding 1 mL of iced cold water, the sample was shaken and the organic phase below was analyzed. Of the prepared sample, 1 μL was injected through an auto–sampler and analyzed with a flame ionization detector (FID).

The analysis conditions were as follows: a 30 m × 0.25 mm DB–FFAP capillary column, hydrogen 40 mL/min as the carrier gas, air zero flow 450 mL/min and high purity nitrogen gas 45 mL/min as the makeup flow, inlet temperature 250 °C, detector temperature 250 °C, and the oven held at 90 °C for 5 min, heated to 220 °C with a heating rate of 20 °C/min, and then held at 220 °C for 7 min.

For analysis of bio-hydrogen productivity, the gas produced in the serum bottle was collected through a 1 mL syringe (Hamilton Company, NV, USA) and injected into the GC. Hydrogen measurement was performed through a thermal conductivity detector (TCD) with a 10 m × 0.3 mm CP–Molsieve 5A capillary column. The analysis conditions were as follows: high purity N_2 gas (99.999%) was used as the carrier gas, the inlet temperature was 100 °C, the detector temperature was 250 °C, and the oven was held at 80 °C.

HPLC was used to measure metabolic product changes in the bio-hydrogen production. A 1 mL sample was centrifuged for 5 min at 4 °C and 13,000 rpm. Then, the supernatant was filtered through a PTFE membrane filter with 0.45 μm pore size. The HPLC analysis conditions were as follows: 5 mM H_2SO_4 as a mobile phase, 60 °C for column temperature, 300 mm × 7.8 mm Aminex HPX-87H ion excursion column, 55 °C for detector temperature, refractive index detector, 25 μL of injection volume, 0.6 mL/min of flow rate, and 55 min of running time.

The measured hydrogen and biogas were converted to values at standard pressure and temperature before being applied to the modified Gompertz model below [25]:

$$\mathrm{H} = P \times \exp\{-\exp[((Rm \times e))/P \times (\lambda - \mathrm{t}) + 1]\} \quad (4)$$

H is the estimated hydrogen production (mL), *P* is the hydrogen production potential (mL), *Rm* is the maximum hydrogen production rate (mL/h), λ is the lag phase (h), t is the time (h), and *e* is 2.7182 [25].

3. Result and Discussion

3.1. Extracted Oil from SCGs

In general, oil extraction from oilseed includes press extraction and organic solvent extraction. The organic solvent extraction method was selected because SCGs are not suitable for the press extraction method as they are finely ground. Solvents used for oil extraction include *n*–hexane, ethanol, methanol, pentane, acetone, and isopropanol [26,27]. According to Al–Hamamre et al. and Pichai et al., various solvents were used to extract SCGs oil, and relatively high yields were found when *n*–hexane was used [26–28]. According to Efthymiopoulos et al., the amount of SCGs and *n*–hexane and extraction time used were 25 g, 200 mL, and 1.5 h, respectively [29].

The total weight of the cellulose filter and SCGs was 30.05 g (Table 1). The total weight of the dried cellulose filter and SCGs after oil extraction was 26.43 g, and the SCGs weight variation was 3.62 g. The mixture of *n*–hexane and SCGs oil was fractionally distilled, and the concentrated SCGs oil was 4.03 g. There was a difference of 0.41 g between the concentrated SCGs oil and the SCGs weight variation. It was found that about 10.4% of *n*–hexane in SCGs oil remained even after fractional distillation (data not shown). Therefore, it is important to select strains that can operate efficiently considering that *n*–hexane is known to be toxic to a relatively large number of bacterial species [30].

Table 1. Conditions of spent coffee grounds oil and yield.

Extraction Conditions of SCGs Oil		Fractional Distillation of Extracted SCGs Oil		Oil Yield	
Hexane	200 mL	SCGs oil	4.03 g	Filter + SCGs	26.43 g
SCGs	25.00 g	Operation time	0.5 h	Weight of SCGs extract	3.62 g
Cellulose filter	5.05 g	Temperature	50 °C	Residual *n*–hexane	0.41 g
Operation time	1.5 h	Rotation	50 rpm	Oil yield	14.46%

The SCGs oil extraction yield was 14.4%, and the fat content in the SCGs was 14.9% based on the dry weight (Table 2). Based on these figures, the oil extraction rate was 97.0%. The results of analyzing the components of SCGs oil through GC/MS and GC showed that the SCGs consisted of 34.1% palmitic acid, 16.8% stearic acid, 10.3% oleic acid, and 38.8% linoleic acid (Supplementary Figure S1). Other studies found that oil accounted for 10–15% of SCGs [29,31]. One of those studies showed that SCGs oil consists of palmitic acid, stearic acid, oleic acid, and linoleic acid [29], which are the same components as for the SCGs used in the present experiment. There was a difference in oil content and composition ratio, and this is believed to be because of different coffee production areas. A large amount of coffee oil was extracted and then sterilized and refrigerated for use in the next experiment. According to previous studies conducted by Al-Hamamreet et al. and others, 14.7% [26] and 15.28% [27] were extracted for 30 min using the Soxhlet extractor. Similar to this study, it was shown that the entirety of the oil in SCGs could be recovered. In addition, although there was a difference in content of fatty acids in SCGs oil at various studies, most of them were palmitic acid, linolcic acid, oleic acid, and stearic acid, showing that they were similar to the results of this study [26].

Table 2. Chemical composition of spent coffee grounds.

Analysis Items	Results	Units
Carbohydrates	62.7	g/100 g
Proteins	12.8	g/100 g
Fats	13.5	g/100 g
Ash	1.6	g/100 g
Moisture	9.4	g/100 g

3.2. *Hydrolysis of Oil Extracted-SCGs*

For hydrogen production experiments, hydrolysis was performed using OESCGs. The OESCGs were hydrolyzed under conditions derived from previous experiments (S/L ratio 10% (*w*/*w*), H_2SO_4 1.0% (*w*/*w*), 130 °C, and 1 h). The hydrolysis results of general SCGs and OESCGs are shown in Figure 1, and the total sugar concentration of the OESCGH was 32.26 g/L (Figure 1A). This is about 15.4% higher than that for the SCGH (27.96 g/L).

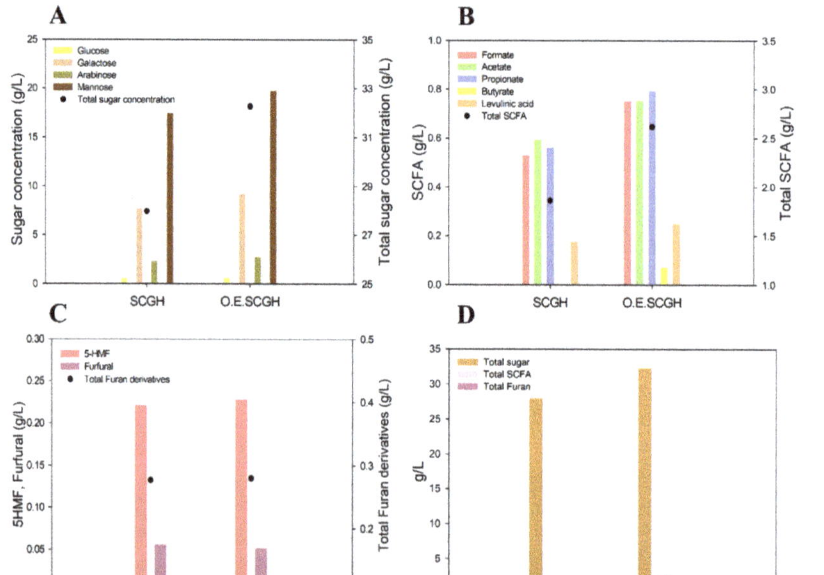

Figure 1. Comparison of spent coffee grounds hydrolysate (SCGH) and oil–extracted spent coffee grounds hydrolysate (OESCGH). (**A**) sugar composition; (**B**) SCFA composition; (**C**) furan derivatives composition; and (**D**) composition of sugar, SCFA and furan derivatives.

In addition, this was similar to the amount when the S/L ratio was 20% (*w*/*w*) and H_2SO_4 1.0% (*w*/*w*) (35.93 g/L), which is thought to be because of an increase in carbohydrate content after oil is extracted. The carbohydrate content of dried SCGs was 69.2% (Table 2), while that of OESCGs was 80.8% (=69.2/[100 − 14.4] × 100).

Comparing the sugar concentration with other studies, Obruca et al. reported that the total sugar concentration was 50.1 g/L under the conditions of 1% H_2SO_4, 121 °C, 90 min, and 15% of the S/L ratio [9]. Hudeckova, Helena, et al. showed 23.86 g/L under the conditions of 2.7% H_2SO_4, 121 °C, 15 min, and 10% S/L ratio [32]. It was confirmed that the sugar recovery rate varies depending on H_2SO_4 concentration, reaction time, and S/L ratio. It was verified that our experimental conditions recovered a relatively high concentration of sugar even that H_2SO_4 and S/L ratios were less than the previous experiments.

Furthermore, the monosaccharide recovery rate compared with the carbohydrate content of SCGs (69.2%) was 40.4%, and that of the OESCGs (80.8%) was 39.9%. It was also found that the ratio (3.1:6.9) of galactose and mannose, which are the main sugars of the SCGH, remained the same as that of the general SCGs hydrolysis (3.1:6.9). These results show that hydrolysates of the same composition with higher sugar concentrations can be obtained through oil extraction.

The concentration of short chain fatty acids (SCFAs) tended to be increased, similar to the sugar concentration, with the SCFA concentration of the OESCGH at 2.61 g/L, higher than that of the SCGH (1.86 g/L), whereas the concentrations of the furan derivatives were 0.27 g/L and 0.28 g/L, respectively, with no increasing tendency (Figure 1B,C). This result is similar to that of the previous hydrolysis for each S/L ratio where SCFAs tended to increase as the S/L ratio increased.

It is known that the concentration of furan derivatives that inhibits the growth of microorganisms is low, and it has been shown that bacterial growth is inhibited above 0.5 g/L 5-hydroxyl methyl furfural or above 2.0 g/L furfural [33,34]. In the SCGs hydrolysate, furan derivatives were produced at a concentration that did not cause inhibition of microbial growth, so the post-treatment process could be omitted. The above inhibitors generated during biomass hydrolysis are removed through physico-chemical processes such as adsorption using powdered activated carbon. This leads to an increase in process costs. Such a post-treatment process is not required for the SCGH; therefore, it provides process advantages and may be used as an ideal substrate.

3.3. Bio-Hydrogen Production Using Oil-Extracted SCGs Hydrolysate

Hydrogen production experiments were conducted using the SCGs hydrolysates prepared above, and hydrogen productivity determined for the SCGs hydrolysates from general SCGs and OESCGs (Figure 2). When using the OESCGH, the lag time was shorter and the bio-hydrogen production was higher than when using the SCGH (Figure 2A–C). Lag time was 4.8 h and 6.1 h, respectively, and hydrogen production was 181.19 mL and 136.58 mL, respectively, in OESCGH and SCGH media. However, there was no significant difference in hydrogen production yield.

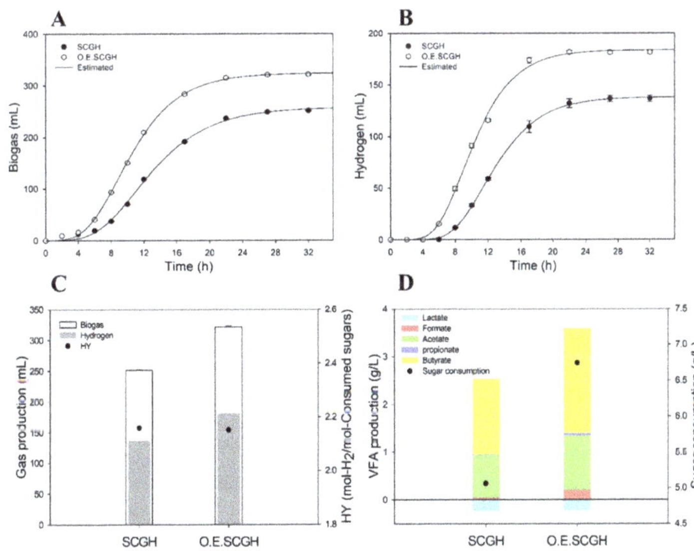

Figure 2. Comparison of dark fermentation productivities based on SCGH, OESCGH media (10 g/L of sugar concentration). (**A**) biogas profile; (**B**) hydrogen profile; (**C**) production amount of biogas and hydrogen, and hydrogen yield; and (**D**) VFA production and sugar consumption.

There was a large difference in VFA production and sugar consumption between SCGH and OESCGH (Figure 2D). In OESCGH media, the consumed sugar concentration was 6.77 g/L, 33.0% higher than in SCGH media (5.09 g/L). VFA production increased with OESCGH (3.58 g/L) by 40.9% compared with SCGH (2.54 g/L). Although OESCGH media showed a fast lag time, high production of hydrogen and VFA, and high sugar consumption, it seems that the yield was similar to that of the SCGH media and this was because of the input amount of the hydrolysate.

When producing RCM–based media, each hydrolysate was added so that the total sugar concentration was 10 g/L. Accordingly, relatively less was added for the volume of OESCGH because OESCGH had a high sugar concentration. Although sugar concentrations were the same in each medium, there were differences in the amounts of SCFA, furan derivatives, and H_2SO_4. One of these factors is believed to have caused the difference in productivity. When SCGH and OESCGH were put in the RCM media in the same volume, there was no significant difference in hydrogen production, sugar consumption, or lag time (Figure 3). Accordingly, the difference in productivity seems to be because of the concentration of sulfate ion from the H_2SO_4 and buffer capacity. In summary, OESCGH has a high sugar concentration, which can induce cost reductions and play a good role in increasing productivity. In addition, it shows a high yield compared to various biomass used in the production of bio–hydrogen, although there may be differences depending on the fermentation conditions (Table 3). It is estimated due to the different composition of hydrolystates, because most lignocellulose biomass has glucose, xylose, and arabinose as the main sugar components, but SCGs hydrolysates are mainly composed of mannose and galactose.

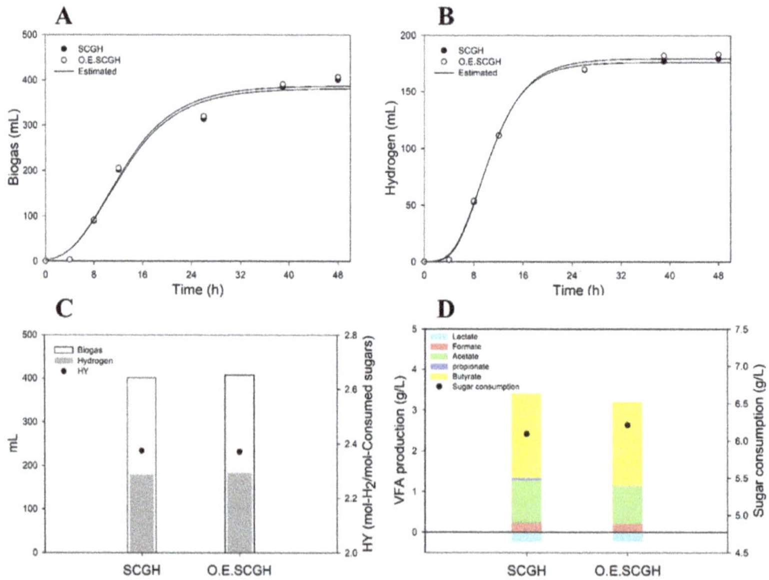

Figure 3. Comparison of dark fermentation productivities based on SCGH, OESCGH media (same volumes of hydrolysates in media [30% (*v/v*)]). (**A**) biogas profile; (**B**) hydrogen profile; (**C**) production amount of biogas and hydrogen, and hydrogen yield; and (**D**) VFA production and sugar consumption.

Table 3. Comparisons of hydrogen productivities using various biomass.

Biomass	Carbohydrates in the Hydrolysate	Organism	H_2 Yield (mol-H_2/mol-Consumed or Added Sugar)	References
Sugarcane bagasse	Glucose, Xylose, Arabinose	*Clostridium butyricum*	1.73	[14]
Rice straw	Xylose	*Clostridium butyricum* CGS5	0.76	[15]
Oil palm empty fruit bunch	Glucose, Xylose	*Enterobacter* sp. KBH6958	1.68	[16]
Corn stover	Glucose, Xylose, Arabinose	*Thermoanaerobacterium thermosaccharolyticum* W16	2.24	[17]
Oil-extracted spent coffee grounds	Glucose, Galactose, Mannose, Arabinose	*Clostridium butyricum*	2.37	This study

3.4. Nitrogen Sources and C/N Ratio Balance Comparison

Generally, the PHA accumulation pathway is triggered when the carbon source is sufficient, but nitrogen sources are limited [35]. Considering that SCGs oil exists as micelles during fermentation and is utilized by the β-oxidation pathway, selection of proper nitrogen sources is important. Thus, six different nitrogen sources (ammonium chloride, yeast extract, beef extract, peptone, soybean extract, and urea) were evaluated at concentrations of 0.1% and 0.5%, along with 2% of SCGs oil in M9 minimal media. Ammonium chloride, peptone, and urea showed high cell growth, in the range of 2.6~3.1 g/L of CDW, compared with the other nitrogen sources; most of the nitrogen sources showed similar PHA production in the range of 0.42~0.51 g/L at 0.1% concentration, except yeast extract (Figure 4A). While ammonium chloride and beef extract showed higher cell growth at 3.0 g/L and 3.4 g/L, respectively, compared with the other nitrogen sources, at 0.5% concentration beef extract provided slightly better production of PHA at 0.6 g/L compared with 0.5 g/L of PHA produced with ammonium chloride (Figure 4B). Yeast extract showed a negative effect on PHA production, while it was confirmed that *P. resinovorans* produced PHA with ammonium chloride even without a complex media component.

To investigate the C/N ratio balance when the SCGs oil is used as a carbon source, ammonium chloride was used with concentrations in the range of 0.01 to 1%. Cell growth and PHA production increased up to 0.5% ammonium chloride, reaching 3.0 g/L and 0.5 g/L, respectively (Figure 5). However, with 1% ammonium chloride, cell growth and PHA production were both reduced dramatically. It may be that excessive nitrogen exposure triggered nitrogen regulatory proteins that had a negative effect on cell growth metabolism, or there may be an effect on the β-oxidation pathway in *Pseudomonas* species [36,37].

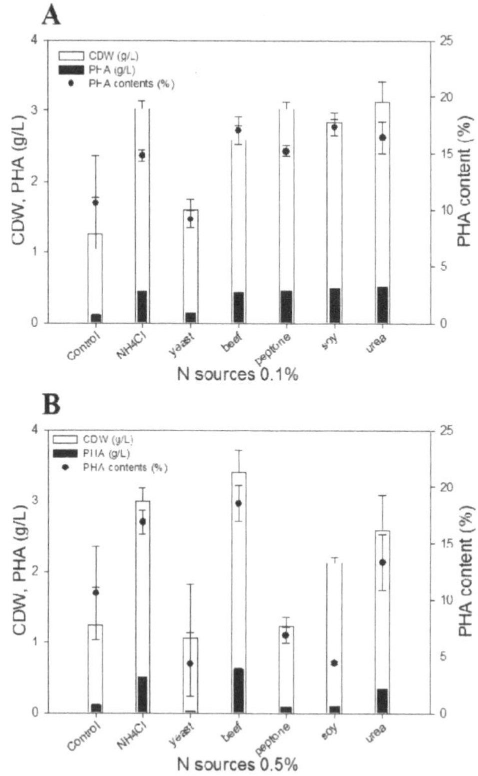

Figure 4. Nitrogen sources comparison for PHA production. Six different nitrogen sources were evaluated at 0.1% (**A**) and 0.5% (**B**) concentration; 2% coffee oil was used in the culture as a constant.

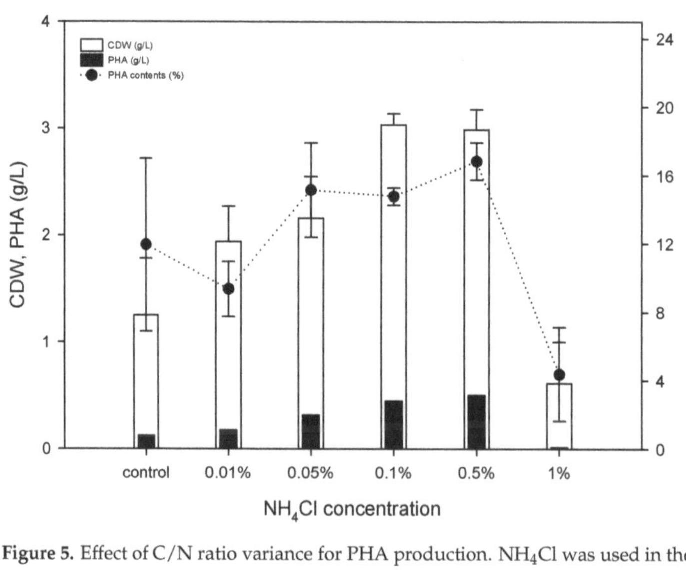

Figure 5. Effect of C/N ratio variance for PHA production. NH_4Cl was used in the range of 0.01 to 1%, and 2% of coffee oil was used in the culture as a constant.

3.5. Fed-Batch Culture for PHA Production

Application of a fed-batch strategy to increase PHA productivity was proved by many other related studies [38–40]. Therefore, *P. resinovorans* was cultured with an optimized C/N balance on a flask scale, and cell growth and PHA production monitored (Figure 6A). Cell growth during the fermentation reached 2.7 g/L after 24 h and 7.9 g/L after 72 h. While PHA production tended to increase steadily, increasing to 0.75 g/L after 24 h and 2.4 g/L after 72 h, PHA content was maintained at around 30%. It may be that knock-out of PHA-degrading enzymes such as *phaZ* or optimization of the PHA accumulation pathway is required in order to increase the PHA content further.

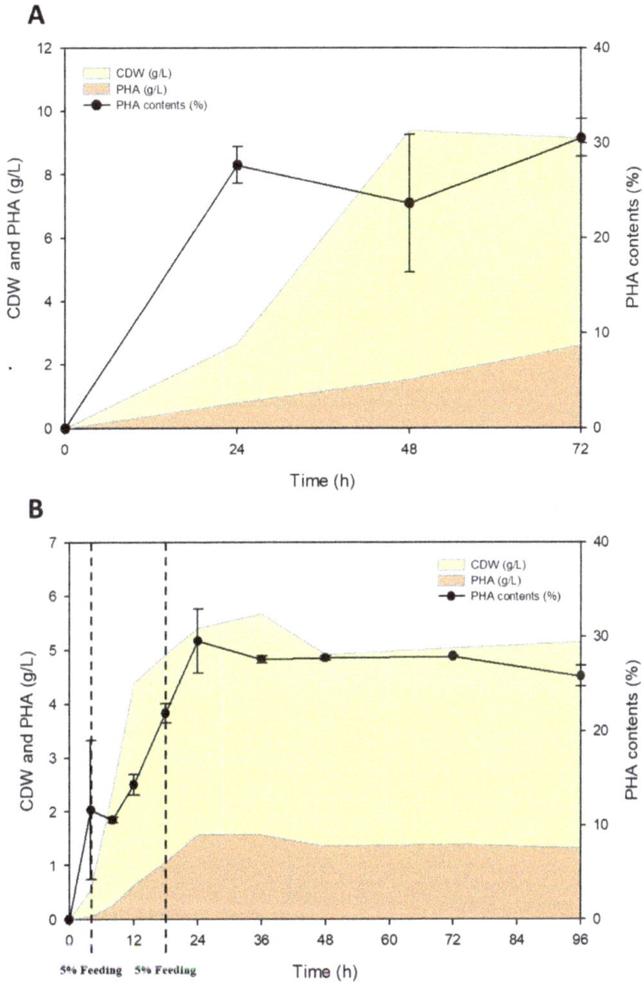

Figure 6. 10 L scale jar fermentation for PHA production. (**A**) *P. resinovorans* was cultured with 2% of SCGs oil initially added as the sole carbon source on a 250 mL flask scale; and (**B**) *P. resinovorans* was cultured with 2% of SCGs oil initially added as the sole carbon source and 5% of SCGs oil fed sequentially after 4 and 18 h during the cultivation on a 10 L jar fermenter scale.

The fed-batch culture was performed through pulse feeds on a 10 L scale jar fermenter. In particular, considering the long lag phase of *P. resinovorans* with the flask scale, 5% SCGs oil was added after 4 h to boost cell growth, and the same concentration of SCGs oil was then added at 18 h. From the results, it was confirmed that the lag phase was much shorter than with the flask scale, and CDW reached 4.4 g/L at 12 h and 5.4 g/L at 24 h of the fermentation (Figure 6B). In addition, PHA production increased at the same time as the growth phase, increasing from 0.6 g/L at 12 h to 1.6 g/L at 24 h. The PHA content increased to 29.5% within 24 h and remained constant. It is believed that the reason for the decrease in the lag phase was because of pH and DO adjustment.

Compared with previously-reported results of PHA production using SCGs oil, it found carbon sources to be effective in mcl-PHA production (Table 4). However, produced amount mcl-PHA is relatively low than PHB, it is estimated that development of strain in genetic or metabolic engineering level is required to improve PHA production.

Table 4. Production of various PHA using SCGs oil.

Organism	Polymer	CDW (g/L)	PHA Contents (%)	PHA Concentration (g/L)	References
Cupriavidus necator H16	PHB	55.0	89.1	49.4	[41]
Cupriavidus necator DSM 428	PHB	16.7	78.4	13.1	[42]
Ralstonia eutropha Re2133	P(3HB-co-3HHx)	0.93	69	0.64	[43]
Pseudomonas resinovorans	Mcl-PHA	5.4	29.5	1.6	This study

3.6. Physical Properties of Produced PHA

Differential Scanning Calorimetry (DSC) was used to investigate the thermal properties of mcl-PHA, including the glass transition temperature (Tg), melting temperature (Tm), crystallization temperature (Tc), and melting enthalpy (ΔHm). The results obtained for the scl-co-mcl-PHA were Tg = −59.79 °C and Tc = 86.8 °C with ΔHm = 38.6 J/g, and Tm = 11.239 °C with ΔHm = 41.2 J/g (Supplementary Figure S2).

The thermal characteristics of our mcl-PHA differed both in Tm and Tg from those of the previously-reported mcl-PHA from *Pseudomonas* sp. PAMC28620, composed of poly(25.5% 3HO-co-52.1% 3HD-co-5.7% 3HdD-co-16.7% 3HtD) with Tm = 172.8 °C, Tg = 3.99 °C, and Tc = 54.61 °C [44]. The thermal properties also differed from those of the mcl-PHA from *Pseudomonas* sp. MPC6. Composed of poly(89.5% 3HB-co-1.8% 3HHx-co-3.3% 3HO-co-4.4% 3HD-co-1.1% 3HdD) with Tm = 163.5 °C, Tg = 2.3 °C, and Tc = 46.0 °C [45].

As highly crystalline polymers have limited application in industrial and medical fields [46], the mcl-PHA from *P. resinovorans* may have potential in applications that require a sticky and relatively low-temperature modeling polymer, as well as one that is biodegradable.

The molecular weight of the obtained PHA was characterized by Gel Permeation chromatography (GPC), which demonstrated a retention time peak start at 14.07 min, peak maximum at 15.86 min, and peak end at 18.93 min (Supplementary Figure S3). The mcl-PHA had average values, with a number average molecular weight (Mn) of 60,424, weight average molecular weight (Mw) of 114,093, Z-average (Mz) of 177,428, and viscosity average molar mass (Mv) of 105,309. The high polydispersity index value was because of the various monomer unit compositions and breakdown of the polymer in the sample preparation steps. The physical properties of the mcl-PHA produced by *P. resinovorans* differed from PHAs produced by other species [47]. Therefore, we hypothesize that the mcl-PHA has potential applications in the biomedical field or similar industries.

3.7. Mass Balance of Spent Coffee Grounds to Bio-Hydrogen and PHA

The mass balance was calculated on a carbohydrate basis to confirm the chemical oxygen demand (COD) reduction of SCGs (Figure 7). The 738 g COD of carbohydrates of the initial SCGs was reduced to 444 g COD in the residue. Then, hydrogen and PHA were produced. Hydrogen was 279 mL/g COD-consumed sugar, which showed higher productivity than the 115 mL–CH_4/g COD using SCG hydrolysates reported in another anaerobic fermentation study [48]. In addition, the amount of PHA produced was 14 g/ kg SCG oil, which is a relatively low amount.

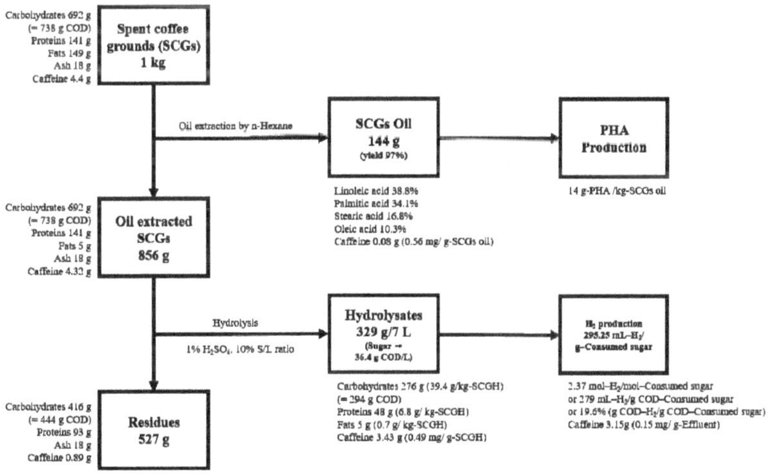

Figure 7. Mass balance of coffee grounds to bio-hydrogen and PHA.

It is believed that this amount could be increased later through optimization of the culture process and production process. One of the important constituents in coffee is caffeine. Gokulakrishnan et al. reported that caffeine concentrations above 2.5 mg/mL can affect microbial growth [49]. Accordingly, it is necessary to reduce the amount of caffeine in SCGs. As shown by the mass balance, 80% of caffeine was removed in the residues.

However, after dark fermentation for hydrogen production, caffeine in effluent was 0.15 mg/g, which is about 6.5 times higher than the caffeine concentration of 23 mg/L at a sewage treatment plant in Seoul [50]. Caffeine did not cause a decrease in productivity during the hydrogen production process, but the amount of caffeine decomposition was very small. Therefore, it is necessary to establish a caffeine adsorption and removal facility in the post-treatment process when the operation is on an industrial scale.

4. Conclusions

This study produced eco–friendly energy and biodegradable plastics to induce the valorization of SCGs that are produced in large quantities every year. For bioplastic production, 97% oil was extracted from total oil in the SCGs. OESCGs were hydrolyzed and showed a sugar concentration of 32.26 g/L. OESCGH exhibited a low inhibitor concentration, and the post–treatment process could be omitted. Hydrogen production was 15% higher than that of SCGH. This shows that the process of extracting oil is more advantageous in the next utilization step. The extracted oil was used as a substrate for PHA production. *P. resinovorans*, which can use *n*–hexane remaining in the oil, produced 5.6 g/L of CDW and 29.7% of PHA content at 24 h in a fed–batch culture. In summary, it was shown that waste disposal costs can be reduced by performing both processes to increase the utilization of SCGs that were previously only hydrolyzed or had oil extracted. In addition, hydrogen and PHA were produced as useful resources. These results show that fossil fuel–energy and petroleum–based plastics can be replaced and enable the proposal of an environmental resource circulation model.

Supplementary Materials: The following supporting information can be downloaded at: https://www.mdpi.com/article/10.3390/polym15030681/s1, Figure S1: Qualitative and quantitative analysis of spent coffee grounds oil (A) GC/MS graph of qualitative analysis (B) composition of fatty acids in SCGs oil; Figure S2: Differential scanning calorimetry results of mcl-PHA; Figure S3: Gel permeation chromatography result of mcl-PHA.

Author Contributions: Conceptualization, B.-J.K., J.-M.J. and J.-J.Y.; methodology, S.K.B., D.-H.K. and Y.-H.Y.; validation, B.-J.K., J.-M.J., S.J. and J.-J.Y.; investigation, J.-M.J., D.-H.K. and S.J.; data curation, B.-J.K. and J.-M.J.; writing—original draft preparation, B.-J.K. and J.-M.J.; writing—review and editing, S.K.B., D.-H.K. and Y.-H.Y.; supervision, J.-J.Y.; project administration, J.-J.Y.; funding acquisition, J.-J.Y. All authors have read and agreed to the published version of the manuscript.

Funding: This work was supported by the National Research Foundation of Korea (NRF) grant funded by the Korean government (Ministry of Science and ICT or MSIT; NRF-2020R1A2C2102381) and R&D Program of MOTIE/KEIT [grant number 20018072].

Institutional Review Board Statement: Not applicable.

Informed Consent Statement: Not applicable.

Data Availability Statement: Not applicable.

Conflicts of Interest: The authors declare no conflict of interest.

References

1. ICO (International Coffee Organization, London, UK). Coffee Production by Exporting Countries. 2021. Available online: https://www.ico.org/ (accessed on 12 December 2022).
2. Korea Cumstoms Service Home Page. Import and Export Trade Statistics of Coffee (by HS Code [0901]). 2021. Available online: https://unipass.customs.go.kr/ets/index_eng.do (accessed on 30 November 2022).
3. Mantell, C.L. *Solid Wastes: Origin, Collection, Processing, and Disposal*; John Wiley & Sons: Hoboken, NJ, USA, 1975.
4. Tun, M.M.; Raclavská, H.; Juchelková, D.; Růžičková, J.; Šafář, M.; Štrbová, K.; Gikas, P. Spent coffee ground as renewable energy source: Evaluation of the drying processes. *J. Environ. Manag.* **2020**, *275*, 111204. [CrossRef] [PubMed]
5. Kim, D.-H.; Jo, I.-S.; Kang, B.-J.; Lee, B.-D.; Kumar, S.; Kim, S.-H.; Yoon, J.-J. Evaluation of bio-hydrogen production using rice straw hydrolysate extracted by acid and alkali hydrolysis. *Int. J. Hydrogen Energy* **2022**, *47*, 37385–37393. [CrossRef]
6. Tsai, Y.-C.; Du, Y.-Q.; Yang, C.-F. Anaerobic biohydrogen production from biodetoxified rice straw hydrolysate. *J. Taiwan Inst. Chem. Eng.* **2021**, *123*, 134–140. [CrossRef]
7. Park, J.-H.; Yoon, J.-J.; Park, H.-D.; Kim, Y.J.; Lim, D.J.; Kim, S.-H. Feasibility of biohydrogen production from Gelidium amansii. *Int. J. Hydrogen Energy* **2011**, *36*, 13997–14003. [CrossRef]
8. Kwon, E.E.; Yi, H.; Jeon, Y.J. Sequential co-production of biodiesel and bioethanol with spent coffee grounds. *Bioresour. Technol.* **2013**, *136*, 475–480. [CrossRef]
9. Obruca, S.; Benesova, P.; Petrik, S.; Oborna, J.; Prikryl, R.; Marova, I. Production of polyhydroxyalkanoates using hydrolysate of spent coffee grounds. *Process Biochem.* **2014**, *49*, 1409–1414. [CrossRef]
10. López-Linares, J.C.; García-Cubero, M.T.; Coca, M.; Lucas, S. Efficient biobutanol production by acetone-butanol-ethanol fermentation from spent coffee grounds with microwave assisted dilute sulfuric acid pretreatment. *Bioresour. Technol.* **2021**, *320*, 124348. [CrossRef]
11. Mata, T.M.; Martins, A.A.; Caetano, N.S. Bio-refinery approach for spent coffee grounds valorization. *Bioresour. Technol.* **2018**, *247*, 1077–1084. [CrossRef]
12. do Sul, J.A.I.; Costa, M.F. The present and future of microplastic pollution in the marine environment. *Environ. Pollut.* **2014**, *185*, 352–364. [CrossRef]
13. Kemp, L.; Xu, C.; Depledge, J.; Ebi, K.L.; Gibbins, G.; Kohler, T.A.; Rockström, J.; Scheffer, M.; Schellnhuber, H.J.; Steffen, W. Climate Endgame: Exploring catastrophic climate change scenarios. *Proc. Natl. Acad. Sci. USA* **2022**, *119*, e2108146119. [CrossRef]
14. Pattra, S.; Sangyoka, S.; Boonmee, M.; Reungsang, A. Bio-hydrogen production from the fermentation of sugarcane bagasse hydrolysate by Clostridium butyricum. *Int. J. Hydrogen Energy* **2008**, *33*, 5256–5265. [CrossRef]
15. Lo, Y.C.; Lu, W.C.; Chen, C.Y.; Chang, J.S. Dark fermentative hydrogen production from enzymatic hydrolysate of xylan and pretreated rice straw by *Clostridium butyricum* CGS5. *Bioresour. Technol.* **2010**, *101*, 5885–5891. [CrossRef] [PubMed]
16. Abdul, P.M.; Jahim, J.M.; Harun, S.; Markom, M.; Lutpi, N.A.; Hassan, O.; Balan, V.; Dale, E.B.; Nor, M.T.M. Effects of changes in chemical and structural characteristic of ammonia fibre expansion (AFEX) pretreated oil palm empty fruit bunch fibre on enzymatic saccharification and fermentability for biohydrogen. *Bioresour. Technol.* **2016**, *211*, 200–208. [CrossRef] [PubMed]
17. Cao, G.; Ren, N.; Wang, A.; Lee, D.-J.; Guo, W.; Liu, B.; Yujie, F.; Qingliang, Z. Acid hydrolysis of corn stover for biohydrogen production using *Thermoanaerobacterium thermosaccharolyticum* W16. *Int. J. Hydrogen Energy* **2009**, *34*, 7182–7188. [CrossRef]

18. Kourmentza, C.; Plácido, J.; Venetsaneas, N.; Burniol-Figols, A.; Varrone, C.; Gavala, H.N.; Reis, M.A. Recent advances and challenges towards sustainable polyhydroxyalkanoate (PHA) production. *Bioengineering* **2017**, *4*, 55. [CrossRef] [PubMed]
19. Kourmentza, C.; Kornaros, M. Biotransformation of volatile fatty acids to polyhydroxyalkanoates by employing mixed microbial consortia: The effect of pH and carbon source. *Bioresour. Technol.* **2016**, *222*, 388–398. [CrossRef]
20. López, N.I.; Pettinari, M.J.; Nikel, P.I.; Méndez, B.S. Polyhydroxyalkanoates: Much more than biodegradable plastics. *Adv. Appl. Microbiol.* **2015**, *93*, 73–106.
21. Wang, S.; Chen, W.; Xiang, H.; Yang, J.; Zhou, Z.; Zhu, M. Modification and potential application of short-chain-length polyhydroxyalkanoate (SCL-PHA). *Polymers* **2016**, *8*, 273. [CrossRef]
22. Ciesielski, S.; Możejko, J.; Przybyłek, G. The influence of nitrogen limitation on mcl-PHA synthesis by two newly isolated strains of *Pseudomonas* sp. *J. Ind. Microbiol. Biotechnol.* **2010**, *37*, 511–520. [CrossRef]
23. Jeon, J.-M.; Park, S.-J.; Son, Y.-S.; Yang, Y.-H.; Yoon, J.-J. Bioconversion of Mixed Alkanes to Polyhydroxyalkanoate by *Pseudomonas resinovornas*: Upcycling of Pyrolysis Oil from Waste-Plastic. *Polymers* **2022**, *14*, 2624. [CrossRef]
24. Bhatia, S.K.; Gurav, R.; Choi, T.-R.; Jung, H.-R.; Yang, S.-Y.; Song, H.-S.; Kim, Y.-G.; Yoon, J.-J.; Yang, Y.-H. Effect of synthetic and food waste-derived volatile fatty acids on lipid accumulation in *Rhodococcus* sp. YHY01 and the properties of produced biodiesel. *Energy Convers. Manag.* **2019**, *192*, 385–395. [CrossRef]
25. Wang, J.; Wan, W. Factors influencing fermentative hydrogen production: A review. *Int. J. Hydrogen Energy* **2009**, *34*, 799–811. [CrossRef]
26. Somnuk, K.; Eawlex, P.; Prateepchaikul, G. Optimization of coffee oil extraction from spent coffee grounds using four solvents and prototype-scale extraction using circulation process. *Agric. Nat. Resour.* **2017**, *51*, 181–189. [CrossRef]
27. Al-Hamamre, Z.; Foerster, S.; Hartmann, F.; Kröger, M.; Kaltschmitt, M. Oil extracted from spent coffee grounds as a renewable source for fatty acid methyl ester manufacturing. *Fuel* **2012**, *96*, 70–76. [CrossRef]
28. Pichai, E.; Krit, S. Optimization of solid-to-solvent ratio and time for oil extraction process from spent coffee grounds using response surface methodology. *ARPN J. Eng. Appl. Sci.* **2015**, *10*, 7049–7052.
29. Efthymiopoulos, I.; Hellier, P.; Ladommatos, N.; Kay, A.; Mills-Lamptey, B. Effect of solvent extraction parameters on the recovery of oil from spent coffee grounds for biofuel production. *Waste Biomass Valorization* **2019**, *10*, 253–264. [CrossRef]
30. He, S.; Ni, Y.; Lu, L.; Chai, Q.; Yu, T.; Shen, Z.; Yang, C. Simultaneous degradation of n-hexane and production of biosurfactants by *Pseudomonas* sp. strain NEE2 isolated from oil-contaminated soils. *Chemosphere* **2020**, *242*, 125237. [CrossRef]
31. Phimsen, S.; Kiatkittipong, W.; Yamada, H.; Tagawa, T.; Kiatkittipong, K.; Laosiripojana, N.; Assabumrungrat, S. Oil extracted from spent coffee grounds for bio-hydrotreated diesel production. *Energy Convers. Manag.* **2016**, *126*, 1028–1036. [CrossRef]
32. Hudeckova, H.; Neureiter, M.; Obruca, S.; Fruhauf, S.; Marova, I. Biotechnological conversion of spent coffee grounds into lactic acid. *Lett. Appl. Microbiol.* **2018**, *66*, 306–312. [CrossRef]
33. Kim, D.-H.; Yoon, J.-J.; Kim, S.-H.; Park, J.-H. Effect of conductive material for overcoming inhibitory conditions derived from red algae-based substrate on biohydrogen production. *Fuel* **2021**, *285*, 119059. [CrossRef]
34. Haroun, B.M.; Nakhla, G.; Hafez, H.; Nasr, F.A. Impact of furfural on biohydrogen production from glucose and xylose in continuous-flow systems. *Renew. Energy* **2016**, *93*, 302–311.
35. Jeon, J.M.; Brigham, C.J.; Kim, Y.H.; Kim, H.J.; Yi, D.H.; Kim, H.; Rha, C.; Sinskey, A.J.; Yang, Y.H. Biosynthesis of poly(3-hydroxybutyrate-co-3-hydroxyhexanoate) (P(HB-co-HHx)) from butyrate using engineered *Ralstonia eutropha. Appl. Microbiol. Biotechnol.* **2014**, *98*, 5461–5469. [CrossRef] [PubMed]
36. Hervas, A.B.; Canosa, I.; Little, R.; Dixon, R.; Santero, E. NtrC-dependent regulatory network for nitrogen assimilation in *Pseudomonas putida. J. Bacteriol.* **2009**, *191*, 6123–6135. [CrossRef] [PubMed]
37. Hervas, A.B.; Canosa, I.; Santero, E. Transcriptome analysis of *Pseudomonas putida* in response to nitrogen availability. *J. Bacteriol.* **2008**, *190*, 416–420. [CrossRef] [PubMed]
38. Borrero-de Acuna, J.M.; Rohde, M.; Saldias, C.; Poblete-Castro, I. Fed-Batch mcl- Polyhydroxyalkanoates Production in *Pseudomonas putida* KT2440 and DeltaphaZ Mutant on Biodiesel-Derived Crude Glycerol. *Front. Bioeng. Biotechnol.* **2021**, *9*, 642023. [CrossRef]
39. Ochoa, S.; Garca, C.; Alcaraz, W. Real-time optimization and control for polyhydroxybutyrate fed-batch production at pilot plant scale. *J. Chem. Technol. Biotechnol.* **2020**, *95*, 3221–3231. [CrossRef]
40. Blunt, W.; Gaugler, M.; Collet, C.; Sparling, R.; Gapes, D.J.; Levin, D.B.; Cicek, N. Rheological Behavior of High Cell Density *Pseudomonas putida* LS46 Cultures during Production of Medium Chain Length Polyhydroxyalkanoate (PHA) Polymers. *Bioengineering* **2019**, *6*, 93. [CrossRef]
41. Obruca, S.; Petrik, S.; Benesova, P.; Svoboda, Z.; Eremka, L.; Marova, I. Utilization of oil extracted from spent coffee grounds for sustainable production of polyhydroxyalkanoates. *Appl. Microbiol. Biotechnol.* **2014**, *98*, 5883–5890. [CrossRef]
42. Cruz, M.V.; Paiva, A.; Lisboa, P.; Freitas, F.; Alves, V.D.; Simões, P.; Barreiros, S.; Reis, M.A. Production of polyhydroxyalkanoates from spent coffee grounds oil obtained by supercritical fluid extraction technology. *Bioresour. Technol.* **2014**, *157*, 360–363.
43. Bhatia, S.K.; Kim, J.H.; Kim, M.S.; Kim, J.; Hong, J.W.; Hong, Y.G.; Kim, H.J.; Jeon, J.M.; Kim, S.H.; Ahn, J.; et al. Production of (3-hydroxybutyrate-co-3-hydroxyhexanoate) copolymer from coffee waste oil using engineered *Ralstonia eutropha. Bioprocess Biosyst. Eng.* **2018**, *41*, 229–235. [CrossRef]

44. Sathiyanarayanan, G.; Bhatia, S.K.; Song, H.S.; Jeon, J.M.; Kim, J.; Lee, Y.K.; Kim, Y.G.; Yang, Y.H. Production and characterization of medium-chain-length polyhydroxyalkanoate copolymer from Arctic psychrotrophic bacterium *Pseudomonas* sp. PAMC 28620. *Int. J. Biol. Macromol.* **2017**, *97*, 710–720. [CrossRef] [PubMed]
45. Orellana-Saez, M.; Pacheco, N.; Costa, J.I.; Mendez, K.N.; Miossec, M.J.; Meneses, C.; Castro-Nallar, E.; Marcoleta, A.E.; Poblete-Castro, I. In-Depth Genomic and Phenotypic Characterization of the Antarctic Psychrotolerant Strain *Pseudomonas* sp. MPC6 Reveals Unique Metabolic Features, Plasticity, and Biotechnological Potential. *Front. Microbiol.* **2019**, *10*, 1154. [CrossRef] [PubMed]
46. Luef, K.P.; Stelzer, F.; Wiesbrock, F. Poly(hydroxy alkanoate)s in Medical Applications. *Chem. Biochem. Eng. Q* **2015**, *29*, 287–297. [CrossRef] [PubMed]
47. Choi, T.R.; Park, Y.L.; Song, H.S.; Lee, S.M.; Park, S.L.; Lee, H.S.; Kim, H.J.; Bhatia, S.K.; Gurav, R.; Choi, K.Y.; et al. Fructose-Based Production of Short-Chain-Length and Medium-Chain-Length Polyhydroxyalkanoate Copolymer by Arctic *Pseudomonas* sp. B14-6. *Polymers* **2021**, *13*, 1398. [CrossRef] [PubMed]
48. Shen, M.-Y.; Huang, Z.-H.; Kuo, Y.-T.; Hsu, J.; Yang, H.-W.; Peng, C.-Y.; Chu, C.-Y. Improvement of gaseous bioenergy production from spent coffee grounds Co-digestion with pulp wastewater by physical/chemical pretreatments. *Int. J. Hydrogen Energy* **2022**, *47*, 40664–40671. [CrossRef]
49. Gokulakrishnan, S.; Chandraraj, K.; Gummadi, S.N. Microbial and enzymatic methods for the removal of caffeine. *Enzym. Microb. Technol.* **2005**, *37*, 225–232. [CrossRef]
50. Choi, K.; Kim, Y.; Park, J.; Park, C.K.; Kim, M.; Kim, H.S.; Kim, P. Seasonal variations of several pharmaceutical residues in surface water and sewage treatment plants of Han River, Korea. *Sci. Total Environ.* **2008**, *405*, 120–128. [CrossRef]

polymers

MDPI

Article

The Modification of Poly(3-Hydroxybutyrate-co-4-hydroxybutyrate) by Melt Blending

Minki Jo [1], Yunjae Jang [2], Eunhye Lee [2], Sooan Shin [2] and Ho-Jong Kang [1,*]

1 Department of Polymer Science and Engineering, Dankook University, 152 Jukjeon-ro, Suji-gu, Yongin-si 16890, Gyeonggi-do, Korea; mc0281@naver.com
2 CJ Cheiljedang Corporation, 55, Gwanggyo-ro 42beon-gil, Yeongtong-gu, Suwon-si 16495, Gyeonggi-do, Korea; yunjae.jang@cj.net (Y.J.); eunhye.lee@cj.net (E.L.); sooanshin@gmail.com (S.S.)
* Correspondence: hjkang@dankook.ac.kr

Abstract: Crystalline and noncrystalline poly(3-hyroxybutylate-co-4-hyroxybutylate) (P(3HB-co-4HB)) were melt blended to obtain mixtures of P(3HB-co-4HB) copolymers. The mixtures and P(3HB-co-4HB) copolymers of different 4HB contents were compared to study the effect of 4HB content on the properties of the copolymers and mixtures. P(3HB-co-4HB) copolymer mixtures, having various 4HB content, have been successfully made by melt blending instead of bacterial biosynthesis. In the case of copolymers, they were noncrystalline when the 4HB content was over 16%, while the P(3HB-co-4HB) mixtures at the same 4HB content were crystalline. The mixtures had a higher glass transition temperature, suggesting that their chain mobility is relatively low compared with the copolymer having the same 4HB content. Due to this effect, the mixture is expected to have a higher melt viscosity and a lower loss tangent to exhibit better melt processing properties. The mechanical properties of the mixtures show a similar behavior to the copolymers in that the tensile strength and the modulus decreases and elongation at the break increases with an increase in the 4HB content.

Keywords: P(3HB-co-4HB); 4HB contents; mixtures; thermal properties; rheological properties; mechanical properties

Citation: Jo, M.; Jang, Y.; Lee, E.; Shin, S.; Kang, H.-J. The Modification of Poly(3-Hydroxybutyrate-co-4-hydroxybutyrate) by Melt Blending. *Polymers* **2022**, *14*, 1725. https://doi.org/10.3390/polym14091725

Academic Editors: Shashi Kant Bhatia and Jens-Uwe Sommer

Received: 31 March 2022
Accepted: 20 April 2022
Published: 23 April 2022

1. Introduction

Polyhydroxyalkanoate (PHA) is a biodegradable polyester produced by microorganisms and has the benefit of being degradable by microorganisms under a diverse environment [1–6] contrary to the chemically manufactured polylactic acid (PLA) [7,8] and polybutylene succinate (PBS) [9,10]. Poly 3-hydroxybutyrate(P3HB), a typical PHA, is highly crystalline and its crystallization rate is very slow due to its stereoregular chemical structure. Therefore, the spherulites formed are very large and spherulite/spherulite interfaces are formed where fracture occurs, making it brittle [11–14]. Degradation occurs easily above the processing temperature of 180 °C, giving it the drawback of a very narrow melt processing window [15,16].

To resolve the drawback, control of the crystallization rate and degradation properties by melt blending with diverse biodegradable polymers such as PLA and PBS, etc., is being carried out [17–19]. Utilization of the chemical structure of P(3HB) to induce reactions such as carboxylation, halogenation, hydroxylation, grafting, epoxidation, etc., by adding diverse chemicals to control the properties through structural changes such as branching and crosslinking, etc., is possible [20–24]. Along with this method, the most widely used method to control the properties of P(3HB) is bacterial synthesis of diverse copolymers. Representative copolymers are P(3HB-co-3HV) [25,26], P(3HB-co-HDD) [27], etc., and recently research has focused on P(3HB-co-4HB) [28–32].

When 4HB is added to the biosynthesis of P(3HB), a P(3HB-co-4HB) copolymer is formed and the flexibility of the 4HB segment decreases the brittle character of P(3HB) along with decrease in the glass transition temperature, decrease in the melting temperature, and decrease in the crystallization rate and crystallinity. When the 4HB content exceeds 20%, the crystallization of P(3HB-co-4HB) is difficult [33], and as a result, modulus and tensile strength decreases and elongation at break increases. Furthermore, the melt processibility decreases significantly due to the decrease in melt viscosity and increase in the loss tangent. Besides the effect of the 4HB segment on the physical properties of the P(3HB-co-4HB) copolymer, the 4HB segment causes an enhancement of thermal stability of P(3HB). The thermal degradation of P(3HB-co-4HB) occurs due to the random unzipping degradation of 3HB, similar to P(3HB), however, 4HB is hardly activated for degradation due to the methylene group to stop unzipping degradation [34]. The cyclization and transesterification of degraded 4HB also results in the improvement of thermal stability of P(3HB-co-4HB).

To resolve the problem of crystallinity change and the decrease in melt processibility with an increase in 4HB content in P(3HB-co-4HB), crystalline P(3HB-co-4HB) and noncrystalline P(3HB-co-4HB) is melt blended to prepare mixtures. The change in the crystallization behavior and the melt processibility with an increase in 4HB content in the mixture is compared with P(3HB-co-4HB) of similar 4HB content.

2. Experimental

2.1. Materials and Mixture Preparation

The copolymer of 3-hyroxybutyrate and 4-hyroxybutyrate, poly(3-hyroxybutylate-co-4-hyroxybutylate), was provided by CJ Cheiljedang and their 4HB content and molecular weights are shown in Table 1. To enhance the properties of P(3HB-co-4HB) copolymer, crystalline P(3HB-co-10% 4HB) and noncrystalline P(3HB-co-53.7% 4HB) were melt blended at composition ratios of 9/1–4/6 in an internal mixer (Haake, Rheomix600, Vreden, Germany). To minimize degradation during the melt blending process, they were blended at 140 °C and 20 rpm for 10 min. All P(3HB-co-4HB) samples were processed in the same way to give the same thermal history. The P(3HB-co-4HB) and P(3HB-co-4HB) mixtures thus obtained were put in 1T thick 150 mm × 150 mm mold in a compression molding machine (QMESYS, QM900A, Anyang, Korea) and kept at 140 °C for 2 min, pressure raised to 8 MPa, then quenched in a water tank at 25 °C for 2 min, to prepare the 1T thick films.

Table 1. 4HB Contents and molecular weight (M_W) of P(3HB-co-4HB) used in this study.

Name	4HB (%)	M_W (k)
P(3HB-co-2.1% 4HB)	2.1	200
P(3HB-co-10% 4HB)	10	600
P(3HB-co-16% 4HB)	16	1000
P(3HB-co-35.6% 4HB)	35.6	580
P(3HB-co-53.7% 4HB)	53.7	695

2.2. The 4HB Contents of P(3HB-co-4HB) Mixtures

^{1}H NMR (Jeol, Jeol400, Tokyo, Japan) was used to determine the 4HB content of the P(3HB-co-10% 4HB) and P(3HB-co-53.7% 4HB) mixtures. The NMR samples were prepared by dissolving $CDCl_3$. The 4.1 ppm peak due to the two hydrogens on the number 8 carbon (4HB) and the 1.25 ppm peak due to the three hydrogens on the number 4 carbon (3HB) were compared to calculate the relative presence of the 3HB and 4HB units and calculate the 4HB content.

2.3. Thermal Properties of P(3HB-co-4HB) Mixtures

Differential scanning calorimeter (TA, Q20, New Castle, DE, USA) was used to study the effect of 4HB content on the thermal properties of the P(3HB-co-4HB) and P(3HB-co-4HB) mixtures. The effect of 4HB content on the changes in the crystallization temperature and enthalpy in the cooling stage and the glass transition temperature, melting temperature, and the enthalpy of melting in the second heating stage were evaluated by scanning in the range −50~250 °C at the heating and cooling rates of 20 °C/min.

2.4. Rheological Properties of P(3HB-co-4HB) Mixtures

The rheological properties of P(3HB-co-4HB) and their mixtures were evaluated on a rotational rheometer (TA, AR200EX, New Castle, DE, USA) with 25 mm diameter samples in the oscillation mode at angular frequencies of 0.1~600 rad/s and the % strain set at 0.1, 3, and 5 at 140 °C, 160 °C, and 180 °C, respectively, to obtain the changes in the complex viscosity (η^*) and loss tangent (tanδ) with frequency change. Using obtained zero shear viscosity (η_0), the curves of $\ln(\eta_0)$ versus 1/T were plotted and the dependence of P(3HB co-4HB) and their mixtures shear viscosity on temperature can be described by the Arrhenius equation.

$$\ln(\eta_0) = A \exp(Ea/RT) \tag{1}$$

where A is constant related to melt viscosity, and Ea is the activation energy for viscous flow.

2.5. Mechanical Properties of P(3HB-co-4HB) Mixtures

The tensile strength, Young's modulus and elongation at break were measured on a tensile tester (LLOYD, LR-30K, West Sussex, UK) with 1T thick 10 mm × 50 mm samples gripping the top and bottom 10 mm on the jig and extending at 50 mm/min until break.

3. Results and Discussion

3.1. 4HB Contents of P(3HB-co-4HB) Mixtures

Figure 1 shows the ^{1}H NMR spectra of the P(3HB-co-4HB) mixtures of different blend ratios. The peak areas of the 4.1 ppm and 1.25 ppm due to 4HB and 3HB units, respectively, change with the composition of the mixture. The 4HB contents calculated from the ^{1}H NMR peak areas are shown in Figure 2 along with the line for the theoretical 4HB content calculated from the blend ratios. The 4HB contents calculated from the NMR data conform to the theoretical values calculated from the blend ratios and P(3HB-co-4HB) mixtures, having 14.4–45% 4HB content that could be prepared. Conformance of the data and the theoretical values suggest that there is no thermal degradation or chemical change in the melt blending process changing the 4HB or 3HB contents. However, only our ^{1}H NMR results did not convince us that there was no chemical change in the melt blending, and we may need further study to find out the light cross-linking in the copolymer networks. At this point, the P(3HB-co-4HB) mixtures of different 4HB contents can be obtained by melt blending, instead of by controlling the 4HB feed in bacterial biosynthesis. From the point of view of the industrial mass production of P(3HB-co-4HB), it may be easier and more useful to control the 4HB content in P(3HB-co-4HB) by melt blending, instead of by bacterial biosynthesis, which involves a complex process. In the copolymer, crystallizable 4HB exists in the main chain of crystalline P(3HB-co-4HB), whereas in the mixtures, a blended noncrystalline P(3HB-co-4HB), having 4HB segments, coexists as a co-domain in the crystalline P(3HB-co-4HB) domain. This difference of micro and macro structure, between the biosynthesized copolymer and their mixture, is expected to have a different effect on the physical properties of P(3HB-co-4HB). Our study focused on the change of physical properties of P(3HB-co-4HB) mixtures by melt blending and compared this with those of bacterial biosynthesis.

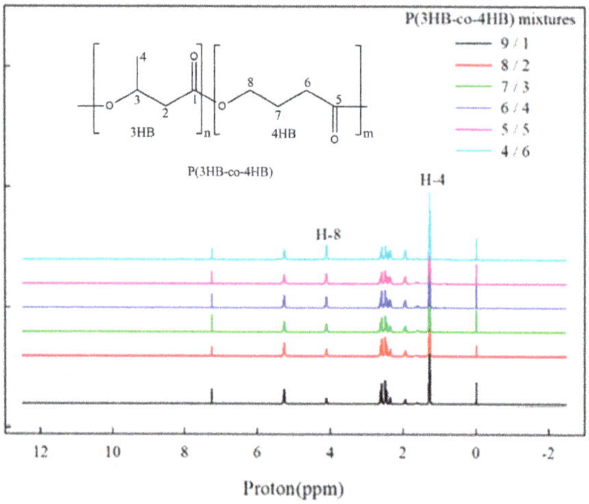

Figure 1. ^{1}H NMR spectra of P(3HB-co-4HB) mixtures.

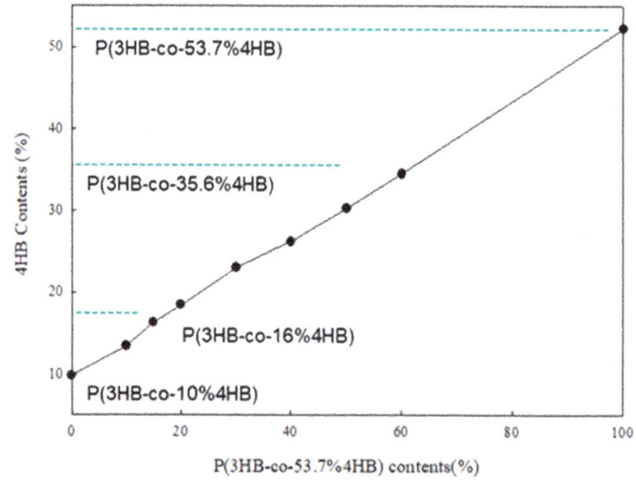

Figure 2. The 4HB contents of P(3HB-co-4HB) mixtures calculated from NMR data and the theoretical line calculated from the composition.

3.2. Thermal Properties of P(3HB-co-4HB) Mixtures

The secondary heating DSC thermograms of P(3HB-co-4HB) and P(3HB-co-4HB) mixtures are shown in Figure 3. The P(3HB-co-4HB) copolymer is crystalline only up to 16% 4HB, while the P(3HB-co-4HB) mixtures of 4HB contents 14.4–45% all show cold crystallization and melting peaks. P(3HB-co-4HB) is a random copolymer and the chain randomness increases with the increase in the 4HB content to hinder the crystallization, and it becomes noncrystalline P(3HB-co-4HB) at 4HB contents above 16%. However, in the case of blends of crystalline P(3HB-co-4HB) and noncrystalline P(3HB-co-4HB), the increase in 4HB content does not affect the randomness of the copolymer chain and the crystallization of the crystalline P(3HB-co-10% 4HB) is not influenced by the presence of the noncrystalline P(3HB-co-53.7% 4HB) resulting in the appearance of the cold crystallization and melting peaks with change in only the peak areas. This demonstrates that 4HB content, which influences the flexibility and biodegradability of P(3HB-co-4HB) copolymers [35], can be high while retaining the crystallinity and mechanical properties in the case of P(3HB-co-

4HB) mixtures from melt blending. Multiple melting peaks are founded in both copolymer and their mixtures. This can be explained by micro-phase separation behavior. The random P(3HB-co-4HB) copolymer may have two phases, a 3HB rich area and a 4HB rich area, each having a different melting temperature. The difference of chain flexibility results in a high melting temperature for the 3HB rich area, and a low melting temperature for the 4HB rich area.

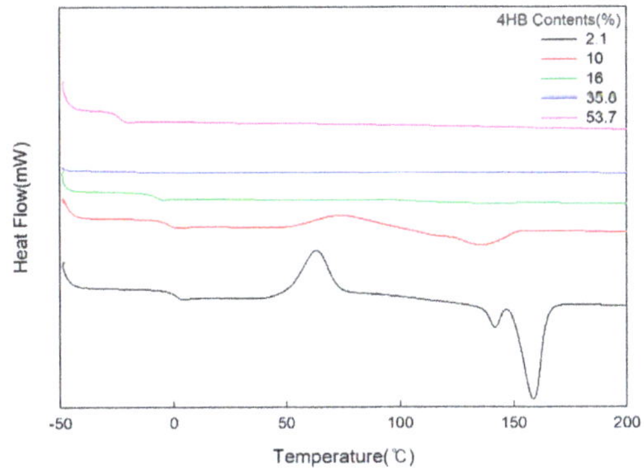

(**a**)

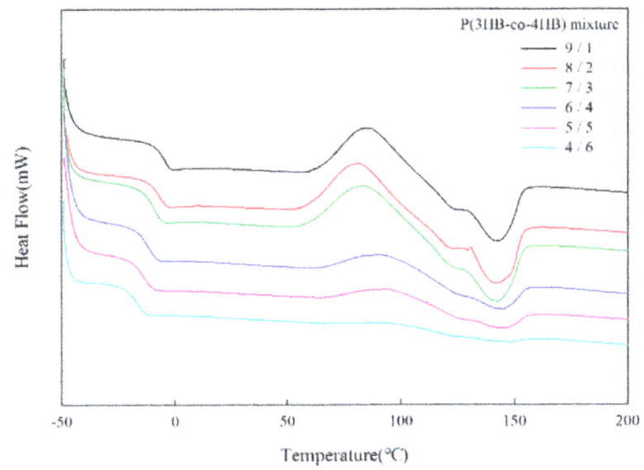

(**b**)

Figure 3. The DSC thermograms of: (**a**) P(3HB-co-4HB) and (**b**) P(3HB-co-4HB) mixtures (cooling rate/heating rate; 20 °C/min).

Figure 4 shows the effect of 4HB content on the glass transition temperature, crystallization temperature and enthalpy, melting temperature and enthalpy of P(3HB-co-4HB) and P(3HB-co-4HB) mixtures. The P(3HB-co-4HB) mixtures prepared from copolymers of different 4HB contents show a single glass transition as a result of their miscibility, and as the 4HB content increases with increase in the composition of flexible P(3HB-co-53.7% 4HB), a decrease in the glass transition temperature occurs. Compared to a copolymer with similar 4HB content, a higher glass transition temperature can be seen in the mixture as the flexibility of the mixture physically blended with the P(3HB-co-53.7% 4HB) chain is lower than that of the copolymer chain with the corresponding amount of 4HB units.

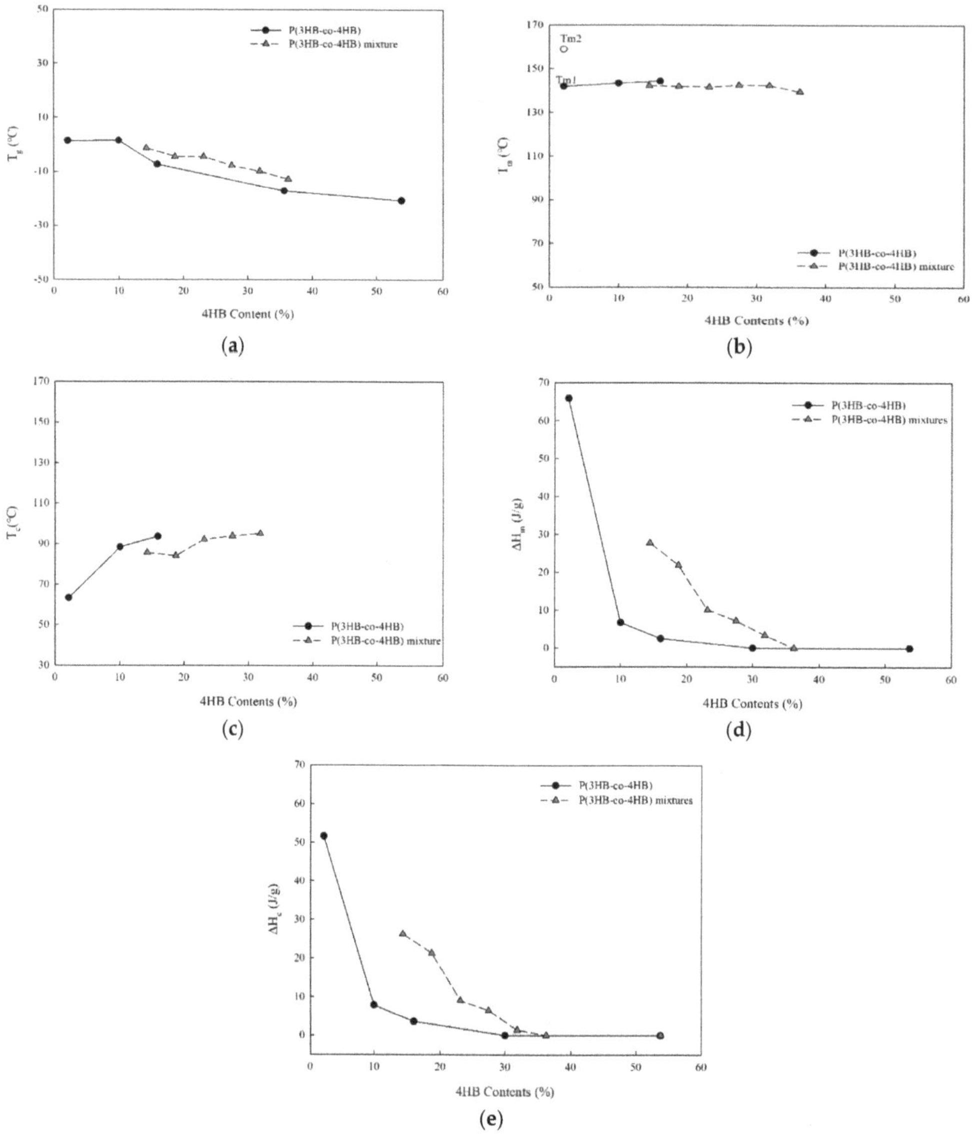

Figure 4. Thermal properties of P(3HB-co-4HB) and P(3HB-co-4HB) mixtures; (**a**) glass transition temperature; (**b**) melting temperature; (**c**) crystallization temperature; (**d**) melting enthalpy; (**e**) crystallization enthalpy.

The melting points of mixtures in Figure 4b exhibit negligible change in 4HB content and the melting peak occurs at, or even above, the 4HB content of 30%, while it does not appear in the P(3HB-co-35.6% 4HB) copolymer. This suggests that the increase in the 4HB content in the copolymer increases the randomness of the copolymer, while the increase in the 4HB content in the mixture through physical mixing does not greatly affect the crystallization of the P(3HB-co-10% 4HB) in the mixture. The appearance of the melting peak is apparently due to the crystallization in the cooling stage. Crystallization temperature of the mixtures in Figure 4c are similar to the copolymer of equivalent 4HB content and crystallization enthalpy and melting enthalpy are much higher, with crystallization occurring at 4HB contents even greater than 30% (Figure 4d,e). Therefore, crystalline P(3HB-co-4HB) mixtures of higher 4HB content can be prepared, while crystalline P(3HB-co-4HB) copolymer cannot be obtained at higher 4HB content.

3.3. Rheological Properties of P(3HB-co-4HB) Mixtures

The complex viscosity at 140 °C of P(3HB-co-4HB) and their mixtures are shown in Figure 5. The complex viscosity of P(3HB-co-4HB) mixtures decrease with the shear rate, which is representative of non-Newtonian behavior similar to the P(3HB-co-4HB) copolymers. The complex viscosity is related with chain flexibility and molecular weight. The complex viscosity of P(3HB-co-16% 4HB), having a higher molecular weight and 4HB content compared to P(3HB-co-10% 4HB), is lower than that of P(3HB-co-10% 4HB). This means that the effect of chain flexibility on complex viscosity is dominant and controls the complex viscosity of P(3HB-co-4HB), when compared with the effect of the molecular weight. Increasing the 4HB content in the copolymer resulted in lowering of the complex viscosity in both the copolymer and their mixtures, however, the viscosity change in the 4HB content at a low shear rate is smaller compared to the copolymers in the case of copolymer mixtures of higher P(3HB-co-53.7% 4HB) composition, and thus have higher 4HB contents. In the case of copolymer mixtures, the viscosity decreases due to the physical melt blending with P(3HB-co-53.7% 4HB) having a relatively low viscosity, however, the decrease in 4HB content is smaller compared to in P(3HB-co-4HB) copolymers where the 4HB units are in the main chain.

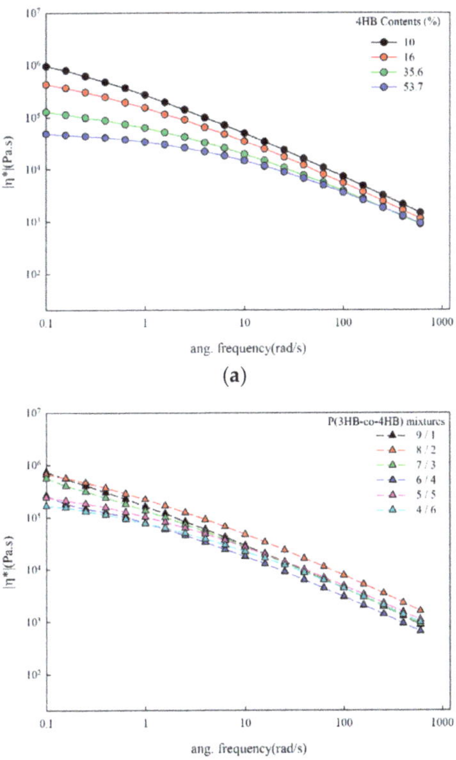

Figure 5. Changes in the complex viscosity with shear rate at 140 °C: (**a**) P(3HB-co-4HB); (**b**) P(3HB-co-4HB) mixtures.

The changes in the viscosity at 0.1 rad/s with 4HB content of P(3HB-co-4HB) copolymers and P(3HB-co-4HB) copolymer mixtures in Figure 6 show that, for same 4HB contents, the P(3HB-co-4HB) mixtures exhibit a higher viscosity than the P(3HB-co-4HB) copolymers. It seems that difference in viscosity between the copolymer and their mixture is more pronounced at a high processing temperature, due to the high chain mobility of 4HB in the copolymer. Changes in the activation energy with 4HB content in P(3HB-co-4HB) copolymers and P(3HB-co-4HB) copolymer mixtures, as shown in Figure 7, show that the activation energy is highest in the case of P(3HB-co-10% 4HB), where the main chain is relatively stiff and decreases as the content of the relatively flexible 4HB in the copolymer increases. Activation energy in the Arrhenius equation reflects the effect of temperature on the mobility of chain. As high activation energy reflects difficult chain motion, the data shows that the increase in 4HB content in P(3HB-co-4HB) increases the chain flexibility. The activation energy of the P(3HB-co-4HB) copolymer mixture also decreases with an increase in the composition of the relatively flexible P(3HB-co 53.7% 4HB). Although the copolymer exhibits a significant decrease in the activation energy at 16% 4HB content, the copolymer mixture exhibits a linear decrease with 4HB content, suggesting that the chain flexibility may be affected by whether 4HB and 3HB units are chemically connected or physically mixed.

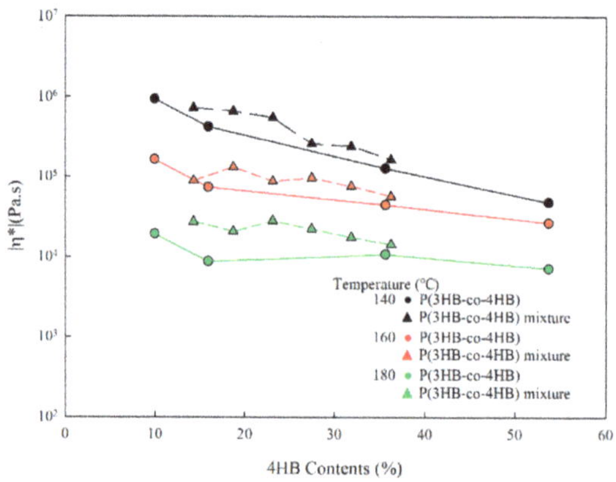

Figure 6. Changes in the complex viscosity of P(3HB-co-4HB) and P(3HB-co-4HB) mixtures with 4HB content at 0.1 rad/s.

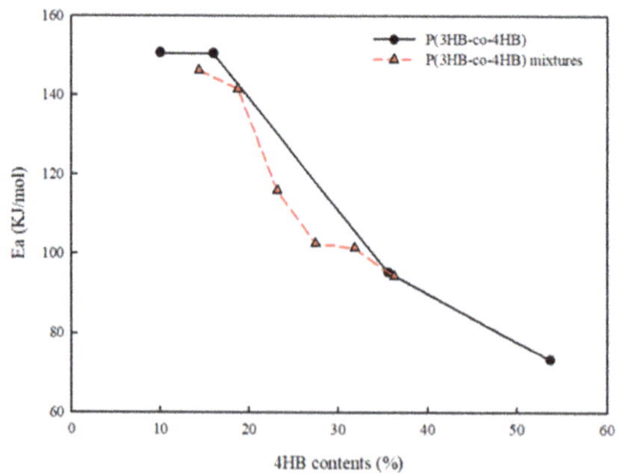

Figure 7. Changes in the activation energy of P(3HB-co-4HB) and P(3HB-co-4HB) mixtures with various 4HB content.

The loss tangent of P(3HB-co-4HB) copolymers and P(3HB-co-4HB) copolymer mixtures at 140 °C, shown in Figure 8, both increase with an increase in 4HB content. The P(3HB-co-4HB) with high 4HB content shows much less significant frequency dependence of the loss tangent. This indicated that P(3HB-co-4HB) with a high 4HB content has a higher elasticity. It is well known that the melt strength is related to the elasticity of polymers. The loss tangent is known to be inversely proportional to the melt strength [36,37]. The increase in 4HB content increases the chain flexibility and thus decreases the melt strength to decrease the melt processibility, where pressure is applied to form the shape. As can be seen in Figure 8, the loss tangent values of the mixtures are similar to P(3HB-co-10% 4HB), close to 1.0, although there is a slight difference depending on the composition. The lowest loss tangent was obtained with 9/1 composition in the mixtures. The changes in the loss tangent with the increase in the 4HB content is much smaller in the copolymer mixtures compared with the copolymers allowing minimal change in processibility in the case of the copolymer mixtures at the same 4HB content, suggesting that using the copolymer mixtures instead of copolymers of the same 4HB content, is more advantageous for melt processibility.

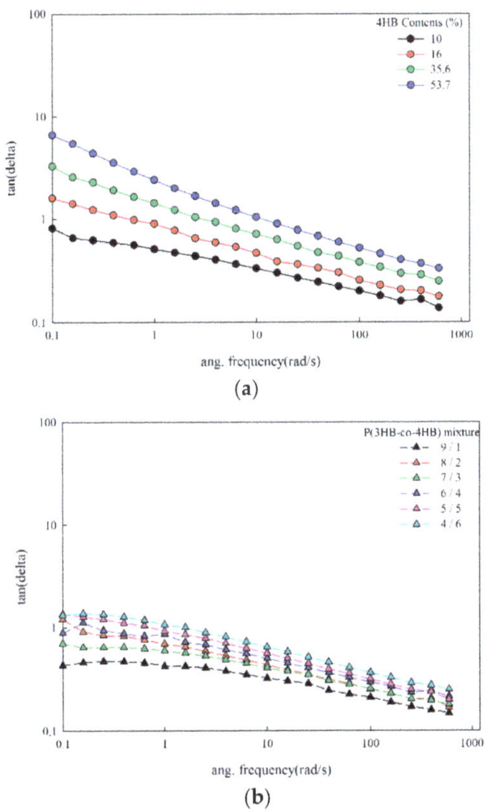

Figure 8. Changes in loss tangent with shear rate at 140 °C: (**a**) P(3HB-co-4HB); (**b**) P(3HB-co-4HB) mixtures.

Figure 9 shows the change in loss tangent with 4HB content at shear rate 0.1 rad/s for P(3HB-co-4HB) and the copolymer mixtures at a different processing temperature. The lower the processing temperature and the lower the 4HB content, the more the loss tangent decreases to result in higher melt strengths and the copolymer mixtures show a lower loss tangent compared to the copolymers, irrespective of the processing temperature or 4HB content. The change in the loss tangent with the increase in 4HB content or processing temperature can be minimized by using copolymer mixtures. This is due to the higher chemically bonded 4HB content of the P(3HB-co-4HB) copolymers, which results in a high increase in tanδ and affect the processibility, while the increase in 4HB content of P(3HB-co-4HB) mixtures, through the physical mixing of P(3HB-co-10% 4HB) and P(3HB-co-53.7% 4HB), does not affect the processibility of P(3HB-co-10% 4HB). From this result, it is found that the enhancement of melt processability could be obtained using P(3HB-co-4HB) mixtures at a low processing temperature.

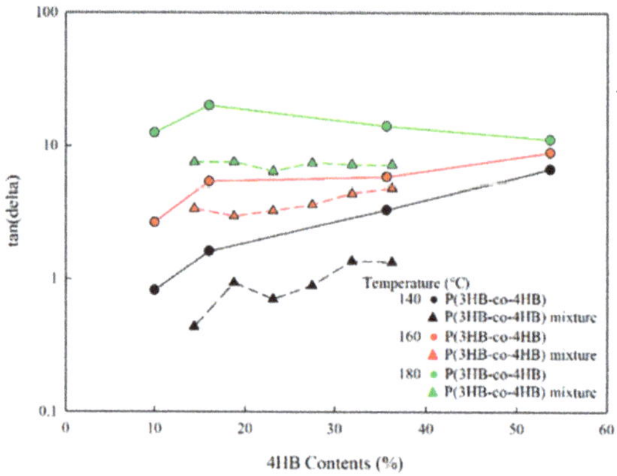

Figure 9. Changes in the loss tangent of P(3HB-co-4HB) and P(3HB-co-4HB) mixtures with 4HB content at 0.1 rad/s.

3.4. Mechanical Properties of P(3HB-co-4HB) Mixtures

Figure 10 shows the effect of 4HB content on the Young's modulus, tensile strength, and elongation at break of P(3HB-co-4HB) copolymers and P(3HB-co-4HB) copolymer mixtures. The Young's modulus and tensile strength decrease with the increase in 4HB content and the elongation at break increases to a certain 4HB content, then decreases. It seems that the change in the mechanical properties of P(3HB-co-4HB) and mixtures as a function of 4HB content is due to the flexibility of the 4HB chain in both crystalline P(3HB-co-4HB) and noncrystalline P(3HB-co-4HB). This is observed in both P(3HB-co-4HB) copolymers and P(3HB-co-4HB) copolymer mixtures. Tensile strength and Young's modulus have an intimate relationship with crystallinity. As the 4HB content increases, crystallinity of both copolymers and copolymer mixtures decrease and spherulites from the crystallization of the 3HB segments become smaller, along with the relatively low crystallization rate of the 4HB segments, which also make the spherulites smaller to decrease the interfacial area of impingement and thus increase the elongation at break. The elongation at break of copolymers decrease in the 4HB contents above 30%, and that of the copolymer mixtures decrease in the 4HB contents above 25%. Above a certain composition, the increase in the P(3HB-co-53.7% 4HB) composition does not increase the elongation at break, but decreases it. The optimum composition for lowering the brittle character of P(3HB-co-10% 4HB) in this study was 7/3, however, for increasing the elongation at break, increasing the 4HB content in P(3HB-co-4HB) copolymer was more effective compared to increasing it in the copolymer mixtures. This suggests that the influence on the elongation at break of 4HB in the copolymer mixtures is lower than that in the copolymers, where they are chemically bound. The processibility of P(3HB-co-4HB) could be enhanced by preparing copolymer mixtures, however, there is a limitation in enhancing the mechanical properties.

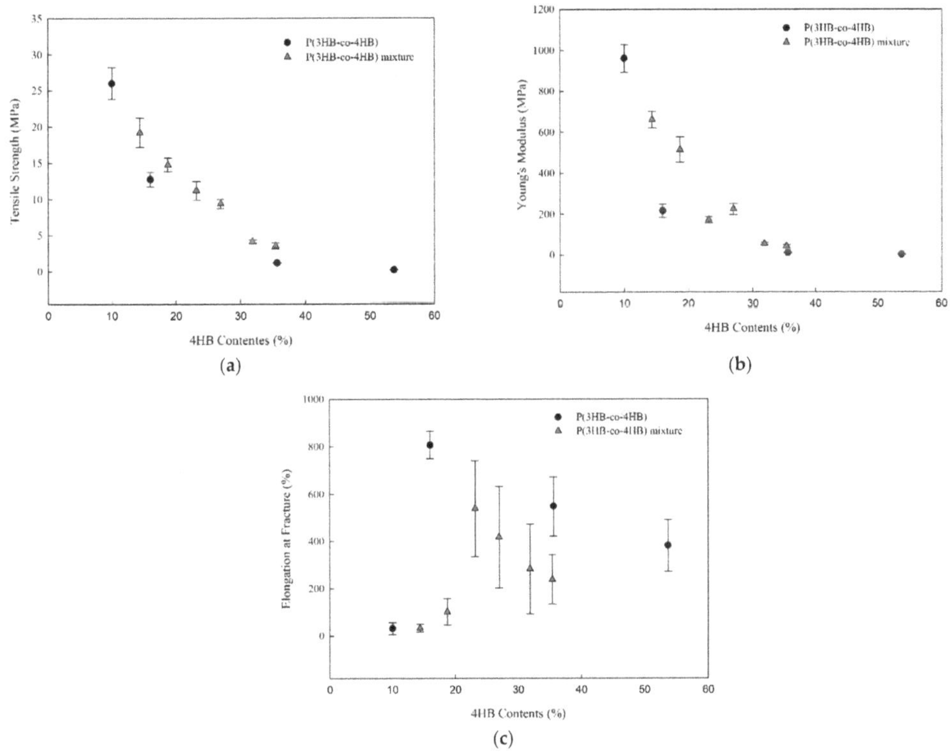

Figure 10. Effect of 4HB content on the mechanical properties of P(3HB-co-4HB) mixtures: (**a**) tensile strength; (**b**) Young's modulus; (**c**) elongation at break.

4. Conclusions

The change in the properties, with variation in the 4HB content of the P(3HB-co-4HB) copolymers and the mixtures of crystalline and noncrystalline P(3HB-co-4HB) copolymers, have been studied. It is found that crystalline P(3HB-co-4HB), with high 4HB content, can be prepared by melt blending. An increase in the 4HB content of P(3HB-co-4HB) copolymers increases the flexibility of the polymer chains to decrease the glass transition temperature and also increases the randomness to decrease the crystallinity. The P(3HB-co-4HB) copolymer mixtures show a higher glass transition temperature and also maintain crystallinity compared to copolymers of the same 4HB content. An increase in the 4HB content of P(3HB-co-4HB) copolymers decreases processibility due to a decrease in the melt viscosity, and an increase in the loss tangent. While in the case of P(3HB-co-4HB), the decrease in the copolymer mixtures' melt viscosity is relatively smaller, and the increase in the loss tangent is also smaller, such that a P(3HB-co-4HB) copolymer mixture exhibiting a minimal decrease in the processibility with an increase in 4HB content can be prepared. The P(3HB-co-4HB) copolymers and mixtures of crystalline and noncrystalline P(3HB-co-4HB) both exhibited a decrease in tensile strength and Young's modulus with an increase in 4HB content, however, an adequate increase in the 4HB content increases the flexibility of the polymer chains to increase the elongation at break.

Author Contributions: Conceptualization, H.-J.K.; investigation, M.J., E.L. and Y.J.; writing—original draft, M.J. and H.-J.K.; funding acquisition, E.L. and S.S. All authors have read and agreed to the published version of the manuscript.

Funding: This research received no external funding.

Institutional Review Board Statement: Not applicable.

Informed Consent Statement: Informed consent was obtained from all subjects involved in the study.

Data Availability Statement: Data are in the authors' possession.

Conflicts of Interest: The authors declare no conflict of interest.

References

1. Kanesawa, Y.; Tanahashi, N.; Doi, Y. Enzymatic degradation of microbial poly(3-hydroxyalkanoates). *Polym. Degrad. Stab.* **1994**, *45*, 179–185. [CrossRef]
2. Chen, G.-Q. (Ed.) Introduction of bacterial plastics PHA, PLA, PBS, PE, PTT, and PPP. In *Plastics from Bacteria: Natural Functions and Applications;* Springer: Berlin/Heidelberg, Germany, 2010; Volume 14, pp. 1–16.
3. Norhafini, H.; Huong, K.H.; Amirul, A.A. High PHA density fed-batch cultivation strategies for 4HB-rich P(3HB-co-4HB) copolymer production by transformant Cupriavidus malaysiensis USMAA1020. *Int. J. Biol. Macromol.* **2019**, *125*, 1024–1032. [CrossRef] [PubMed]
4. Chanprateep, S.; Buasri, K.; Muangwong, A.P. Utiswannakul, Biosynthesis and biocompatibility of biodegradable poly(3-hydroxybutyrate-co-4-hydroxybutyrate). *Polym. Degrad. Stab.* **2010**, *95*, 2003–2012. [CrossRef]
5. Chen, G.Q. A microbial polyhydroxyalkanoates (PHA) based bio- and materials industry. *Chem. Soc. Rev.* **2009**, *38*, 2434–2446. [CrossRef] [PubMed]
6. Ong, S.Y.; Chee, J.Y.; Sudesh, K. Degradation of Polyhydroxyalkanoate (PHA): A Review. *J. Sib. Fed. Univ. Biol.* **2017**, *10*, 211–225. [CrossRef]
7. Raquez, J.M.; Habili, Y.; Murariu, M.; Dubois, P. Polylactide (PLA)-based nanocomposites. *Prog. Polym. Sci.* **2013**, *38*, 1504–1542. [CrossRef]
8. Lim, L.T.; Auras, R.; Rubino, M. Processing technologies for poly(lactic acid). *Prog. Polym. Sci.* **2008**, *33*, 820–852. [CrossRef]
9. Liu, L.F.; Yu, J.Y.; Cheng, L.D.; Yang, X.J. Biodegradability of poly(butylene succinate) (PBS) composite reinforced with jute fibre. *Polym. Degrad. Stab.* **2009**, *94*, 90–94. [CrossRef]
10. Xu, D.J.; Guo, B.H. Poly(butylene succinate) and its copolymers: Research, development and industrialization. *Biotechnol. J.* **2010**, *5*, 1149–1163. [CrossRef]
11. El-Hadi, A.; Schnabel, R.; Straube, E.; Müller, G.; Henning, S. Correlation between degree of crystallinity, morphology, glass temperature, mechanical properties and biodegradation of poly(3-hydroxyalkanoate) PHAs and their blends. *Polym. Test.* **2002**, *21*, 665–674. [CrossRef]
12. Erich, M.; Hannes, G.; Maximilian, L. PHB-bio based and biodegradable replacement for PP: A Review. *Nov. Tech. Nutr. Food Sci.* **2018**, *2*, 206–209.
13. Savenkova, L.; Gercberga, Z.; Nikolaeva, V.; Dzene, A.; Bibers, I.; Kalnin, M. Mechanical properties and biodegradation characteristics of PHB-based films. *Process. Biochem.* **2000**, *35*, 573–579. [CrossRef]
14. Patricia, P.S.O.; Pereira, F.V.; Santos, M.C.; Souza, P.P.; Roa, J.P.B.; Orefice, R.L. Increasing the elongation at break of polyhydroxybutyrate biopolymer: Effect of cellulose nanowhiskers on mechanical and thermal properties. *J. Appl. Polym. Sci.* **2013**, *127*, 3613–3621. [CrossRef]
15. Panaitescu, D.M.; Popa, M.S.; Raditoiu, V.; Frone, A.N.; Sacarescu, L.; Gabor, A.R.; Teodorescu, M. Effect of calcium stearate as a lubricant and catalyst on the thermal degradation of poly (3-hydroxybutyrate). *Int. J. Biol. Macromol.* **2021**, *190*, 780–791. [CrossRef] [PubMed]
16. Pachekoski, W.M.; Dalmolin, C.; Agnelli, J.A.M. The Influence of the Industrial Processing on the Degradation of Poly(hidroxybutyrate)—PHB. *Macromol. Res.* **2013**, *16*, 327–332. [CrossRef]
17. Koyama, N.; Doi, Y. Morphology and biodegradability of a binary blend of poly ((R)-3-hydroxybutyric acid) and poly ((R, S)-lactic acid). *Can. J. Microbiol.* **1995**, *41*, 316–322. [CrossRef]
18. Wang, Y.X.; Huang, G.C.; Luo, L.Q. Thermal Properties of Poly(Lactic Acid) Modified Poly(3-Hydroxybutyrate-co-4-Hydroxbutyrate). *Mater. Sci. Forum* **2016**, *852*, 677–685.
19. Kennouche, S.; Moigne, N.L.; Kaci, M.; Quantin, J.C.; Caro-Bretelle, A.S.; Delaite, C.; Lopez-Cuesta, J.M. Morphological characterization and thermal properties of compatibilized poly(3-hydroxybutyrate-co-3-hydroxyvalerate) (PHBV)/poly(butylene succinate) (PBS)/halloysite ternary nanocomposites. *Eur. Polym. J.* **2016**, *75*, 142–162. [CrossRef]
20. Bian, Y.; Han, L.; Han, C.; Lin, H.; Zhang, H.; Bian, J.; Dong, L. Intriguing crystallization behavior and rheological properties of radical-based crosslinked biodegradable poly (3-hydroxybutyrate-co-4-hydroxybutyrate). *CrystEngComm* **2014**, *16*, 2702–2714. [CrossRef]
21. Li, M.; Li, Z.Q.; Xu, J.; Wei, D.S.; Zhu, H.W.; Li, D. Thermal property, morphology, mechanical and rheological properties of a modified bio-polymers prepared by blending poly (3-hydrobutyrate-co-4-hydrobutyrate) with chain extenders. *Open J. Adv. Mater. Res.* **2011**, *152*, 924–930. [CrossRef]
22. Xiang, H.; Chen, Z.; Zheng, N.; Zhang, X.; Zhu, L.; Zhou, Z.; Zhu, M. Melt-spun microbial poly (3-hydroxybutyrate-co-3-hydroxyvalerate) fibers with enhanced toughness: Synergistic effect of heterogeneous nucleation, long-chain branching and drawing process. *Int. J. Biol. Macromol.* **2019**, *122*, 1136–1143. [CrossRef] [PubMed]

23. Fei, B.; Chen, C.; Chen, S.; Peng, S.; Zhuang, Y.; An, Y.; Dong, L. Crosslinking of poly [(3-hydroxybutyrate)-co-(3-hydroxyvalerate)] using dicumyl peroxide as initiator. *Polym. Int.* **2004**, *53*, 937–943. [CrossRef]
24. Rupp, B.; Ebner, C.; Rossegger, E.; Slugovc, C.; Stelzer, F.; Wiesbrock, F. UV-induced crosslinking of the biopolyester poly (3-hydroxybutyrate)-co-(3-hydroxyvalerate). *Curr. Green Chem.* **2010**, *12*, 1796–1802. [CrossRef]
25. Bhubalan, K.; Lee, W.H.; Loo, C.Y.; Yamamoto, T.; Tsuge, T.; Doi, Y.; Sudesh, K. Controlled biosynthesis and characterization of poly (3-hydroxybutyrate-co-3-hydroxyvalerate-co-3-hydroxyhexanoate) from mixtures of palm kernel oil and 3HV-precursors. *Polym. Degrad. Stab.* **2008**, *93*, 17–23. [CrossRef]
26. Loo, C.Y.; Sudesh, K. Biosynthesis and native granule characteristics of poly (3-hydroxybutyrate-co-3-hydroxyvalerate) in Delftia acidovorans. *Int. J. Biol. Macromol.* **2007**, *40*, 466–471. [CrossRef] [PubMed]
27. Liu, Q.; Luo, G.; Zhou, X.R.; Chen, G.Q. Biosynthesis of poly (3-hydroxydecanoate) and 3-hydroxydodecanoate dominating polyhydroxyalkanoates by β-oxidation pathway inhibited Pseudomonas putida. *Metab. Eng.* **2011**, *13*, 11–17. [CrossRef] [PubMed]
28. Kim, J.S.; Lee, B.H.; Kim, B.S. Production of poly (3-hydroxybutyrate-co-4-hydroxybutyrate) by Ralstonia eutropha. *Biochem. Eng. J.* **2005**, *23*, 169–174. [CrossRef]
29. Saito, Y.; Nakamura, S.; Hiramitsu, M.; Doi, Y. Microbial synthesis and properties of poly (3-hydroxybutyrate-co-4-hydroxybutyrate). *Polym. Int.* **1996**, *39*, 169–174. [CrossRef]
30. Valentin, H.E.; Dennis, D. Production of poly (3-hydroxybutyrate-co-4-hydroxybutyrate) in recombinant Escherichia coli grown on glucose. *J. Biotechnol.* **1997**, *58*, 33–38. [CrossRef]
31. Al-Kaddo, K.B.; Mohamad, F.; Murugan, P.; Tan, J.S.; Sudesh, K.; Samian, M.R. Production of P(3HB-co-4HB) copolymer with high 4HB molar fraction by Burkholderia contaminans Kad1 PHA synthase. *Biochem. Eng. J.* **2020**, *153*, 107394–107400. [CrossRef]
32. Yun, O.Y.; Min, X.; Li, Y. Properties analysis of biodegradable material P(3HB-co-4HB). *Open J. Adv. Mater. Res.* **2011**, *380*, 168–172.
33. Wen, X.; Lu, X.; Peng, Q.; Zhu, F.; Zheng, N. Crystallization behaviors and morphology of biodegradablepoly(3-hydroxybutyrate-co-4-hydroxybutyrate). *J. Therm. Anal. Calorim.* **2012**, *109*, 959–966. [CrossRef]
34. Omura, T.; Gato, T.; Maehara, A.; Kimura, S.; Abe, H.; Iwata, T. Thermal degradation behavior of poly [(R)-3-hydroxybutyrate-co-4-hydroxybutyrate]. *Polym. Degrad. Stab.* **2021**, *183*, 109460. [CrossRef]
35. Doi, Y.; Segawa, A.; Kunioka, M. Biosynthesis and characterization of poly(3-hydroxybutyrate-co-4-hydroxybutyrate) in Alcaligenes eutrophus. *Int. J. Biol. Macromol.* **1990**, *12*, 106–111. [CrossRef]
36. Yang, S.Z.; Madbouly, S.A.; Schrader, J.A.; Grewell, D.; Kessler, M.R.; Graves, W.R. Processing and characterization of bio-based poly (hydroxyalkanoate)/poly(amide) blends: Improved flexibility and impact resistance of PHA-based plastics. *J. Appl. Polym. Sci.* **2015**, *132*, 42209–42218. [CrossRef]
37. Harrison, G.M.; Melik, D.H. Application of degradation kinetics to the rheology of poly(hydroxyalkanoates). *J. Appl. Polym. Sci.* **2006**, *102*, 1794–1802. [CrossRef]

polymers

MDPI

Article

Supercritical CO_2 Foaming of Poly(3-hydroxybutyrate-*co*-4-hydroxybutyrate)

Tao Zhang [1], Yunjae Jang [2], Eunhye Lee [2], Sooan Shin [2] and Ho-Jong Kang [1,*]

1 Department of Polymer Science and Engineering, Dankook University, 152 Jukjeon-ro, Suji-gu, Yongin-si 16890, Gyeonggi-do, Korea; taozhang1214@gmail.com

2 CJ Cheiljedang Corporation, 55 Gwanggyo-ro, 42 beon-gil, Yeongtong-gu, Suwon-si 16495, Gyeonggi-do, Korea; yunjae.jang@cj.net (Y.J.); eunhye.lee@cj.net (E.L.); sooanshin@gmail.com (S.S.)

* Correspondence: hjkang@dankook.ac.kr

Abstract: The supercritical carbon dioxide foaming characteristics of the biodegradable polymer poly(3-hydroxybutyrate-co-4-hydroxybutyrate) (P(3HB-*co*-4HB)) are studied for environmentally friendly packaging materials. The effect of the 4HB composition of the P(3HB-*co*-4HB) copolymers on the foaming conditions such as pressure and temperature is studied and the density and the expansion ratio of the resulting P(3HB-*co*-4HB) foam are together evaluated. The increase in the 4HB content reduces the crystallinity and tan δ value of P(3HB-*co*-4HB) required for the growth of the foam cells. Therefore, the foaming temperature needs to be lower to retain a suitable tan δ value of P(3HB-*co*-4HB) for foaming. It was found that P(3HB-*co*-4HB) with less crystallinity showed better formability and cell uniformity. However, foaming is not possible regardless of the foaming temperature when the 4HB content of P(3HB-*co*-4HB) is over 50%, due to the high tan δ value. A lower foam density and higher expansion ratio can be obtained with crystalline P(3HB-*co*-4HB) of low 4HB content, compared with non-crystalline P(3HB-*co*-4HB) of high 4HB content. The expansion ratio of P(3HB-*co*-4HB) foams can be increased slightly by using a chain extender, due to the lowing of crystallinity and tan δ. This is most effective in the case of P(3HB-*co*-4HB), whose 4HB content is 16%.

Keywords: poly(3-hydroxybutyrate-*co*-4-hydroxybutyrate); supercritical carbon dioxide; P(3HB-*co*-4HB) foams; expansion ratio; chain extender

Citation: Zhang, T.; Jang, Y.; Lee, E.; Shin, S.; Kang, H.-J. Supercritical CO_2 Foaming of Poly(3-hydroxybutyrate-*co*-4-hydroxybutyrate). *Polymers* **2022**, *14*, 2018. https://doi.org/10.3390/polym14102018

Academic Editor: Alberto Romero García

Received: 28 March 2022
Accepted: 10 May 2022
Published: 15 May 2022

1. Introduction

Polyhydroxyalkanoates (PHAs) are biodegradable polymers produced by microorganisms and have very wide applicability as an alternative material for synthetic polymers, which cause environmental pollution as they are degraded under diverse earth environments, contrary to other biodegradable polymers [1,2]. Due to their excellent biocompatibility and in vivo degradation properties, their application is being extended to medical materials [3]. The developed PHA products are melt processed films [4,5] or foams [6] for environmentally friendly packing material and research is being carried out regarding their application to medical materials, such as biodegradable sutures from electrospinning [7] and scaffolds for tissue engineering from 3D printing [8], etc.

Poly(3-hydroxybutyrate) (P3HB) was the first biodegradable polymer to be studied among the PHAs, but had the limitation of having a narrow window for melt processing, as thermal degradation occurs very easily [9]. Furthermore, its crystallization rate is very low, resulting in large spherulites with the interface among these resulting in brittleness and the crystallization continually occurs during solidification, making the stable production of molded products difficult [10]. To overcome these problems, blending with biodegradable polymers that are compatible with P3HB, such as polylactic acid (PLA) and polybutylene succinate (PBS) [11,12], modification of the structure by branching or crosslinking through the addition of diverse additives [13,14], and the bacterial synthesis of copolymers from biodegradable monomers [15] have been suggested. The comonomers widely used with 3-hydroxybutyrate in the study of PHA copolymers are 3-hydroxyvalerate to prepare poly(3-hydroxybutyrate-*co*-3-hydroxyvalerate) [16], 4-hydroxybutyrate to prepare poly(3-hydroxybutyrate-*co*-4-hydroxybutyrate) [17], and hydroxyhexanoate to prepare poly(3-hydroxybutyrate-*co*-hydroxyhexanoate) [18].

Due to the higher price of PHAs compared with commodity polymers, they have been considered for use in high priced medical materials, but the decrease in production costs through mass production is resulting in higher possibilities of extending its application to packaging material [19,20]. As a relatively smaller amount of PHA is needed to prepare foamed material with a porous structure, the development of packaging material from PHA using foaming agents is being carried out [21]. Polymer foams can be made by the addition of diverse chemical foaming agents [22,23], using environmentally friendly supercritical fluid in melt extrusion [24,25], and the introduction of supercritical fluid to beads in the solid state to prepare foam beads [26]. Obtained foam beads can be molded by chest injection molding [27]. There are two typical processes to produce bead foam, continuous process and batch process using foam extrusion and autoclave, respectively. In the autoclave bead, polymer beads are saturated with a physical blowing agent, and pressure in the autoclave is suddenly released to expand the beads. Bead foaming is an appropriate processing method for semi-crystalline polymers, such as polypropylene (PP) [28], polyurethane (TPU) [29], and PLA [30]. The use of carbon dioxide supercritical fluid to maintain the environmentally friendly nature of biodegradable polymers is reported mostly in PLA foams [31–35], and in the case of PHA, the melt extrusion foaming of PHBV with supercritical CO_2 is being carried out [36,37]. However, research on the foaming of P3HB has not been reported due to the lack of foamability. It is necessary to develop an eco-friendly supercritical CO_2 process using biodegradable P(3HB-*co*-4HB) with improved chain flexibility.

The effect of 4HB content in the bead foam process using supercritical CO_2 on the processing conditions and the resulting structure of the foams is studied. The effect of foaming temperature and pressure on the cell structure is studied in detail by evaluating the foam density, expansion ratio, etc. to explore the possibilities of application of these foams to packaging material. The effect of the addition of chain extending agents on the foam structure is also studied.

2. Experimental

The P(3HB-*co*-4HB) copolymer used in this study is a biodegradable polymer produced by CJ Cheiljedang, among which those with a 4HB content of 10–16% are confirmed to be crystalline and those with a 4HB content of 30 or 53.7% are confirmed to be non-crystalline. The chemical structure and 4HB content, along with molecular weights and heat of melting, are shown in Figure 1 and Table 1, respectively. A multifunctional styrene-acrylic oligomer (Joncryl, ADR 4370) purchased from BASF was used as a chain extender for the required control of the viscoelasticity of P(3HB-*co*-4HB) in the supercritical CO_2 foaming process.

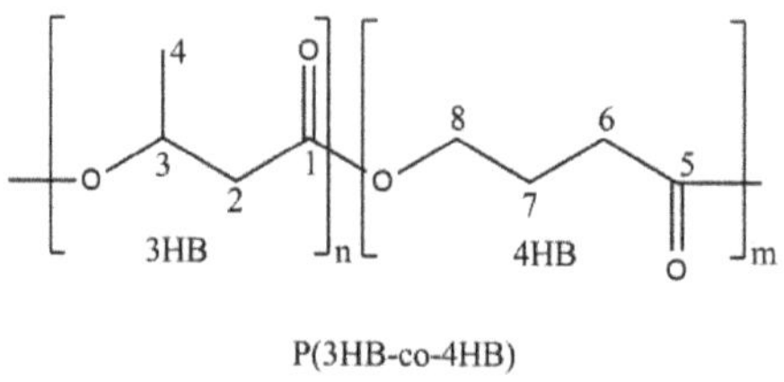

Figure 1. Chemical structure of P(3HB-*co*-4HB).

Table 1. Characteristics of P(3HB-co-4HB) used in this study.

Sample Code [Polymer]	4HB Content (%)	M_w (k)	ΔH_m (J/g)	T_{m1}(°C)	T_{m2}(°C)	T_c(°C)
PHA10 [P(3HB-*co*-10% 4HB)]	10	600	23.97	129.96	145.92	75.82
PHA10 with ADR(5phr)	10	600	9.65	113.59	141.96	72.53
PHA10 [P(3HB-*co*-16% 4HB)]	16	1000	3.99	115.82	155.59	59.67
PHA10 with ADR(5phr)	16	1000	2.54	113.55	154.96	56.84
PHA10 [P(3HB-*co*-30% 4HB)]	30	687	-	-	-	-
PHA10 [P(3HB-*co*-50% 4HB)]	53.7	901	-	-	-	-

P(3HB-*co*-4HB) of different 4HB contents were melt processed in an internal mixer (Haake, Rheomix600, Verden, Germany) at 20 rpm for 10 min at 140 °C, minimizing the thermal degradation to prepare samples of the same thermal history. When a 5 phr chain extender was added, the melt processing was carried out at 180 °C, which is the temperature at which the chain extender can react. The obtained P(3HB-*co*-4HB) samples were compression molded in a compression molding machine (QMESYS, QM900, Anyang, Korea) by heating at 140 °C for 2 min, raising the pressure to 8 MPa, and molded for 2 min, then quenched in water to obtain 1T thick 150 mm × 150 mm sheets. The sheets were cut into 5 mm × 5 mm pieces to use as beads for foaming.

The tan δ of P(3HB-*co*-4HB) was evaluated by measuring 1 mm × 2.5 mm sheets on a dynamic mechanical thermal analyzer (TA, DMA Q800/2980, New Castle, DE, USA) in the 30–140 °C range, at the oscillation frequency of 1 MHz.

The P(3HB-*co*-4HB) foam beads were prepared by adding 10 g of beads to a lab-made autoclave, as shown in Figure 2, diffusing supercritical CO_2 into P(3HB-*co*-4HB) for 60 min at the supercritical condition of CO_2 of 50–135 °C and 70–100 bar, then releasing the pressure abruptly to atmospheric pressure.

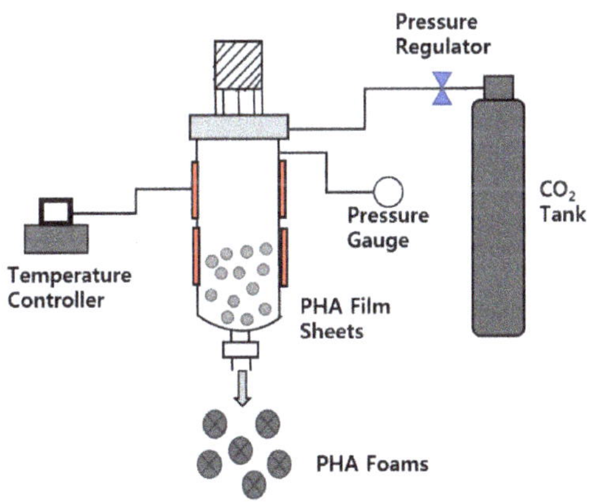

Figure 2. Schematic of supercritical CO_2 assisted foaming process.

The structure of the P(3HB-*co*-4HB) foams was evaluated with SEM pictures, taken on a scanning electron microscope (Hitachi, SEM S-5200, Hitachi, Tokyo, Japan) and by measuring the density before foaming (ρ_p) and the density after foaming (ρ_f) from which the expansion ratio (Φ) was calculated according to the following Equation (1).

$$\Phi = \frac{\rho_p}{\rho_f} \quad (1)$$

Cell density (N_f) refers to the number of cells (n) per cubic centimeter (A) in a foamed sample and can be defined by the following Equation (2).

$$N_f = \left(\frac{n}{A}\right)^{\frac{3}{2}} \text{X } \Phi \tag{2}$$

To measure the resilience rate of the crystalline P(3HB-*co*-10% 4HB) foam beads and noncrystalline P(3HB-*co*-30% 4HB) foam beads, the thickness of the foam beads was first determined as T_0, then after compressing the foam beads to 50% its thickness and releasing the pressure after 30 min, the thickness was measured as T; the resilience rate was calculated according to the following Equation (3).

$$R = \frac{T}{T_0} \times 100\ (\%) \tag{3}$$

3. Results and Discussion

3.1. Foamability of P(3HB-co-4HB) Copolymer

The SEM micrographs that show the possibilities of cell formation in the foams prepared in the range 50–130 °C at 90 bar for different 4HB contents are shown in Figure 3. As can be observed, the temperature at which the cells develop depends on the 4HB content. In the case of P(3HB-*co*-10% 4HB), whose relative crystallinity is highest, (23.97 J/g) as shown in Table 1, cell structure develops when processed at 130 °C, while in the case of P(3HB-*co*-16% 4HB), whose relatively crystallinity is lower (3.99 J/g), cell structure develops when processed at 100–110 °C. P(3HB-*co*-30% 4HB) is non-crystalline and the cell structure develops at a lower foaming temperature of 50–80 °C, but cell structure does not develop in the case of P(3HB-*co*-50% 4HB) in the range studied.

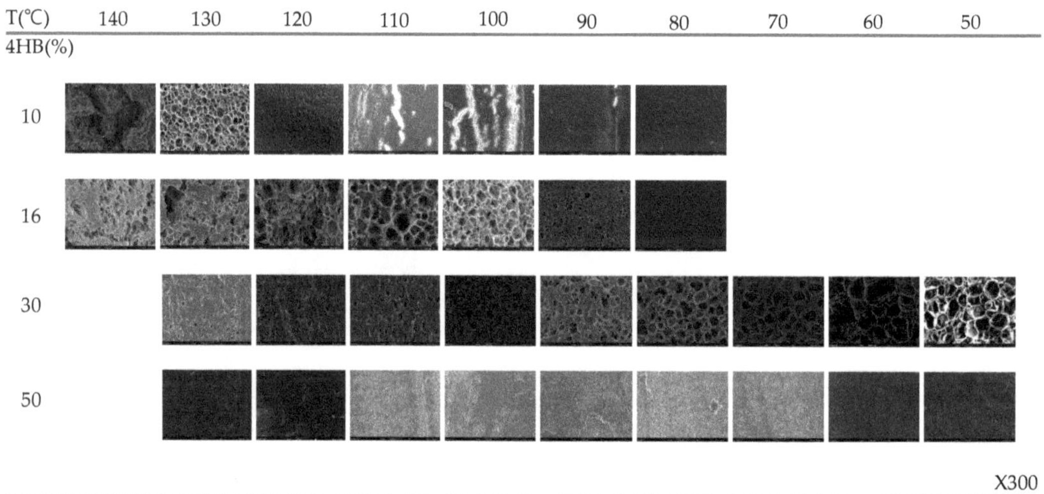

Figure 3. *Cont.*

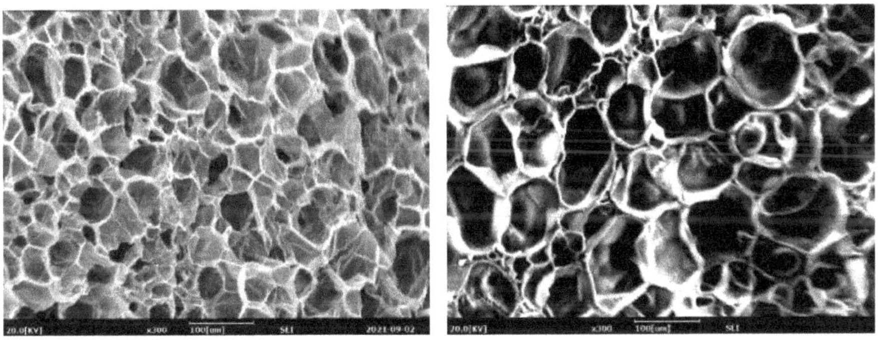

Figure 3. SEM micrographs of P(3HB-*co*-4HB) with different 4HB content foamed at 90 bar.

In supercritical foaming, the supercritical fluid diffuses into the polymer to create nuclei and they grow to form cells [38]. Nucleation and cell growth are affected by the crystalline/non-crystalline structure of the polymer and viscoelasticity, specifically, reasonable crystallinity and tan δ value under the foaming conditions are required for nucleation and continuous growth of the cell. In crystalline P(3HB-*co*-4HB), poor solubility and diffusivity of CO_2 in crystallites caused the lowering of nucleation of the cell and prohibits cell growth, which resulted in less formability.

The change in tan δ with temperature for the P(3HB-*co*-4HB) is shown in Figure 4. The tan δ value and crystallinity are closely related to the chain mobility in the solid state. An increase in the tan δ value is observed with an increase in temperature and 4HB content. The tan δ is the ratio of loss modulus to storage modulus and shows a drastic change at various transition temperatures. The drastic change in the tan δ with temperature can be observed as the temperature reaches the melting transition temperature of crystalline P(3HB-*co*-4HB) or the rubber transition temperature of non-crystalline P(3HB-*co*-4HB). The drastic change occurs at around 130 °C and 110 °C for crystalline P(3HB-*co*-10% 4HB) and P(3HB-*co*-16% 4HB), respectively, as their melting temperature depends on the 4HB content. On the other hand, the change can be observed above 50 °C and 30 °C for non-crystalline P(3HB-*co*-30% 4HB) and P(3HB-*co*-50% 4HB), respectively, where the mobility of the non-crystalline chains start to increase. The tan δ value increases with 4HB content in P(3HB-*co*-4HB) and the change in the tan δ value also increases at the respective transition temperatures. In the case of the P(3HB-*co*-4HB) copolymers, the chain mobility of 4HB is greater than that of 3HB, resulting in a higher tan δ value and a more sensitive change with temperature. P(3HB-*co*-50% 4HB) exhibits a higher tan δ value than other P(3HB-*co*-4HB) copolymers regardless of the temperature, due to the too much chain flexibility of the 4HB chain. The adequate tan δ value required for cell growth in supercritical fluid foaming can be indirectly assessed by the tan δ value reflecting the polymer's viscoelasticity. As observed in Figure 3, for copolymers with a 4HB content up to 30%, the tan δ values below the transition temperature where the chains show a drastic change in their mobility are very similar, with the values being around 0.1. Therefore, foaming is expected to be possible below the transition where chain mobility is increased, which is 130 °C for P(3HB-*co*-10% 4HB), 110 °C for P(3HB-*co*-16% 4HB), and below 80 °C for P(3HB-*co*-30% 4HB). However, P(3HB-*co*-50% 4HB) exhibits a tan δ value above 0.3, regardless of the temperature reflecting the extremely chain mobility; thus, cell growth in the foaming process is not possible. As observed in Figure 3, in the case of P(3HB-*co*-10% 4HB) of low 4HB content, the temperature range where foaming is possible is very narrow around 130 °C, as the stiffness of the chains is high, resulting in a drastic decrease in the mobility of chains. In the case of P(3HB-*co*-16% 4HB) where the flexibility of the chains is increased with an increase in 4HB content, foaming is possible in a relatively broader temperature range of 100–110 °C, and P(3HB-*co*-30% 4HB), which has the highest chain mobility, has the widest window where foaming is possible of 50–90 °C. Therefore, it is evident that in non-crystalline P(3HB-*co*-4HB) an adequate tan δ related to chain mobility above the glass transition temperature is an important factor in foaming.

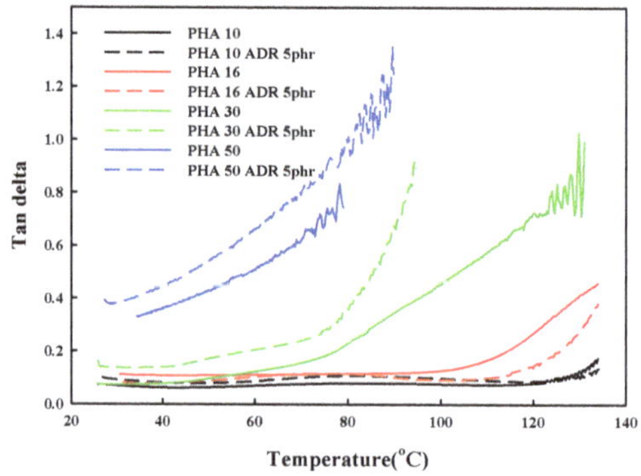

Figure 4. DMA thermograms of various P(3HB-*co*-4HB) with and without ADR.

3.2. Characteristics of P(3HB-co-4HB) Foams

The cell structure of P(3HB-*co*-4HB) foams shown in Figure 3 generally shows thinner cell walls, which hinder the formation of closed cells and result in an open cell structure compared with thermoplastic polyurethane (TPU) foam [39], which is highly elastic, and the cell size is also very irregular. However, the non-crystalline P(3HB-*co*-30% 4HB) foam exhibits thicker walls and more regular size in the closed cell, compared with crystalline P(3HB-*co*-4HB). This can be explained by the crystallinity and their morphology affecting the foaming temperature. In crystalline P(3HB-*co*-4HB), the chain flexibility to foam the cell can be obtained just under the melting temperature, and then the strength for cell nucleation and growth is rapidly lowered with a further increase in temperature, which resulted in uncontrolled cell growth and cell coalescence. However, cell growth and stabilization in non-crystalline P(3HB-*co*-4HB) are more stable because the forming could be archived in a broad range of temperatures with high elastic properties above the glass transition temperature. The cell structure is related to the resiliency against applied pressure. Figure 5 shows the resilience rate of the crystalline P(3HB-*co*-10% 4HB) foam and noncrystalline P(3HB-*co*-30% 4HB) foam. The resilience of P(3HB-*co*-30% 4HB), which is noncrystalline due to the high 4HB content, is greater than that of crystalline P(3HB-*co*-10% 4HB). This is due to the higher elasticity of P(3HB-*co*-30% 4HB) from its higher 4HB content and also due to the thicker cell walls and more regular cell structure. The resilience rate of P(3HB-*co*-30% 4HB) foamed at different temperatures, as shown in Figure 5b and showed much higher resilience for foams prepared at lower temperatures. This shows that the elastic properties of P(3HB-*co*-30% 4HB), along with open/close cell structure, the thickness of the cell walls, and cell size, which in turn depends on an expansion ratio, has a close relationship with resiliency. These results suggest that the possibility of using crystalline P(3HB-*co*-4HB) foams as an environmentally friendly packaging material is higher than that of using it for industrial structural material, which requires elasticity and resiliency to recover to its original form after compression, as it has an open cell structure compared with TPU foams. On the other hand, the non-crystalline P(3HB-*co*-30% 4HB) foams prepared at low temperatures with 100% resilience rates suggest the possibility of developing it as an industrial structural material.

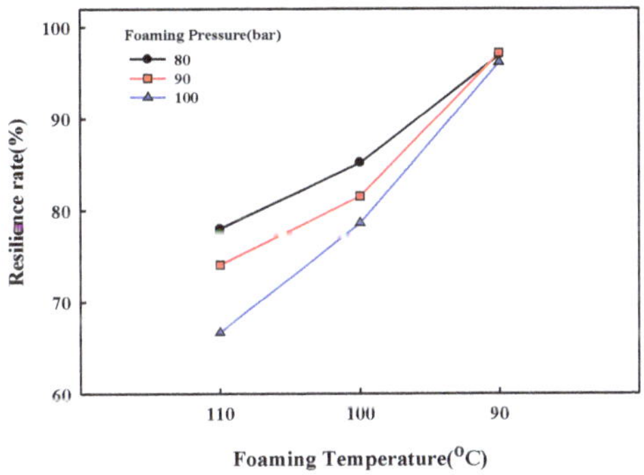

(a)

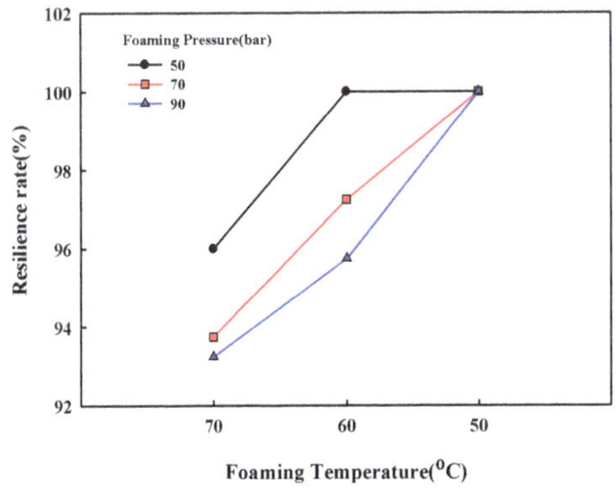

(b)

Figure 5. Resilience characteristics of P(3HB-*co*-4HB) foams; (**a**) crystalline P(3HB-*co*-16% 4HB); (**b**) non-crystalline P(3HB-*co*-30% 4HB).

Figure 6 shows the effect of foaming temperature on the density and the expansion rates calculated from density measurements before and after the expansion of P(3HB-*co*-4HB) beads of different 4HB content. The different foaming characteristic of P(3HB-*co*-30% 4HB) of different 4HB content was confirmed by the SEM micrographs in Figure 3. P(3HB-*co*-50% 4HB) with the highest 4HB content does not show a significant change in density over the range studied, suggesting that cell growth from the diffused supercritical CO_2 is not possible, due to its high tan δ value. The growth of cells in P(3HB-*co*-30% 4HB) at the foaming temperatures of 50–80 °C due to adequate tan δ and chain mobility can be confirmed. The cell growth of crystalline P(3HB-*co*-10% 4HB) and P(3HB-*co*-16% 4HB) at low temperatures is difficult due to the crystalline structure and lack of chain mobility, but near the melting temperature, chain mobility occurs and the tan δ becomes adequate for foaming to result in cell growth. The highest expansion ratios are obtained at 130 °C and 110 °C for P(3HB-*co*-10% 4HB) and P(3HB-*co*-16% 4HB), respectively. However, the uniformity of cells is very poor, due to the rapid decrease in strength to maintain cells during the stabilization step. The expansion ratio of non-crystalline P(3HB-*co*-30% 4HB) is below 6, while that of crystalline P(3HB-*co*-10% 4HB) and P(3HB-*co*-16% 4HB) is over 20. These P(3HB-*co*-4HB) foams have different expansion ratios, suggesting that they can be used to prepare foams for different applications.

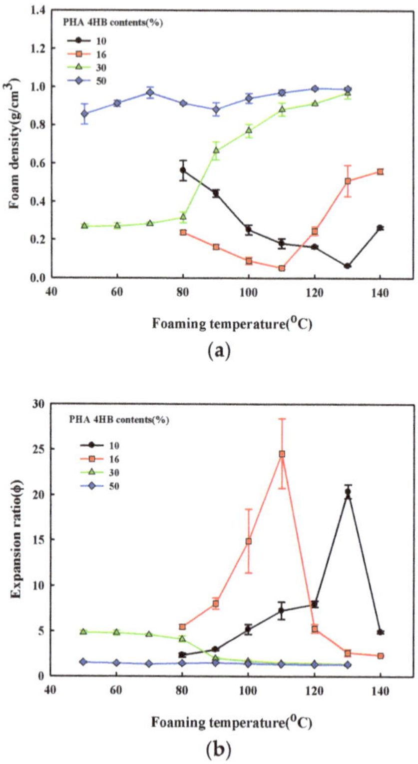

Figure 6. Effect of foaming temperature on the structure of the P(3HB-*co*-4HB) bead foamed at 90 bar; (**a**) foam density; (**b**) expansion ratio.

The effect of foaming pressure on foam formation of P(3HB-*co*-4HB) is shown in Figure 7. The foaming temperature was set at the temperature at which the maximum expansion ratio could be obtained from the respective P(3HB-*co*-4HB) at 90 bar. The cell size increases with the increase in pressure. The foam density and expansion ratio of the P(3HB-*co*-4HB) copolymers shown in Figure 8 show a decrease in the foam density and an increase in the expansion ratio with an increase in pressure. An increase in the foaming pressure allows more diffusion of supercritical CO_2 into P(3HB-*co*-4HB) to form a greater number of nuclei in P(3HB-*co*-4HB) beads, which results in a decrease in the foam density and an increase in the expansion ratio on cell growth when the pressure is relieved. This effect of pressure is more evident in the higher 4HB content P(3HB-*co*-4HB), showing that diffusion of the supercritical CO_2 is easier in the non-crystalline P(3HB-*co*-4HB) than the crystalline P(3HB-*co*-4HB).

3.3. Effect of Chain Extender on Foam Structures

SEM micrographs that show the effect of the addition of a chain extender on the supercritical CO_2 foaming are shown in Figure 9. Although the addition of a chain extender to crystalline P(3HB-*co*-10% 4HB) increases cell size when foamed at 130 °C, it does not affect the temperature window where foaming is possible, while in the case of P(3HB-*co*-16% 4HB), the temperature window appears at a higher temperature and is broadened from 110–100 °C to 120–130 °C. Figure 8b shows that although foaming of non-crystalline P(3HB-*co*-30% 4HB) is possible in the temperature range 50–90 °C in the absence of a chain extender, it is not when a chain extender is added. It was found that the degree of crystallinity of P(3HB-*co*-4HB) modified by a chain extender decreased dramatically, as shown in Table 1. This indicated that the crystalline structural change and the chain flexibility took place by chemical modification. As a result, the tan δ decreases observed in Figure 4 are caused by a structural change in the formation of chain branching or crosslinking from the reaction of epoxy groups of the chain extender Joncryl ADR 4370, containing epoxy groups, and the 3HB hydroxy groups in the crystalline P(3HB-*co*-4HB) of low 4HB content, resulting in a decrease in tan δ. However, in the case of P(3HB-*co*-4HB) of high 4HB content, viscosity decrease occurs instead of the aforementioned increase and decreases tan δ to decrease the foamability. Figure 10 shows the effect of a chain extender on the foam density and expansion ratio of the foam. The addition of the chain extender does not affect the foaming conditions in the case of P(3HB-*co*-10% 4HB) due to relatively high crystallinity, but a general increase in the expansion ratio with a decrease in foam density occurs. However, in the case of P(3HB-*co*-16% 4HB), the change in tan δ with the change in the chain structure allows foaming at temperatures above 110 °C and the foam density remains low above 110 °C to allow foaming at 120–130 °C, where foaming was not possible in the absence of a chain extender and an expansion ratio of around 20 can be obtained. Therefore, the control of the foaming properties with a chain extender is only possible in crystalline P(3HB-*co*-16% 4HB) with reasonably low crystallinity and tan δ value.

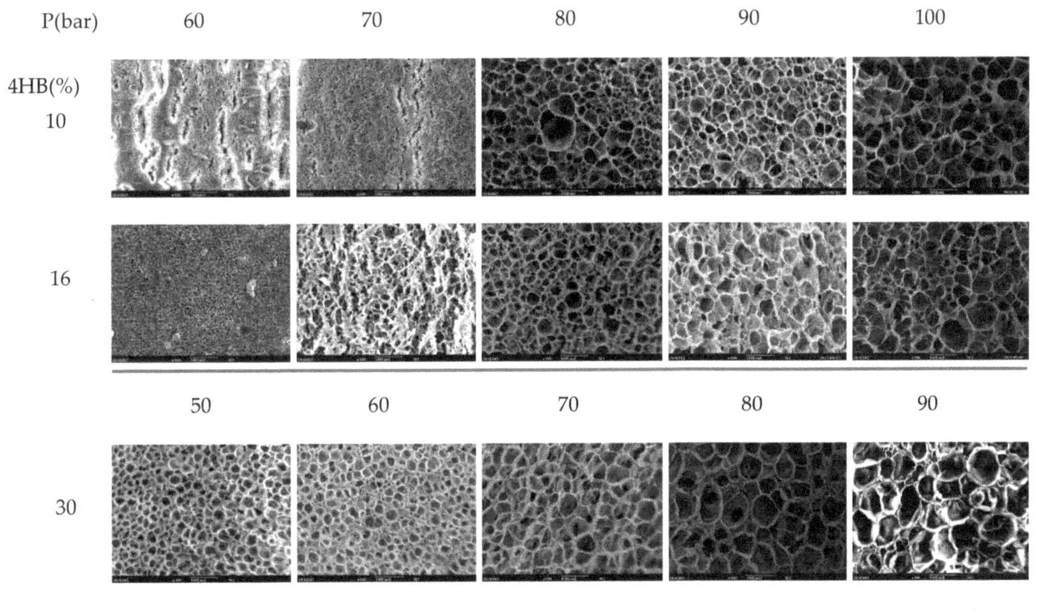

Figure 7. SEM micrographs of P(3HB-*co*-4HB) foamed at various foaming pressure. Foaming temperatures of P(3HB-*co*-10% 4HB), P(3HB-*co*-16% 4HB), and P(3HB-*co*-30% 4HB) are 130 °C, 100 °C, 50 °C, respectively.

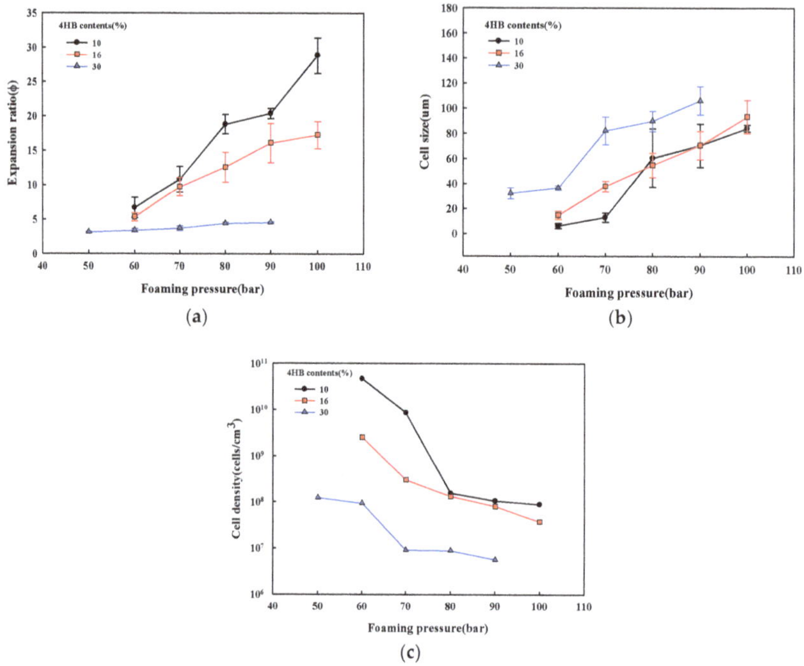

Figure 8. Effect of forming pressure on the structure of the P(3HB-*co*-4HB) foams. Foaming temperatures of P(3HB-*co*-10% 4HB), P(3HB-*co*-16% 4HB), and P(3HB-*co*-30% 4HB) are 130 °C, 100 °C, 50 °C, respectively; (**a**) expansion ratio; (**b**) cell size; (**c**) cell density.

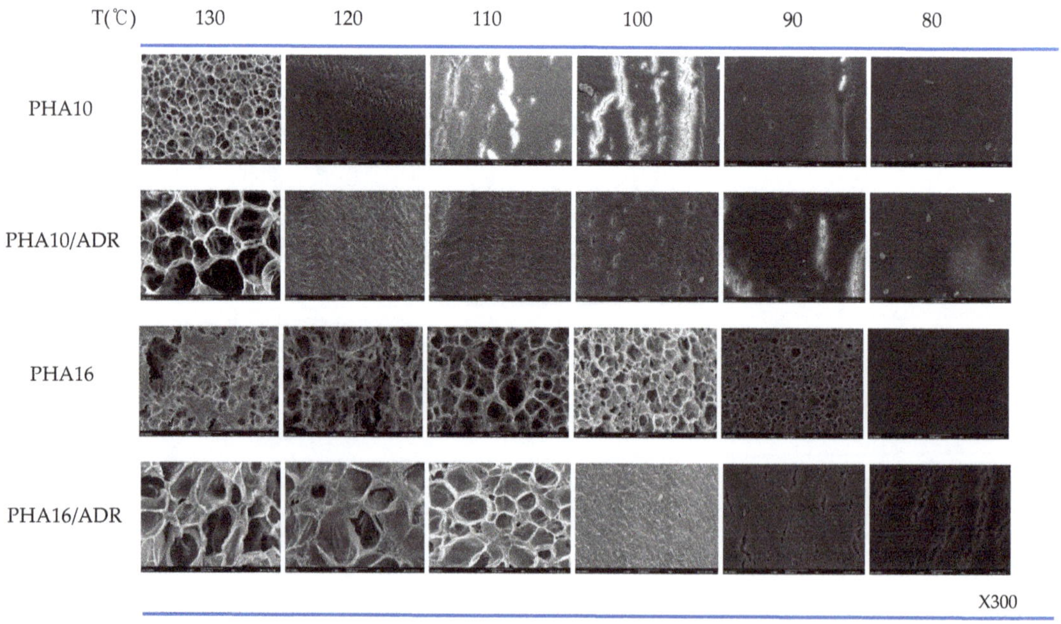

Figure 9. SEM micrographs of foamed crystalline P(3HB-*co*-4HB) with and without chain extender.

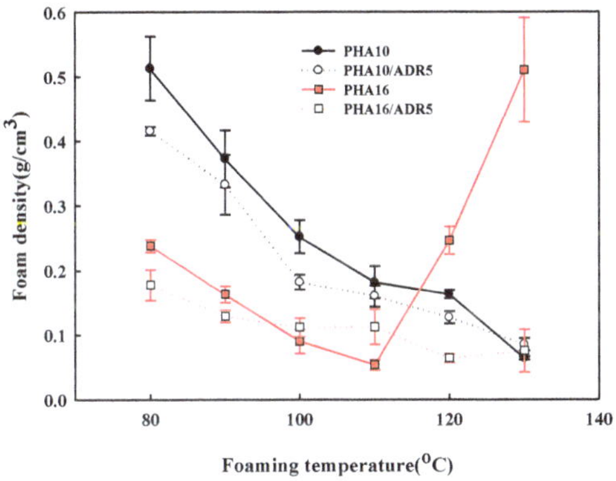

(**a**)

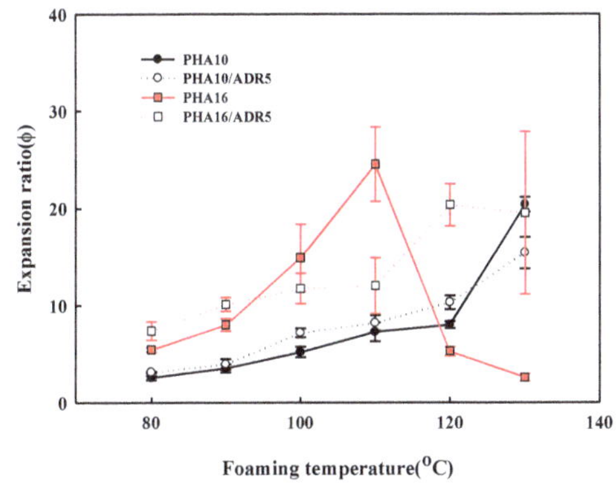

(**b**)

Figure 10. Effect of chain extender on the structure of the P(3HB-*co*-4HB) foams; (**a**) foam density; (**b**) expansion ratio.

4. Conclusions

Biodegradable P(3HB-*co*-4HB) of different 4HB contents and foam beads processed with supercritical carbon dioxide was used to study the effects of 4HB content on the foaming conditions and foam properties. It was found that various foam structures for packaging and industrial materials were obtained by controlling the 4HB content in P(3HB-*co*-4HB), which resulted in changes in the crystallinity and viscoelastic properties of P(3HB-*co*-4HB). The low 4HB content P(3HB-*co*-4HB) that maintained a crystalline structure can be foamed just under the melting temperature, but the range of the foaming temperature becomes broader with an increase in the 4HB content. The expansion ratio is around 20 and an open cell structure is formed. The non-crystalline P(3HB-*co*-4HB) of high 4HB content has a relatively wide window of foaming around the temperature at which chain mobility occurs. The formed foam has thicker cell walls, expansion ratios of about 6, and the cell size is regular, resulting in enhanced resiliency against compression, compared with crystalline P(3HB-*co*-4HB). The chain extender changes the chain structure of crystalline P(3HB-*co*-16% 4HB), which changes the crystallinity and chain mobility and, thus, changes the temperature at which foaming is possible and the cell structure. To enhance the physical properties of P(3HB-*co*-4HB) foams, future studies for the blends and composites using P(3HB-*co*-4HB) may be considered.

Author Contributions: Conceptualization, H.-J.K.; investigation, T.Z., E.L. and Y.J.; writing—original draft, T.Z. and H.-J.K.; funding acquisition, E.L. and S.S. All authors have read and agreed to the published version of the manuscript.

Funding: CJ Cheiljedang.

Institutional Review Board Statement: Not applicable.

Informed Consent Statement: Informed consent was obtained from all the subjects involved in the study.

Data Availability Statement: Data are in the authors' possession.

Conflicts of Interest: The authors declare no conflict of interest.

References

1. Chen, G.Q. A microbial polyhydroxyalkanoates (PHA) based bio- and materials industry. *Chem. Soc. Rev.* **2009**, *38*, 2434–2446. [CrossRef] [PubMed]
2. Ong, S.Y.; Chee, J.Y.; Sudesh, K. Degradation of Polyhydroxyalkanoate (PHA): A Review. *J. Sib. Fed. Univ. Biol.* **2017**, *10*, 211–225. [CrossRef]
3. Puppi, D.; Pecorini, G.; Chiellini, F. Biomedical Processing of Polyhydroxyalkanoates. *Bioengineering* **2019**, *6*, 0108. [CrossRef] [PubMed]
4. Thellen, C.; Coyne, M.; Froio, D.; Auerbach, M.; Wirsen, C.; Ratto, J.A. A Processing. Characterization and Marine Biodegradation Study of Melt-Extruded Polyhydroxyalkanoate (PHA) Films. *J. Polym. Environ.* **2008**, *16*, 1–11. [CrossRef]
5. Vostrejs, P.; Adamcova, D.; Vaverkova, M.D.; Enev, V.; Kalina, M.; Machovsky, M.; Sourkov, M.; Marovaa, I.; Kovalcik, A. Active biodegradable packaging films modified with grape seeds lignin. *RSC Adv.* **2020**, *10*, 29202–29213. [CrossRef]
6. Cravo, C.; Duarte, A.R.C.; Duarte, C.M.M. Solubility of carbon dioxide in a natural biodegradable polymer: Determination of diffusion coefficients. *J. Supercrit. Fluids* **2007**, *740*, 194–199. [CrossRef]
7. Volova, T.; Goncharov, D.; Sukovatyi, A.; Shabanov, A.; Nikolaeva, E.; Shishatskaya, E. Electrospinning of polyhydroxyalkanoate fibrous scaffolds: Effects on electrospinning parameters on structure and properties. *J. Biomater. Sci. Polym. Ed.* **2013**, *25*, 370–393. [CrossRef]
8. Grande, D.; Ramier, J.; Versace, D.L.; Renard, E.; Langlois, V. Design of functionalized biodegradable PHA-based electrospun scaffolds meant for tissue engineering applications. *New Biotechnol.* **2017**, *37*, 129–137. [CrossRef]
9. Thomas, S.; Shumilova, A.A.; Kiselev, E.G.; Baranovsky, S.V.; Vasiliev, A.D.; Nemtsev, I.V.; Kuzmin, A.P.; Sukovatyi, A.G.; Avinash, R.P.; Volova, T.G. Thermal, mechanical and biodegradation studies of biofiller based poly-3-hydroxybutyrate biocomposites. *Int. J. Biol. Macromol.* **2020**, *155*, 1373–1384. [CrossRef]
10. Che, X.M.; Ye, H.M.; Chen, G.Q. Effects of uracil on crystallization and rheological property of poly(R-3-hydroxybutyrate-co-4-hydroxybutyrate). *Compos. A Appl. Sci. Manuf.* **2018**, *109*, 141–150. [CrossRef]
11. Sun, J.Y.; Shen, J.J.; Chen, S.K.; Cooper, M.A.; Fu, H.B.; Wu, D.M.; Yang, Z.G. Nanofiller Reinforced Biodegradable PLA/PHA Composites: Current Status and Future Trends. *Polymers* **2018**, *10*, 0505. [CrossRef] [PubMed]
12. Parulekar, Y.; Mohanty, A.K. Extruded Biodegradable Cast Films from Polyhydroxyalkanoate and Thermoplastic Starch Blends: Fabrication and Characterization. *Macromol. Mater. Eng.* **2007**, *292*, 1218–1228. [CrossRef]
13. Musiol, M.; Jurczyk, S.; Sobota, M.; Klim, M.; Sikorska, W.; Zieba, M.; Janeczek, H.; Rydz, J.; Kurcok, P.; Johnston, B.; et al. (Bio)Degradable Polymeric Materials for Sustainable Future—Part 3: Degradation Studies of the PHA/Wood Flour-Based Composites and Preliminary Tests of Antimicrobial Activity. *Materials* **2020**, *13*, 2200. [CrossRef] [PubMed]
14. Ventura, H.; Laguna-Gutierrez, E.; Rodriguez-Perez, M.A.; Ardanuy, M. Effect of chain extender and water-quenching on the properties of poly(3-hydroxybutyrate-co-4-hydroxybutyrate) foams for its production by extrusion foaming. *Eur. Polym. J.* **2016**, *85*, 14–25. [CrossRef]
15. Li, S.Y.; Dong, C.L.; Wang, S.Y.; Ye, H.M.; Chen, G.Q. Microbial production of polyhydroxyalkanoate block copolymer by recombinant Pseudomonas putida. *Appl. Microbiol. Biotechnol.* **2010**, *90*, 659–669. [CrossRef]
16. Ahankari, S.S.; Mohanty, A.K.; Misra, M. Mechanical behaviour of agro-residue reinforced poly(3-hydroxybutyrate-co-3-hydroxyvalerate), (PHBV) green composites: A comparison with traditional polypropylene composites. *Compos. Sci. Technol.* **2011**, *71*, 653–657. [CrossRef]
17. Al-Kaddo, K.B.; Mohamad, F.; Murugan, P.; Tan, J.S.; Sudesh, K.; Samian, M.R. Production of P(3HB-co-4HB) copolymer with high 4HB molar fraction by Burkholderia contaminans Kad1 PHA synthase. *Biochem. Eng. J.* **2020**, *153*, 107394–107400. [CrossRef]
18. Tripathil, L.; Wu, L.P.; Chen, J.C.; Chen, G.Q. Synthesis of Diblock copolymer poly-3-hydroxybutyrate -block-poly-3-hydroxyhexanoate [PHB-b-PHHx] by a β-oxidation weakened Pseudomonas putida KT2442. *Microb. Cell. Factories* **2012**, *11*, 44–54. [CrossRef]
19. Vijayendra, S.V.N.; Shamala, T.R. Film forming microbial biopolymers for commercial applications—A review. *Crit. Rev. Biotechnol.* **2014**, *34*, 338–357. [CrossRef]

20. Zhang, J.Q.; Kasuya, K.; Hikima, T.; Takata, M.; Takemura, A.; Lwata, T. Mechanical properties, structure analysis and enzymatic degradation of uniaxially cold-drawn films of poly[(R)-3-hydroxybutyrate-co-4-hydroxybutyrate]. *Polym. Degrad. Stab.* **2011**, *96*, 2130–2138. [CrossRef]
21. Plackett, D.; Siro, I. Polyhydroxyalkanoates (PHAs) for food packaging. *Multifunct. Nanoreinforced Polym. Food Packag.* **2011**, *4*, 498–526.
22. Liao, Q.; Tsui, A.; Billington, S.; Frank, C.W. Extruded Foams from Microbial Poly(3-hydroxybutyrate-co-3-hydroxyvalerate) and Its Blends with Cellulose Acetate Butyrate. *Polym. Eng. Sci.* **2012**, *52*, 1495–1508. [CrossRef]
23. Wright, Z.C.; Frank, C.W. Increasing Cell Homogeneity of Semicrystalline, Biodegradable Polymer Foams with a Narrow Processing Window via Rapid Quenching. *Polym. Eng. Sci.* **2014**, *54*, 2877–2886. [CrossRef]
24. Moigne, N.L.; Sauceau, M.; Benyakhlef, M.; Jemai, R.; Benezet, J.C.; Rodier, E.; Lopez-Cuesta, J.M.; Fages, J. Foaming of poly(3-hydroxybutyrate-co-3-hydroxyvalerate)/organo-clays nano-biocomposites by a continuous supercritical CO_2 assisted extrusion process. *Eur. Polym. J.* **2014**, *61*, 157–171. [CrossRef]
25. Takahashi, S.; Hassler, J.C.; Kiran, E. Melting behavior of biodegradable polyesters in carbon dioxide at high pressures. *J. Supercrit. Fluids* **2012**, *72*, 278–287. [CrossRef]
26. Ke, J.; Zhang, L.; Li, D.L.; Bao, J.B.; Wang, Z.B. Foaming of Poly(3-hydroxybutyrate-co-3-hydroxyvalerate) with Supercritical Carbon Dioxide: Foaming Performance and Crystallization Behavior. *ACS Omega* **2020**, *5*, 9839–9845.
27. Ge, C.B.; Ren, Q.; Wang, S.P.; Zheng, W.G.; Zhai, W.T.; Park, C.B. Steam-chest molding of expanded thermoplastic polyurethane bead foams and their mechanical properties. *Chem. Eng. Sci.* **2017**, *17*, 337–346. [CrossRef]
28. Hossieny, N.; Ameli, A.; Park, C.B. Characterization of Expanded Polypropylene Bead Foams with Modified Steam-Chest Molding. *Ind. Eng. Chem. Res.* **2013**, *52*, 8236–8247. [CrossRef]
29. Jiang, J.J.; Liu, F.; Yang, X.; Xiong, Z.J.; Liu, H.W.; Xu, D.H.; Zhai, W.T. Evolution of ordered structure of TPU in high-elastic state and their influences on the autoclave foaming of TPU and inter-bead bonding of expanded TPU beads. *Polymer* **2021**, *228*, 123872–123884. [CrossRef]
30. Tang, L.Q.; Zhai, W.T.; Zheng, W.G. Autoclave preparation of expanded polypropylene/poly(lactic acid) blend bead foams with a batch foaming process. *J. Cell. Plast.* **2011**, *47*, 429–446. [CrossRef]
31. Gong, P.J.; Zhai, S.; Lee, R.; Zhao, C.X.; Buahom, P.; Li, G.X.; Park, C.B. Environmentally Friendly Polylactic Acid-Based Thermal Insulation Foams Blown with Supercritical CO_2. *Ind. Eng. Chem. Res.* **2015**, *57*, 5464–5471. [CrossRef]
32. Yang, Y.C.; Li, X.Y.; Zhang, Q.Q.; Xia, C.H.; Chen, C.; Chen, X.H.; Yu, P. Foaming of poly(lactic acid) with supercritical CO_2: The combined effect of crystallinity and crystalline morphology on cellular structure. *J. Supercrit. Fluids* **2019**, 145122–145132. [CrossRef]
33. Nofar, M.; Park, C.B. Poly(lactic acid)foaming. *Prog. Polym. Sci.* **2014**, *39*, 1721–1741. [CrossRef]
34. Standau, T.; Zhao, C.J.; Castellon, S.M.; Bonten, C.; Altstadt, V. Chemical Modification and Foam Processing of Polylactide (PLA). *Polymers* **2019**, *11*, 0306. [CrossRef]
35. Szegda, D.; Duangphet, S.; Song, J.; Tarverdi, K. Extrusion foaming of PHBV. *J. Cell. Plast.* **2014**, *50*, 145–162. [CrossRef]
36. Javadi, A.; Srithep, Y.; Clemons, C.C.; Turng, L.S.; Gong, S.Q. Processing of poly(hydroxybutyrate-co-hydroxyvalerate)-based bionanocomposite foams using supercritical fluids. *J. Mater. Res.* **2012**, *27*, 1506–1517. [CrossRef]
37. Frerich, S.C. Biopolymer foaming with supercritical CO_2—Thermodynamics, foaming behaviour and mechanical characteristics. *J. Supercrit. Fluids* **2015**, *96*, 349–358. [CrossRef]
38. Banerjee, R.; Ray, S.S. Foamability and Special Applications of Microcellular Thermoplastic Polymers: A Review on Recent Advances and Future Direction. *Macromol. Mater. Eng.* **2020**, *305*, 2000366–2000414. [CrossRef]
39. Zhang, T.; Lee, S.J.; Yoo, Y.H.; Park, K.H.; Kang, H.J. Compression Molding of Thermoplastic Polyurethane Foam Sheets with Beads Expanded by Supercritical CO_2 Foaming. *Polymers* **2021**, *13*, 0656. [CrossRef]

polymers

MDPI

Article

Microbial Poly(hydroxybutyrate-co-hydroxyvalerate) Scaffold for Periodontal Tissue Engineering

Seubsakul Phuegyod [1,†], Sasivimon Pramual [2,†], Nungnit Wattanavichean [3], Supasuda Assawajaruwan [1], Taweechai Amornsakchai [4], Panithi Sukho [5], Jisnuson Svasti [2], Rudee Surarit [6,7] and Nuttawee Niamsiri [1,*]

1 Department of Biotechnology, Faculty of Science, Mahidol University, Bangkok 10400, Thailand
2 Laboratory of Biochemistry, Chulabhorn Research Institute, Bangkok 10210, Thailand
3 School of Materials Science and Innovation, Faculty of Science, Mahidol University, Nakhon Pathom 73170, Thailand
4 Center of Excellence for Innovation in Chemistry, Department of Chemistry, Faculty of Science, Mahidol University, Nakhon Pathom 73170, Thailand
5 Department of Clinical Sciences and Public Health, Faculty of Veterinary Science, Mahidol University, Nakhon Pathom 73170, Thailand
6 Department of Oral Biology, Faculty of Dentistry, Mahidol University, Bangkok 10400, Thailand
7 Faculty of Dentistry, Siam University, Bangkok 10160, Thailand
* Correspondence: nuttawee.nia@mahidol.ac.th
† These authors contributed equally to this work.

Abstract: In this study, we fabricated three dimensional (3D) porous scaffolds of poly(hydroxybutyrate-co-hydroxyvalerate) with 50% HV content. P(HB-50HV) was biosynthesized from bacteria *Cupriavidus necator* H16 and the in vitro proliferation of dental cells for tissue engineering application was evaluated. Comparisons were made with scaffolds prepared by poly(hydroxybutyrate) (PHB), poly(hydroxybutyrate-co-12%hydroxyvalerate) (P(HB-12HV)), and polycaprolactone (PCL). The water contact angle results indicated a hydrophobic character for all polymeric films. All fabricated scaffolds exhibited a high porosity of 90% with a sponge-like appearance. The P(HB-50HV) scaffolds were distinctively different in compressive modulus and was the material with the lowest stiffness among all scaffolds tested between the dry and wet conditions. The human gingival fibroblasts (HGFs) and periodontal ligament stem cells (PDLSCs) cultured onto the P(HB-50HV) scaffold adhered to the scaffold and exhibited the highest proliferation with a healthy morphology, demonstrating excellent cell compatibility with P(HB-50HV) scaffolds. These results indicate that the P(HB-50HV) scaffold could be applied as a biomaterial for periodontal tissue engineering and stem cell applications.

Keywords: polyhydroxyalkanoates; poly(hydroxybutyrate-co-hydroxyvalerate); tissue engineering; 3D porous scaffolds; human gingival fibroblasts; periodontal ligament stem cells

Citation: Phuegyod, S.; Pramual, S.; Wattanavichean, N.; Assawajaruwan, S.; Amornsakchai, T.; Sukho, P.; Svasti, J.; Surarit, R.; Niamsiri, N. Microbial Poly(hydroxybutyrate-co-hydroxyvalerate) Scaffold for Periodontal Tissue Engineering. *Polymers* **2023**, *15*, 855. https://doi.org/10.3390/polym15040855

Academic Editor: Shashi Kant Bhatia

Received: 4 January 2023
Revised: 31 January 2023
Accepted: 2 February 2023
Published: 9 February 2023

1. Introduction

Periodontitis is a chronic inflammatory oral disease caused by bacteria infection [1]. Typically, the infection destroys periodontal cells including gingival fibroblast, periodontal ligament fibroblasts, and alveolar bone, which are the supporting tissue and bone that hold the tooth. As the disease progresses, more oral tissues are damaged, causing deep pockets, which eventually lead to teeth loss if left untreated [2].

Tissue engineering has been employed to regenerate the lost periodontal tissues and restore both structure and function. In this regard, three dimensional (3D) porous scaffolds represent important components for tissue engineering as a supporting material for cell proliferation or differentiation before being applied to repair the damaged area [3,4]. Scaffolds provide attachment sites and structural guidance for cells that enable them to synthesize appropriate extracellular matrix (ECM) proteins and ultimately proliferate into functional tissues [5]. In addition, the choice of scaffold can be critical as its chemical and physical properties provide guidance cues for the cells to behave appropriately. Scaffold biomaterials for successful tooth regeneration applications should have some requirements such as being biocompatible, biodegradable, and possess mechanical properties that are consistent with the implanted area as well as being used in the appropriate amount and with an accessible volume of porosities for the diffusion of oxygen, cells, and nutrients [6,7]. To date, many polymeric materials have been reported to create biodegradable scaffolds for dental tissue engineering including poly(lactide) (PLA) [8], poly(lactide-co-glycolide) (PLGA) [9,10], and polycaprolactone (PCL) [11–13].

Polyhydroxyalkanoates (PHAs) are aliphatic polyesters synthesized by microorganisms to store excess carbon and energy. Poly(hydroxybutyrate-co-hydroxyvalerate) or P(HB-HV) copolymers are a member of the PHA family [14]. P(HB-HV) has shown great potential for tissue engineering with attractive characteristics of natural origin, biocompatibility, and biodegradability. The properties are adjustable by changing the content of the HV unit. P(HB-HV) are less crystalline, less stiff, and more flexible than the PHB homopolymer due to the incorporation of the HV monomer in the polymer chain [15]. Recently, studies have revealed that different types of scaffolds made with P(HB-HV) demonstrate desirable advantages for tissue engineering. The application of macroporous P(HB-8HV) matrices in the repair of full-thickness cartilage defects in rabbits in vivo was reported by Kose et al. At 8 and 20 weeks after seeding, in vivo results with chondrocyte seeded P(HB-8HV) matrices presented early cartilage formation resembling normal articular cartilage and revealed minimal foreign body reaction. This study also showed that P(HB-8HV) matrices maintained their integrity for 21 days and permitted appropriate gradual degradation and allowed for tissue remodeling to take place [16]. Abazari et al. demonstrated the increased survival rate and insulin-producing cell (IPC) differentiation potential of induced pluripotent stem cells (iPSCs) cultured on a nanofibrous 3D P(HB-5HV) scaffold in comparison with the 2D substrate. iPSCs-P(HB-5HV), as a promising cell-copolymer construct, could potentially be applied in pancreatic tissue engineering applications to diabetic patient treatment [17]. The P(HB-3HV) scaffold was tested for degradation in simulated body fluid (SBF), pH 7.4. After 8-week periods, the P(HB-3HV) scaffolds revealed about 51% weight loss along time due to the high porous structure when compared with the dense and compact films, which showed about a 9% weight loss. Culturing of MC3T3-E1 pre-osteoblast cells on the P(HB-3HV) scaffold samples obtained after 6 weeks of degradation did not lead to the formation of cytotoxic components [18]. In spite of extensive research on P(HB-HV) and their blends as scaffolds for tissue engineering, the HV molar contents of the available commercial P(HB-HV) published are 12 mol% or lower. P(HB-HV) films consisting of various HV content (5–80%) produced by *Paracoccus denitrificans* have been reported to be biocompatible with connective tissue, bone, and dermal fibroblast cells [19]. *Haloferax mediterranei* ES1 produced P(HB-HV) nanofibrous meshes were also shown to be excellent in vitro and to show in vivo biocompatibility with skin tissues [20]. As reported earlier, the more flexible P(HB-HV) with HV contents of 50 mol% can be successfully biosynthesized from bacteria *Cupriavidus necator* H16. This material has already been employed as a drug delivery platform [21–23]. Until now, there have been no studies available in the literature concerning the application of P(HB-50HV) produced by *C. necator* H16 as a scaffold to support cell growth and promote tissue regeneration.

In this study, the 3D porous scaffolds were fabricated from bacterial derived P(HB-50HV) via a particulate leaching method using salt particles as a strategy for the regeneration of periodontal cells. Comparisons were made with scaffold prepared from poly(hydroxybutyrate) (PHB) and poly(hydroxybutyrate-co-12%hydroxyvalerate) (P(HB-12HV)), and the well-established synthetic polycaprolactone (PCL). The scaffolds were characterized with respect to the morphology of the surface and cross section, porosity, mechanical strength, and protein absorption. Subsequently, biological performance of the scaffolds in terms of biocompatibility and cell proliferation was assessed. In this regard, human gingival fibroblasts (HGFs) [10] and periodontal ligament stem cells (PDLSCs) [24] were used since they have been widely studied for the initial evaluation of biomaterials for periodontal tissue engineering applications.

2. Materials and Methods

2.1. Materials

Poly(hydroxybutyrate) (PHB, Mw 3.5×10^5 g/mol), poly(hydroxybutyrate-co-hydroxyvalerate) containing HV content 12 mol% (P(HB-12HV), Mw 2.5×10^5 g/mol), polycaprolactone (PCL, Mw 6.5×10^4 g/mol), and 3-(4,5-Dimethyl-2-thiazolyl)-2,5-diphenyl-2H-tetrazolium bromide (MTT) were obtained from Sigma-Aldrich (St. Louis, MO, USA). Chloroform, methanol, and ethanol were purchased from RCI Labscan (Bangkok, Thailand). Sodium chloride (NaCl) (Ajax Chemicals Ltd., Sydney, Australia) with the particle size range of 425–500 µm was obtained by sieving through an analytical sieve shaker Octagon digital (Endecotts Ltd., London, UK) using two certified sieve sizes with 425 and 500 µm. Dulbecco's modified Eagle medium (DMEM), fetal bovine serum (FBS), and trypsin-EDTA were obtained from Invitrogen (Carlsbad, CA, USA). All chemicals and solutions were used as supplied without further purification. Poly(hydroxybutyrate-co-hydroxyvalerate) with 50% HV content (P(HB-50HV) Mw 1.69×10^6 g/mol) was biosynthesized by in house bacterial cultivation according to a previously described protocol [21].

2.2. Characterization of Polymer Films

Thin polymeric films of PHB, P(HB-12HV), P(HB-50HV), and PCL were prepared by a casting method using 10 mL of 2% (*w*/*v*) polymer stock solution in chloroform on a clean Petri dish. Chloroform was evaporated in a fume hood at room temperature for 24 h. The final thickness of film ranged from 0.05 to 0.10 mm.

The Fourier transform infrared (FTIR) spectra of all PHA and PCL thin films were obtained with a Perkin-Elmer FTIR ATR-FTIR spectrometer (Perkin-Elmer, Spectrum GX FTIR; Shelton, CT, USA). The sample spectra were recorded over 20 scans between 400 and 4000 cm^{-1} wavenumbers at a resolution of 4 cm^{-1}.

The hydrophilicity of the polymeric surfaces was examined by an optical bench-type contact angle goniometry DM-CE1 (Kyowa Interface Science, Niiza, Japan) using a sessile drop method at room temperature.

2.3. Fabrication and Characterization of Scaffolds

Salt-leached scaffolds of PHB, P(HB-12HV), P(HB-50HV), and PCL were fabricated following the established procedure [25]. In brief, the polymer was dissolved in chloroform to prepare a 5% (*w*/*v*) stock solution. The 1 mL polymer solution was then poured on a bed of sieved NaCl particles (with size range of 425–500 µm) in a clean glass vial. The weight ratio of porogen (NaCl) to polymer was set at 9:1. The scaffolds were placed in a fume hood at room temperature for the slow evaporation of chloroform over 2 days followed by repeated rinsing with distilled water to remove any residual salt and air-dried. All scaffolds were prepared as a cylindrical shape with 10 mm diameter and 3 mm height.

The fabricated scaffolds were mounted onto an aluminum stub, gold-coated, and then observed by scanning electron microscopy (SEM, JSM-6360; JEOL Techniques, Tokyo, Japan) with an accelerating voltage of 20 kV for the surface topography and cross section images.

The porosity or void volume fraction V_f (%) of the scaffold was calculated using the following equation:

$$V_f = (1 - (\rho_s/\rho_m)) \times 100$$

where ρ_s is the apparent density of the porous scaffold and ρ_m is the density of the polymer material [26].

2.4. Compressive Mechanical Testing of Scaffolds

The scaffolds were subjected to mechanical measurements under compressive mode in order to determine the compressive stress and compressive modulus (E). The tests were performed at room temperature using a Texture analyzer (TA-XT2i, Stable Micro Systems,

Ltd., Godalming, UK) with a 50 kN load cell at a crosshead speed of 0.1 mm/s [27]. Cylindrical specimens were tested under both dry and wet conditions. The load deformation curves of the samples obtained were converted into stress–strain curves. The compressive stress (MPa) was used to calculate the secant modulus according to the following equation:

$$\text{Compressive modulus at 30\% strain} = \text{Compressive stress (MPa)}/0.3$$

Under the wet condition, the compressive properties of each scaffold were measured in DMEM to mimic the physiological environment [28]. The scaffolds were preconditioned by soaking in the DMEM containing 10% FBS for 24 h at 37 °C. Then, scaffolds were placed in a Petri dish containing fresh media and compressed using a similar setup as above-mentioned. Each reported value was averaged from six independent measurements.

2.5. Evaluation of Protein Absorption on Scaffolds

The protein absorption onto porous scaffolds was determined following a previously published protocol with some modifications [29]. The scaffold sample was cut into equal sizes (10 mm diameter and 3 mm height) and sterilized by soaking in 70% ethanol for 1 h, followed by air drying in a laminar hood. Scaffolds were incubated in 1 mL of DMEM containing 10% FBS for 24 h at 37 °C in a humid atmosphere containing 5% CO_2. Bradford protein assays were performed to determine the residual FBS proteins left in DMEM by using bovine serum albumin (BSA) as a standard [30]. Then, 1 mL of Bradford reagent was added to 100 μL of DMEM solution and incubated for 20 min in the dark. The absorption at 595 nm was measured. The amount of FBS proteins absorbed onto the scaffold could be determined indirectly by subtracting the initial amount of proteins present in DMEM with the residual proteins left in the DMEM solution after removing the scaffold. The absorbed proteins could be reported as % (w/w) proteins absorbed per scaffold.

2.6. Cell Culture

The human gingival fibroblasts (HGFs) and periodontal ligament stem cells (PDLSCs) were originally obtained from American Type Culture Collection (ATCC®). The cells were cultured in DMEM supplemented with 10% (v/v) FBS, 100 units/mL penicillin, and 100 μg/mL streptomycin in an environment of 95% air and 5% CO_2 at 37 °C.

2.7. In Vitro Cell Proliferation Study

The scaffold samples were sterilized with 70% ethanol followed by UV exposure. Each scaffold was then transferred to 48-well plates and washed with phosphate buffered saline (PBS). Prior to seeding the cells, the scaffolds were soaked with 1 mL of fresh cell culture media containing 10% FBS for 3 h at ambient temperature to precondition the scaffolds, as previously described [31]. Thereafter, the preconditioning media were removed and the cells were seeded at a density of 5×10^4 cells/well. Cell cultivations on the scaffolds were carried out over 8 days for the HGF cells and 21 days for the PDLSCs cells, respectively. The cell proliferations were evaluated using the MTT colorimetric assay. For each time point, the scaffolds with cultured cells were washed twice with PBS and transferred into a new clean well. Then, 0.5 mL of MTT solution (1 mg/mL) was added to each well, followed by incubation at 37 °C for 4 h. The excess MTT solution was then removed and formazan crystals that formed in the living cells were dissolved by adding 0.5 mL isopropanol. The liquid solution measured the absorbance at 570 nm using an Epoch microplate spectrophotometer (BioTek Instruments, Inc., Winooski, VT, USA). To visualize nuclear and cytoskeletal morphologies, both the HGF and PDLSC cells were fixed with 2% paraformaldehyde and permeabilized with 0.1% Triton X-100. After washing with PBS, the nuclei were stained with Hoechst 33342 (Invitrogen Corporation, Carlsbad, CA, USA) and the actin filaments were labeled with Alexa Fluor 568 phalloidin solution (Invitrogen Corporation, Carlsbad, CA, USA). The images were collected with a confocal laser scanning microscope (FV10i-DOC; Olympus, Tokyo, Japan).

2.8. Statistical Analysis

Data are expressed as the mean ± SD of three independent experiments. The software package PASW Statistics 18 for Windows (SPSS Inc., Chicago, IL, USA) was used for the statistical analysis. The p-value < 0.05 was considered statistically significant.

3. Results and Discussion

3.1. Characterization of Polymer Films

The FTIR spectra shown in Figure 1 were used to assess the functional groups present in the polymers. The FTIR spectra of the PHB, P(HB-12HV), P(HB-50HV), and PCL polymers are also shown for comparison. Since both PHAs and PCL contain ester bonds, peaks of C=O stretching were observed around 1730 and 1625 cm^{-1} for the PHAs and PCL, respectively. Both the PHA and PCL spectra also showed slightly different C–H stretching and bending, located from 3000 to 2800 cm^{-1} and from 1500 to 1000 cm^{-1} [32]. Although the PHB, P(HB-12HV), and P(HB-50HV) polymers are chemically similar, the differences in the HV composition of the polymers could be distinguished by FTIR spectra. The PHB homopolymer showed characteristic peaks at 1724 cm^{-1} for C=O stretching and 1281 cm^{-1} for C–O stretching [33–35]. Apart from additional peaks at 797 cm^{-1}, responsible for C–H bending, the presence of HV in the P(HB-HV) copolymers could be identified by observing the FTIR peak shifts. A major shift occurred at the C=O stretching region, in which the peak shifted from 1724 cm^{-1} in PHB to 1735 cm^{-1} in P(HB-HV). The greater the change to the higher wavenumber, the higher the %HV monomer in the polymer chain. This phenomenon was also observed in other peaks such as C–O stretching at 1281 cm^{-1} and the C–H stretching region around 3000 cm^{-1}. In addition, several peaks from FTIR can be used to denote the crystallinity state of different PHA polymers. The peaks at 1453, 1380, 1281, 1057, and 826 cm^{-1} shifted to a higher wavenumber when the crystallinity was low [36,37]. Our results showed that there were around five to 10 wavenumber shifts in the mentioned peaks among PHB, P(HB-12HV), and P(HB-50HV). Therefore, the PHAs used in this study were confirmed as having differences in %HV as well as their crystallinity.

The hydrophilicity of a polymer surface is the key parameter affecting cell–material interaction and the adsorption of protein on the polymer surface, which subsequently influence cell behaviors [38]. The results of the water contact angle measurements are summarized in Table 1. All samples showed contact angles of below 90° considering hydrophilic behavior. The highest contact angle value of the PCL film indicated the greater hydrophobicity of PCL than the other PHAs. The contact angle value of P(HB-50HV) was significantly higher than the other PHA films tested ($p < 0.05$). This might be due to more ethyl groups of the HV monomer present in the side chain of the copolymers [39]. Kim et al. reported a water contact angle of 79.5° of the P(HB-60HV) film produced by *Haloferax mediterranei* ES1 [20].

Table 1. Surface hydrophilicity of the polymer films.

Type of Polymers	Water Contact Angle (°)
PHB	70.2 ± 3.5^{a}
P(HB-12HV)	67.9 ± 2.1^{a}
P(HB-50HV)	76.8 ± 1.8^{b}
PCL	81.4 ± 1.8^{c}

(Mean ± SD, $n = 3$, different superscript letters indicate a significant difference at $p < 0.05$).

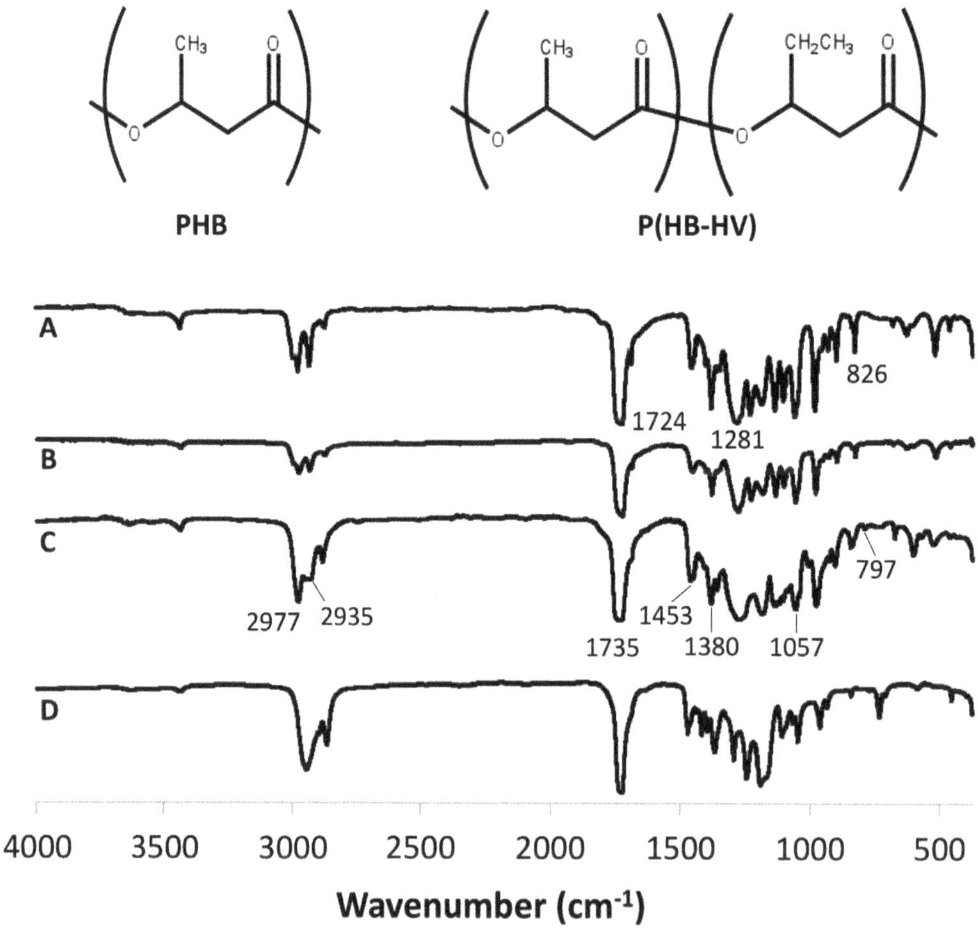

Figure 1. FTIR spectra of the (A) PHB, (B) P(HB-12HV), (C) P(HB-50HV) and (D) PCL polymers.

3.2. *Characterization of Scaffolds*

The PHB, P(HB-12HV), P(HB-50HV), and PCL scaffolds were fabricated via a particulate salt leaching technique. All fabricated porous scaffolds exhibited a high porosity of 90% with a sponge-like appearance (Figure 2). The structure of the pores as well as the surface and cross-sectional topologies of the 3D porous scaffolds were examined using SEM, as shown in Figure 3. All polymeric scaffolds were similar in terms of the surface and cross-sectional topographies that comprised of interconnected open pores throughout the scaffolds. The well-tailored pore sizes ranged between 425 and 500 μm on both the surface and inside the scaffolds, suggesting sufficient surface areas for cell attachment. Furthermore, the pore shape observed was similar to the shape of the imprinted salt crystals. Our results agree with earlier findings for scaffolds with pore sizes of around 400 μm, which are considered suitable for the growth and proliferation of bone cells [40]. In general, the scaffolds were highly porous with interconnected pore networks that facilitate nutrient and oxygen diffusion and waste removal during tissue formation. The interconnected networks between open pores are also important for cellular attachment, proliferation, and migration for tissue vascularization [26,41].

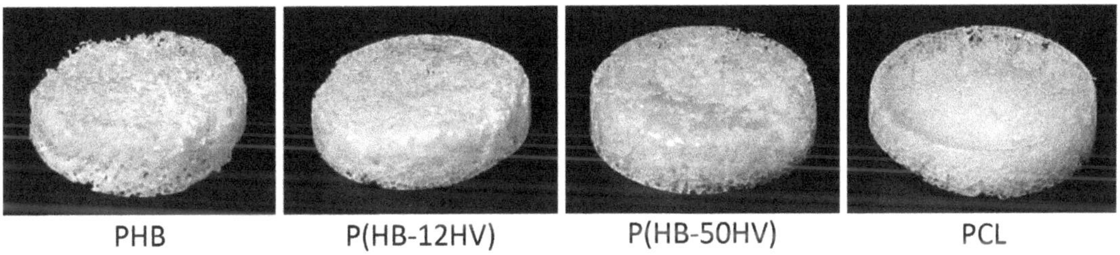

Figure 2. Photographic images of 3D porous scaffolds via the particulate salt leaching technique.

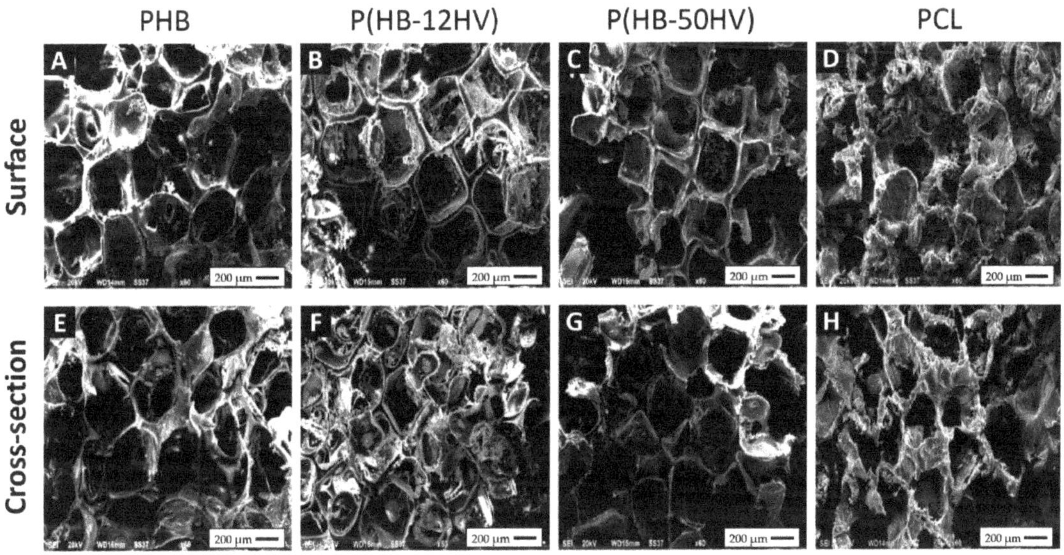

Figure 3. SEM micrographs of representative 3D porous scaffold samples with pore sizes ranging from 425 to 500 μm: PHB scaffold (**A**,**E**), P(HB-12HV) scaffold (**B**,**F**), P(HB-50HV) scaffold (**C**,**G**), and PCL scaffold (**D**,**H**).

3.3. Mechanical Properties of Scaffolds

The typical stress–strain curves obtained from the compressive stress measurement at 30% strain were used to calculate the compressive modulus of all scaffolds. As presented in Figure 4, the compressive modulus values under the dry condition of the PHB, P(HB-12HV), and P(HB-50HV) scaffolds were found to be 0.75 ± 0.02, 0.39 ± 0.08, and 0.04 ± 0.01 MPa, respectively. The compressive modulus of the PCL scaffold at 0.45 ± 0.01 MPa was not significantly different from the scaffold made with P(HB-12HV) polymers. In addition, the lowest compressive modulus was observed in the P(HB-50HV) scaffold. Our results suggest that increasing the HV content in the P(HB-HV) polymer chain at 50% could lead to a significant decrease in the compressive modulus of a 3D porous scaffold while maintaining the same % porosity. Previous studies have reported that an increase of %HV up to 50–60% could cause a lower melting temperature due to a decrease in the crystallinity of the PHA copolymers, resulting in ductile mechanical properties such as higher elongation to break and greater flexibility with a faster degradation rate under specific physiological conditions [42,43]. Among the P(HB-HV) with various %HV contents produced by *P. denitrificans*, film sheets composed of P(HB-HV) with a HV of 53–60 mol% were found to be more flexible and tougher [19]. Here, the P(HB-50HV) scaffold appeared to be the most soft and flexible scaffold in the dry state.

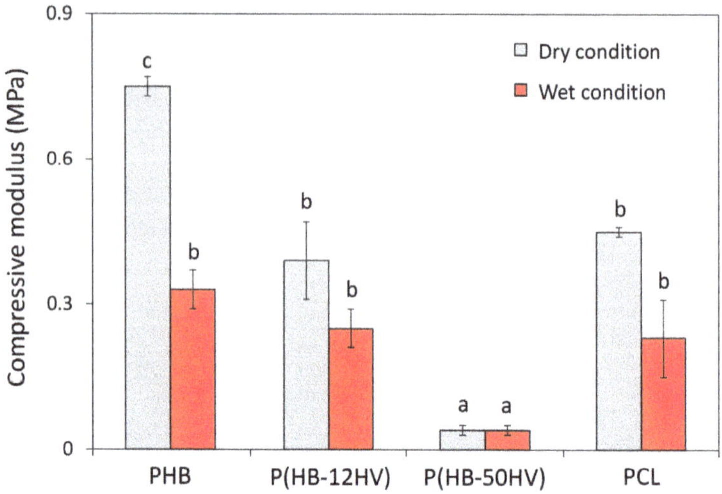

Figure 4. Compressive secant modulus at 30% strain compared between the dry and wet conditions. Data are reported as the average values from six independent scaffolds with standard deviations ($n = 6$). The different letters (a, b, and c) above the bar graph indicate significant differences ($p < 0.05$) between the scaffold materials tested under the same conditions.

The wet-state mechanical properties of 3D porous scaffolds were also investigated in order to determine their compressive behavior in a realistic environment. In Figure 4, the compressive modulus values under wet conditions of the PHB, P(HB-12HV), and PCL scaffolds were found to be approximately in the same levels at 0.33 ± 0.04, 0.25 ± 0.04, and 0.23 ± 0.08 MPa, respectively, which were 62%, 36%, and 49% reduced from the dry condition, respectively. However, there was no change in the compressive modulus of the P(HB-50HV) scaffold under the wet state when compared to the dry state, which still remained at 0.04 ± 0.01 MPa. There was a clear decrease in the mechanical properties from the dry to wet state in all scaffolds, except for P(HB-50HV). Our findings are in line with earlier reports that observed decreased compressive moduli of 3D polymeric scaffolds under wet conditions, which used PBS and cell culture media [27,28]. The water molecules could intersperse and intercalate among the polymer chains that finally spread the polymer chains apart by losing the crystalline network characteristics of the polymer [44]. Notably, there was no discernable difference between the dry and wet conditions on the compressive modulus of the P(HB-50HV) scaffold. One explanation could be that the compressive modulus of P(HB-50HV) at the dry state is already quite low, and that any eventual water plasticizer effect might be too small to be detected.

3.4. Cell Proliferation

Early cell adhesion and proliferation are necessary in developing scaffolds for periodontal regeneration. The cell adhesion ability and proliferation enhancement of HGFs and PDLSCs on 3D porous scaffolds were studied using the MTT assay. The HGF cells were cultured on different scaffolds for 0, 1, 2, 4, 6, and 8 days. In Figure 5A, the HGF cells grew quite slowly during the first 2 days for all types of scaffolds tested. Interestingly, cell numbers at 8 days were significantly the highest for the P(HB-50HV) scaffold, which showed about a 16-fold increase from the start, followed by the P(HB-12HV) and PCL scaffolds (10-fold), PHB scaffold (8-fold), and 2D control surface (6-fold), respectively.

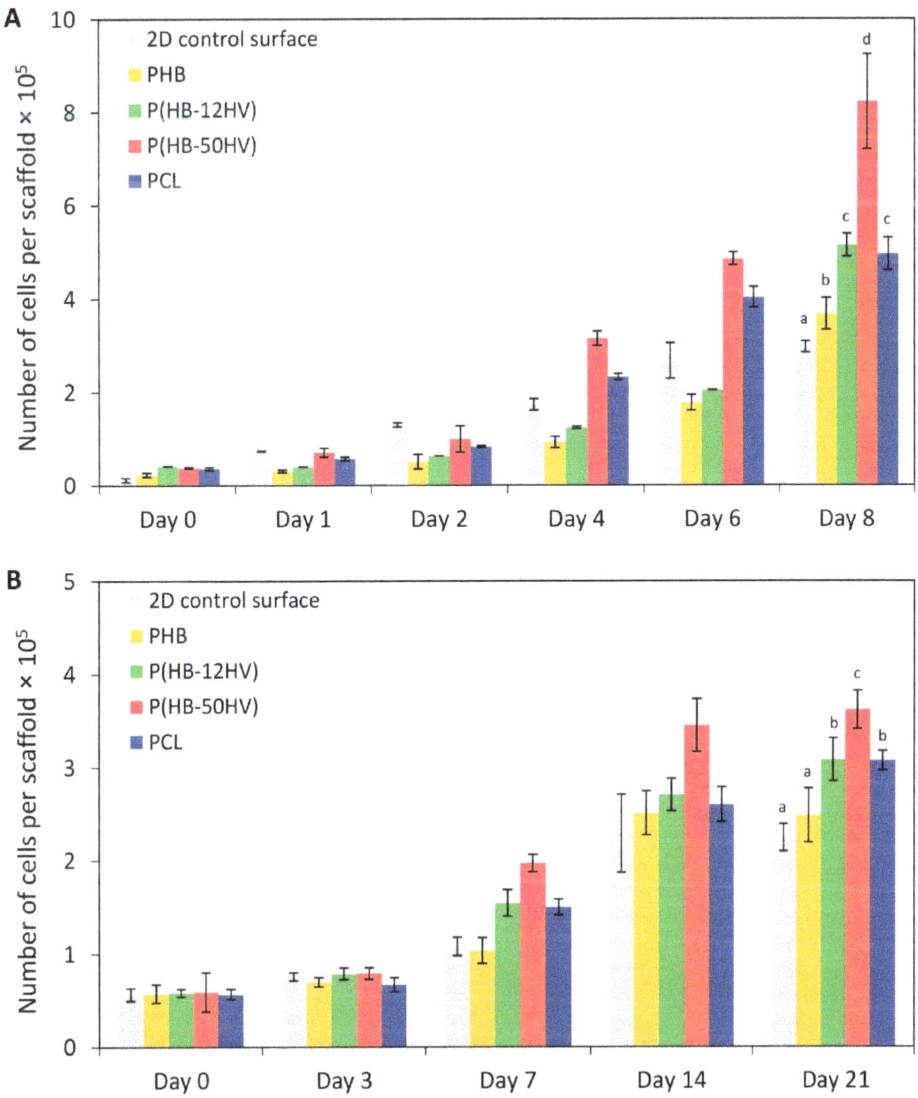

Figure 5. Cell proliferation of (**A**) HGF and (**B**) PDLSC cells grown on different types of 3D porous scaffolds. The values are the means with standard deviation derived from three independent scaffolds ($n = 3$). Bars labeled with different letters (a, b, c, and d) indicate significant differences within the same day ($p < 0.05$).

The proliferation ability of PDLSC cells was investigated at 0, 3, 7, 14, and 21 days, as shown in Figure 5B. Similar cell numbers were found on the 2D control surface and in all scaffolds at day 0. On the last day of the experiment, the highest number of PDLSC cells was significantly increased on the P(HB-50HV) scaffold with a 7-fold increase from the initial cell loaded compared with the P(HB-12HV) and PCL scaffolds (6-fold), and the PHB scaffold and 2D control surface (5-fold).

In this study, it was clear that all PHA scaffolds could support the attachment and proliferation of HGF and PDLSC cells. This is the first report on the cytotoxicity and biocompatibility of P(HB-50HV) produced by *C. necator* H16 as a candidate scaffold for dental tissue engineering. However, the P(HB-50HV) scaffold appeared to be the most suitable to support both HGF and PDLSC cell growth among the three types of PHA scaffolds, and was even better than the conventional PCL scaffold. Since the chemical properties of PHA and PCL scaffolds are quite similar such as the functional group and hydrophilicity, protein absorption on the material surface is known to be one of the important considerations to promoting cell attachment and the growth of anchorage-dependent mammalian cells on a solid substratum [45]. The amount of protein absorbed on the scaffolds' surface were found to be similar, as shown in Figure 6. Thus, the HGF and PDLSC cell proliferation was related to the mechanical properties of the 3D porous scaffolds. Many previous studies have reported that the stiffness of the material has an effect on cell attachment signaling, leading to a difference in the cell proliferation and differentiation [46,47]. These results indicate that the P(HB-50HV) scaffold had the lowest stiffness, which could promote the greatest adhesion and proliferation of HGF and PDLSC cell adhesion and proliferation, and thus should be considered as a suitable material for the tissue engineering of periodontal cells and other soft tissue-like cells. Regarding the scaffold mechanical properties, our results are in line with the finding reported previously that the fabricated PCL scaffolds with lower modulus values than the PLGA scaffolds showed a 2-fold higher growth rate of stromal cells [48].

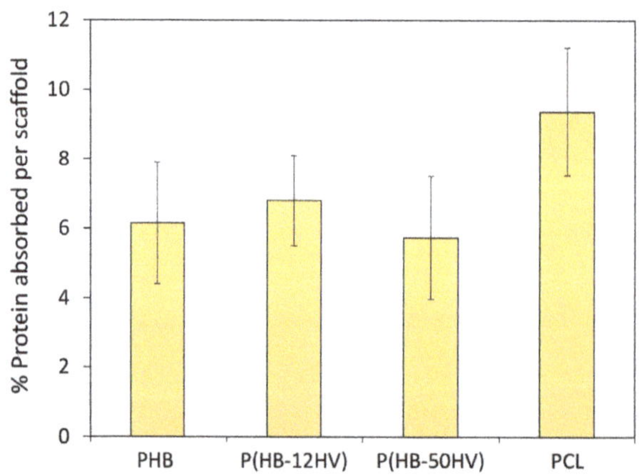

Figure 6. The adsorption of FBS proteins on 3D porous scaffolds. The values are the means with standard deviation derived from three independent scaffolds ($n = 3$).

3.5. Cell Morphology

The morphology of healthy HGF and PDLSC cells cultured on the P(HB-50HV) scaffold was further analyzed by fluorescence staining of the nucleus and F-actin filaments. The round-shape of the HGF cells was observed at day 0 (Figure 7A) followed by cell migration and the development of interconnecting network development by day 2 (Figure 7B). After 8 days of culture, the HGF cells were distributed throughout the entire scaffold with a strong presence of F-actin, resulting in a dense interconnecting network of cells (Figure 7C).

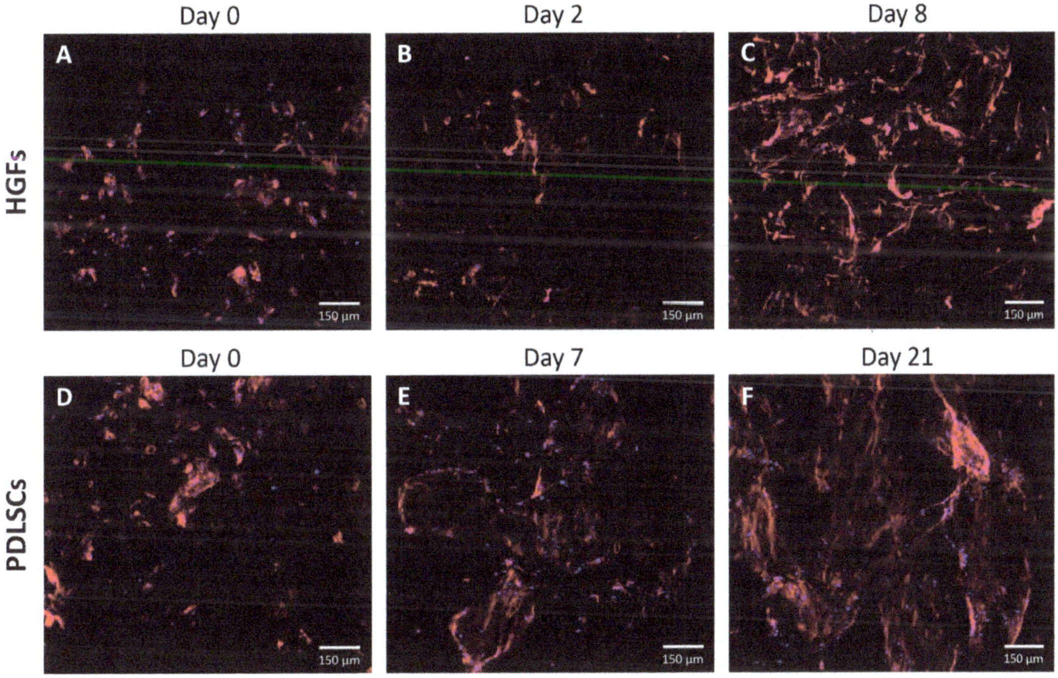

Figure 7. Confocal micrographs of the (**A–C**) HGF and (**D–F**) PDLSC cells cultured on the P(HB-50HV) scaffold. Cell nucleus was stained with Hoechst 33342 (blue). F-actin was stained with Alexa Fluor 568 phalloidin (red).

The initial cell adhesion of PDLSC cells at day 0 were observed to be round-shape (Figure 7D). The PDLSC cells proliferated considerably well on the scaffold surface and gradually progressed to high cell density all over the cultured scaffold from day 7 (Figure 7E) to day 21 (Figure 7F). Importantly, both HGF and PDLSC cell morphology on the P(HB-50HV) scaffold showed both spindle and stellate shapes, which are typically good indicators of healthy fibroblasts [49,50]. The results showed that the biocompatibility of the P(HB-50HV) scaffold has promising potential for periodontal tissue engineering.

The scaffold made from bacterially-derived P(HB-50HV) copolymers developed in this study showed a softness property and displayed a capability to promote good proliferation and the interconnection of periodontal cells including HGF and PDLSC cells. The microbial P(HB-50HV) scaffold is not only effective for normal fibroblast cell regeneration, but also demonstrates good potential to promote stem cell proliferation. With these interesting properties, the P(HB-50HV) scaffold is an attractive material for tissue engineering strategies.

4. Conclusions

In this study, 3D porous scaffolds made from PHB, P(HB-12HV), P(HB-50HV), and PCL polymers were successfully fabricated via the salt leaching method with similar properties in terms of chemical functionality, surface hydrophilicity, surface topography, % porosity, and serum protein absorption. Furthermore, the P(HB-50HV) scaffolds were distinctively different in their compressive modulus by having the lowest stiffness among all of the scaffolds tested. The proliferation of dental cells including HGF and PDLSC cells was investigated with four different types of scaffolds. Interestingly, the P(HB-50HV) scaffold showed the highest proliferation of both HGF and PDLSC cells over all of the PHA scaffolds and the control PCL scaffold. Cells grown on the P(HB-50HV) scaffold had the characteristic of healthy fibroblasts in forming highly dense interconnecting networks. Taken together with the hydrophilicity, softness property, greater cell proliferation, and morphology of dental cells grown on the P(HB-50HV) scaffold, these results confirm the possibility of using a microbial-derived P(HB-50HV) scaffold as a biomaterial for periodontal tissue engineering and stem cell applications.

Author Contributions: Conceptualization, N.N., S.P. (Seubsakul Phuegyod), S.P. (Sasivimon Pramual), J.S. and R.S.; Methodology, S.P. (Seubsakul Phuegyod), S.P. (Sasivimon Pramual), N.W., S.A. and P.S.; Investigation, S.P. (Seubsakul Phuegyod), S.P. (Sasivimon Pramual), N.W., S.A., T.A. and N.N.; Writing—original draft preparation, S.P. (Seubsakul Phuegyod) and S.P. (Sasivimon Pramual); Writing—review and editing, N.N., R.S. and J.S.; Visualization, S.P. (Sasivimon Pramual) and N.N.; Supervision, N.N., R.S. and J.S.; Project administration, N.N.; Funding acquisition, N.N. All authors have read and agreed to the published version of the manuscript.

Funding: This research was supported by the National Science and Technology Development Agency (NSTDA) through the Coordinating Center for Thai Government Science and Technology Scholarship Students (CSTS): A New Researcher Scholarship of CSTS-MOST, Thailand Science Research and Innovation (TSRI), Chulabhorn Research Institute (grant no. 36821/4274353). The research was also partially supported by the Faculty of Science, Mahidol University.

Institutional Review Board Statement: Not applicable.

Informed Consent Statement: Not applicable.

Data Availability Statement: The data presented in this study are available on request from the corresponding author.

Conflicts of Interest: The authors declare no conflict of interest.

References

1. Sedghi, L.M.; Bacino, M.; Kapila, Y.L. Periodontal Disease: The Good, The Bad, and The Unknown. *Front. Cell. Infect. Microbiol.* **2021**, *11*, 1210. [CrossRef] [PubMed]
2. Kwon, T.; Lamster, I.B.; Levin, L. Current Concepts in the Management of Periodontitis. *Int. Dent. J.* **2021**, *71*, 462–476. [CrossRef]
3. Swanson, W.B.; Yao, Y.; Mishina, Y. Novel approaches for periodontal tissue engineering. *Genesis* **2022**, *60*, e23499. [CrossRef] [PubMed]
4. Deng, R.; Xie, Y.; Chan, U.; Xu, T.; Huang, Y. Biomaterials and biotechnology for periodontal tissue regeneration: Recent advances and perspectives. *J. Dent. Res. Dent. Clin. Dent. Prospects.* **2022**, *16*, 1–10. [CrossRef]
5. Eltom, A.; Zhong, G.; Muhammad, A. Scaffold Techniques and Designs in Tissue Engineering Functions and Purposes: A Review. *Adv. Mater. Sci. Eng.* **2019**, *2019*, 3429527. [CrossRef]
6. Chen, S.; Huang, X. Nanomaterials in Scaffolds for Periodontal Tissue Engineering: Frontiers and Prospects. *Bioengineering* **2022**, *9*, 431. [CrossRef]
7. Rahman, S.U.; Nagrath, M.; Ponnusamy, S.; Arany, P.R. Nanoscale and Macroscale Scaffolds with Controlled-Release Polymeric Systems for Dental Craniomaxillofacial Tissue Engineering. *Materials* **2018**, *11*, 1478. [CrossRef]
8. Diomede, F.; Gugliandolo, A.; Scionti, D.; Merciaro, I.; Cavalcanti, M.F.; Mazzon, E.; Trubiani, O. Biotherapeutic Effect of Gingival Stem Cells Conditioned Medium in Bone Tissue Restoration. *Inter. J. Mol. Sci.* **2018**, *19*, 329. [CrossRef]
9. Gangolli, R.A.; Devlin, S.M.; Gerstenhaber, J.A.; Lelkes, P.I.; Yang, M. A Bilayered Poly (Lactic-Co-Glycolic Acid) Scaffold Provides Differential Cues for the Differentiation of Dental Pulp Stem Cells. *Tissue. Eng. Part A* **2019**, *25*, 224–233. [CrossRef] [PubMed]
10. Nan, L.; Zheng, Y.; Liao, N.; Li, S.; Wang, Y.; Chen, Z.; Wei, L.; Zhao, S.; Mo, S. Mechanical force promotes the proliferation and extracellular matrix synthesis of human gingival fibroblasts cultured on 3D PLGA scaffolds via TGF-β expression. *Mol. Med. Rep.* **2019**, *19*, 2107–2114. [CrossRef]
11. Baranowska-Korczyc, A.; Warowicka, A.; Jasiurkowska-Delaporte, M.; Grześkowiak, B.; Jarek, M.; Maciejewska, B.M.; Jurga-Stopa, J.; Jurga, S. Antimicrobial electrospun poly(ε-caprolactone) scaffolds for gingival fibroblast growth. *RSC Adv.* **2016**, *6*, 19647–19656. [CrossRef]
12. Naik, C.; Srinath, N.; Ranganath, M.; Umashankar, D.; Gupta, H. Evaluation of polycaprolactone scaffold for guided bone regeneration in maxillary and mandibular defects: A clinical study. *Nat. J. Maxillofac. Surg.* **2020**, *11*, 207–212. [CrossRef] [PubMed]
13. Azaryan, E.; Hanafi-Bojd, M.Y.; Alemzadeh, E.; Emadian Razavi, F.; Naseri, M. Effect of PCL/nHAEA nanocomposite to osteo/odontogenic differentiation of dental pulp stem cells. *BMC Oral Health* **2022**, *22*, 505. [CrossRef] [PubMed]
14. Raza, Z.A.; Abid, S.; Banat, I.M. Polyhydroxyalkanoates: Characteristics, production, recent developments and applications. *Inter. Biodeterior. Biodegrad.* **2018**, *126*, 45–56. [CrossRef]

15. Ibrahim, M.I.; Alsafadi, D.; Alamry, K.A.; Hussein, M.A. Properties and Applications of Poly(3-hydroxybutyrate-co-3-hydroxyvalerate) Biocomposites. *J. Polymer. Environ.* **2021**, *29*, 1010–1030. [CrossRef]
16. Köse, G.T.; Korkusuz, F.; Ozkul, A.; Soysal, Y.; Ozdemir, T.; Yildiz, C.; Hasirci, V. Tissue engineered cartilage on collagen and PHBV matrices. *Biomaterials* **2005**, *26*, 5187–5197. [CrossRef] [PubMed]
17. Abazari, M.F.; Zare Karizi, S.; Hajati-Birgani, N.; Norouzi, S.; Khazeni, Z.; Hashemi, J.; Shafaghi, L.; Soleimanifar, F.; Mansour, R.N.; Enderami, S.E. PHBV nanofibers promotes insulin-producing cells differentiation of human induced pluripotent stem cells. *Gene* **2021**, *768*, 145333. [CrossRef]
18. Amaro, L.; Correia, D.M.; Martins, P.M.; Botelho, G.; Carabineiro, S.A.C.; Ribeiro, C.; Lanceros-Mendez, S. Morphology Dependence Degradation of Electro- and Magnetoactive Poly(3-hydroxybutyrate-co-hydroxyvalerate) for Tissue Engineering Applications. *Polymers* **2020**, *12*, 953. [CrossRef]
19. Napathorn, S.C. Biocompatibilities and biodegradation of poly(3-hydroxybutyrate-co-3-hydroxyvalerate)s produced by a model metabolic reaction-based system. *BMC Microbiol.* **2014**, *14*, 285. [CrossRef]
20. Kim, H.S.; Chen, J.; Wu, L.P.; Wu, J.; Xiang, H.; Leong, K.W.; Han, J. Prevention of excessive scar formation using nanofibrous meshes made of biodegradable elastomer poly(3-hydroxybutyrate-co-3-hydroxyvalerate). *J. Tissue Eng.* **2020**, *11*, 1–12. [CrossRef]
21. Pramual, S.; Assavanig, A.; Bergkvist, M.; Batt, C.A.; Sunintaboon, P.; Lirdprapamongkol, K.; Svasti, J.; Niamsiri, N. Development and characterization of bio-derived polyhydroxyalkanoate nanoparticles as a delivery system for hydrophobic photodynamic therapy agents. *J. Mater. Sci. Mater. Med.* **2016**, *27*, 40. [CrossRef] [PubMed]
22. Panith, N.; Assavanig, A.; Lertsiri, S.; Bergkvist, M.; Surarit, R.; Niamsiri, N. Development of tunable biodegradable polyhydroxyalkanoates microspheres for controlled delivery of tetracycline for treating periodontal disease. *J. Appl. Polym. Sci.* **2016**, *133*, 44128. [CrossRef]
23. Pramual, S.; Lirdprapamongkol, K.; Svasti, J.; Bergkvist, M.; Jouan-Hureaux, V.; Arnoux, P.; Frochot, C.; Barberi-Heyob, M.; Niamsiri, N. Polymer-lipid-PEG hybrid nanoparticles as photosensitizer carrier for photodynamic therapy. *J. Photochem. Photobiol. B* **2017**, *173*, 12–22. [CrossRef] [PubMed]
24. Song, I.S.; Han, Y.S.; Lee, J.H.; Um, S.; Kim, H.Y.; Seo, B.M. Periodontal Ligament Stem Cells for Periodontal Regeneration. *Curr. Oral Health Rep.* **2015**, *2*, 236–244. [CrossRef]
25. Intranuovo, F.; Gristina, R.; Brun, F.; Mohammadi, S.; Ceccone, G.; Sardella, E.; Rossi, F.; Tromba, G.; Favia, P. Plasma Modification of PCL Porous Scaffolds Fabricated by Solvent-Casting/Particulate-Leaching for Tissue Engineering. *Plasma Process. Polym.* **2014**, *11*, 184–195. [CrossRef]
26. Loh, Q.L.; Choong, C. Three-dimensional scaffolds for tissue engineering applications: Role of porosity and pore size. *Tissue Eng. Part B Rev.* **2013**, *19*, 485–502. [CrossRef]
27. Hutmacher, D.W.; Schantz, T.; Zein, I.; Ng, K.W.; Teoh, S.H.; Tan, K.C. Mechanical properties and cell cultural response of polycaprolactone scaffolds designed and fabricated via fused deposition modeling. *J. Biomed. Mater. Res.* **2001**, *55*, 203–216. [CrossRef]
28. Wu, L.; Zhang, J.; Jing, D.; Ding, J. "Wet-state" mechanical properties of three-dimensional polyester porous scaffolds. *J. Biomed. Mater. Res. A* **2006**, *76*, 264–271. [CrossRef]
29. Aoki, N.; Akasaka, T.; Watari, F.; Yokoyama, A. Carbon nanotubes as scaffolds for cell culture and effect on cellular functions. *Dent. Mater. J.* **2007**, *26*, 178–185. [CrossRef]
30. Bradford, M.M. A rapid and sensitive method for the quantitation of microgram quantities of protein utilizing the principle of protein-dye binding. *Anal. Biochem.* **1976**, *72*, 248–254. [CrossRef]
31. Bäumchen, F.; Smeets, R.; Koch, D.; Gräber, H.G. The impact of defined polyglycolide scaffold structure on the proliferation of gingival fibroblasts in vitro: A pilot study. *Oral Surg. Oral Med. Oral Pathol. Oral Radiol. Endodontol.* **2009**, *108*, 505–513. [CrossRef] [PubMed]
32. Coates, J. Interpretation of Infrared Spectra, A Practical Approach. In *Encyclopedia of Analytical Chemistry*; John Wiley & Sons: Hoboken, NJ, USA, 2006.
33. Shamala, T.R.; Divyashree, M.S.; Davis, R.; Kumari, K.S.; Vijayendra, S.V.; Raj, B. Production and characterization of bacterial polyhydroxyalkanoate copolymers and evaluation of their blends by fourier transform infrared spectroscopy and scanning electron microscopy. *Indian J. Microbiol.* **2009**, *49*, 251–258. [CrossRef]
34. Porras, M.; Cubitto, M.; Villar, M. Quantitative Determination of intracellular PHA in Bacillus megaterium BBST4 strain Using Mid FTIR Spectroscopy. In Proceedings of the XIV SLAP/XII CIP 2014, Porto de Galinhas, Brazil, 12–16 October 2014. [CrossRef]
35. Kumar, M.; Singhal, A.; Verma, P.K.; Thakur, I.S. Production and Characterization of Polyhydroxyalkanoate from Lignin Derivatives by Pandoraea sp. ISTKB. *ACS Omega* **2017**, *2*, 9156–9163. [CrossRef]
36. Xu, J.; Guo, B.H.; Yang, R.; Wu, Q.; Chen, G.-Q.; Zhang, Z.M. In situ FTIR study on melting and crystallization of polyhydroxyalkanoates. *Polymer* **2002**, *43*, 6893–6899. [CrossRef]
37. Isak, I.; Patel, M.; Riddell, M.; West, M.; Bowers, T.; Wijeyekoon, S.; Lloyd, J. Quantification of polyhydroxyalkanoates in mixed and pure cultures biomass by Fourier transform infrared spectroscopy: Comparison of different approaches. *Lett. Appl. Microbiol.* **2016**, *63*, 139–146. [CrossRef] [PubMed]
38. Alves, N.M.; Pashkuleva, I.; Reis, R.L.; Mano, J.F. Controlling Cell Behavior Through the Design of Polymer Surfaces. *Small* **2010**, *6*, 2208–2220. [CrossRef]

39. Yu, D.G.; Lin, W.C.; Lin, C.H.; Yang, M.C. Cytocompatibility and antibacterial activity of a PHBV membrane with surface-immobilized water-soluble chitosan and chondroitin-6-sulfate. *Macromol. Biosci.* **2006**, *6*, 348–357. [CrossRef]
40. Cheng, M.Q.; Wahafu, T.; Jiang, G.F.; Liu, W.; Qiao, Y.Q.; Peng, X.C.; Cheng, T.; Zhang, X.L.; He, G.; Liu, X.Y. A novel open-porous magnesium scaffold with controllable microstructures and properties for bone regeneration. *Sci. Rep.* **2016**, *6*, 24134. [CrossRef]
41. Yazid, F. Scaffold Selection for Tissue Engineering in Dentistry. *Med. Health* **2020**, *15*, 34–53. [CrossRef]
42. Verlinden, R.A.J.; Hill, D.J.; Kenward, M.A.; Williams, C.D.; Radecka, I. Bacterial synthesis of biodegradable polyhydroxyalkanoates. *J. Appl. Microbiol.* **2007**, *102*, 1437–1449. [CrossRef]
43. Sudesh, K.; Abe, H.; Doi, Y. Synthesis, structure and properties of polyhydroxyalkanoates: Biological polyesters. *Prog. Polym. Sci.* **2000**, *25*, 1503–1555. [CrossRef]
44. Passerini, N.; Craig, D.Q. An investigation into the effects of residual water on the glass transition temperature of polylactide microspheres using modulated temperature DSC. *J. Control Release* **2001**, *73*, 111–115. [CrossRef] [PubMed]
45. Cai, S.; Wu, C.; Yang, W.; Liang, W.; Yu, H.; Liu, L. Recent advance in surface modification for regulating cell adhesion and behaviors. *Nanotechnol. Rev.* **2020**, *9*, 971–989. [CrossRef]
46. Yeung, T.; Georges, P.C.; Flanagan, L.A.; Marg, B.; Ortiz, M.; Funaki, M.; Zahir, N.; Ming, W.; Weaver, V.; Janmey, P.A. Effects of substrate stiffness on cell morphology, cytoskeletal structure, and adhesion. *Cell Motil. Cytoskelet.* **2005**, *60*, 24–34. [CrossRef]
47. Sun, M.; Chi, G.; Li, P.; Lv, S.; Xu, J.; Xu, Z.; Xia, Y.; Tan, Y.; Xu, J.; Li, L.; et al. Effects of Matrix Stiffness on the Morphology, Adhesion, Proliferation and Osteogenic Differentiation of Mesenchymal Stem Cells. *Int. J. Med. Sci.* **2018**, *15*, 257–268. [CrossRef] [PubMed]
48. Baker, S.C.; Rohman, G.; Southgate, J.; Cameron, N.R. The relationship between the mechanical properties and cell behaviour on PLGA and PCL scaffolds for bladder tissue engineering. *Biomaterials* **2009**, *30*, 1321–1328. [CrossRef]
49. Guo, Y.; Wang, X.; Wang, C.; Chen, S. In vitro behaviour of human gingival fibroblasts cultured on 3D-printed titanium alloy with hydrogenated TiO_2 nanotubes. *J. Mater. Sci. Mater. Med.* **2022**, *33*, 27. [CrossRef]
50. Sugiura, R.; Hamano, S.; Tomokiyo, A.; Hasegawa, D.; Yoshida, S.; Sugii, H.; Fujino, S.; Adachi, O.; Kadowaki, M.; Yamashita, D.; et al. PAX9 Is Involved in Periodontal Ligament Stem Cell-like Differentiation of Human-Induced Pluripotent Stem Cells by Regulating Extracellular Matrix. *Biomedicines* **2022**, *10*, 2366. [CrossRef]

polymers MDPI

Article

Finding a Benign Plasticizer to Enhance the Microbial Degradation of Polyhydroxybutyrate (PHB) Evaluated by PHB Degrader *Microbulbifer* sp. SOL66

Jang Yeon Cho [1], Su Hyun Kim [1], Hee Ju Jung [1], Do Hyun Cho [1], Byung Chan Kim [1], Shashi Kant Bhatia [1], Jungoh Ahn [2], Jong-Min Jeon [3], Jeong-Jun Yoon [3], Jongbok Lee [4] and Yung-Hun Yang [1,*]

1 Department of Biological Engineering, College of Engineering, Konkuk University, Seoul 05029, Korea
2 Biotechnology Process Engineering Center, Korea Research Institute Bioscience Biotechnology (KRIBB), Cheongju 28116, Korea
3 Green & Sustainable Materials R&D Department, Research Institute of Clean Manufacturing System, Korea Institute of Industrial Technology (KITECH), Cheonan 31056, Korea
4 Department of Biological and Chemical Engineering, Hongik University, Sejong 30016, Korea
* Correspondence: seokor@konkuk.ac.kr; Tel.: + 82-2-450-3936

Abstract: As a biodegradable plastic, polyhydroxybutyrate (PHB) has relatively poor mechanical properties, preventing its wider use. Various plasticizers have been studied to improve the mechanical properties of PHB; however, due to the slow degradation speed in the soil environment and lack of evaluation methods, studies on the degradation of PHB with plasticizers are rarely reported. In this study, by applying *Microbulbifer* sp. SOL66, which is able to degrade PHB very quickly, a benign plasticizer was evaluated with good properties and good degradability, not inhibiting microbial activities. Eight different plasticizers were applied with PHB and *Microbulbifer* sp. SOL66, PHB film containing 10% and 20% tributyl citrate showed significant biodegradability of PHB. It was confirmed that tributyl citrate could increase the speed of PHB degradation by *Microbulbifer* sp. SOL66 by 88% at 1 day, although the degree of degradation was similar after 3 days with and without tributyl citrate. By the analysis of microbial degradation, physical, chemical, and mechanical properties, tributyl citrate was shown not only to improve physical, chemical, and mechanical properties but also the speed of microbial degradation.

Keywords: polyhydroxybutyrate; plasticizer; biodegradation; *Microbulbifer* sp. SOL66

Citation: Cho, J.Y.; Kim, S.H.; Jung, H.J.; Cho, D.H.; Kim, B.C.; Bhatia, S.K.; Ahn, J.; Jeon, J.-M.; Yoon, J.-J.; Lee, J.; et al. Finding a Benign Plasticizer to Enhance the Microbial Degradation of Polyhydroxybutyrate (PHB) Evaluated by PHB Degrader *Microbulbifer* sp. SOL66. *Polymers* **2022**, *14*, 3625. https://doi.org/10.3390/polym14173625

Academic Editor: Helena Felgueiras

Received: 29 July 2022
Accepted: 29 August 2022
Published: 1 September 2022

1. Introduction

In recent years, people have become more aware of environmental pollution and have made efforts to avoid the continuous use of non-degradable conventional plastics [1,2]. Bioplastics are a suitable alternative to disposable products and can be produced and degraded by microorganisms [3]. There are many kinds of bioplastics such as polylactic acid (PLA), poly(butylene adipate-*co*-terephthalate) (PBAT), and polyhydroxybutyrate (PHB) that can substitute non-degradable conventional plastics [4,5]. Since PLA and PBAT have superior properties, they are already being used little by little in various industrial fields such as packaging and agriculture industries [6,7].

In contrast to PLA and PBAT, which are not produced directly by microorganisms, PHB has the advantage that it can be produced directly through fermentation. Several microorganisms have been reported to have PHB accumulation capability, since PHB is a metabolite in the butanoate metabolism of microorganisms [8]. Moreover, PHB accumulators also express PHB depolymerase and can easily degrade PHB compared with PLA and PBAT. However, compared to PLA and PBAT, which are easily observed in daily life, PHB is not frequently observed. The main reason is that PHB is brittle: it does not stretch well and is easily torn, compared with other kinds of bioplastics [9–12].

There are many ways to strengthen PHB properties, but the most commonly used method is to use a plasticizer. A plasticizer is an additive added to change the physical properties of a material. When a plasticizer is added to a polymer such as plastic, the attractive force between the polymer chains is reduced, thereby increasing its flexibility [13]. Many kinds of plasticizers have already been added to non-degradable conventional plastics [14]. Since the addition of plasticizers greatly affects the physical properties of plastics, research on adding various plasticizers to bioplastics is being actively conducted [15,16]. Because most plasticizers are chemical substances with complex structures that are not related to microorganisms, there is a possibility that some plasticizers may adversely affect microorganisms [17]. In this case, the advantages of bioplastics may be weakened by the effect of plasticizers on microorganisms or their biodegradability. Therefore, carrying out research into finding a plasticizer that has good compatibility with bioplastic-degrading microorganisms is important. When carrying out such research, there are limitations of time and place when conducting all kinds of experiments in natural environments such as soil or ocean.

Therefore, in this study, plasticized PHB was prepared by adding various kinds of plasticizers to PHB, and their properties were compared through various analytical instruments such as differential scanning calorimetry (DSC), X-ray diffraction (XRD), thermogravimetric analysis (TGA), and universal testing machine (UTM). The effect of various plasticizers on the *Microbulbifer* sp. SOL66 (already been proven as a superior PHB degrading strain in our laboratory) was studied [18,19]. PHB degradation was also evaluated by studying the difference in the weight loss of PHB added with various kinds of plasticizers. This result suggests which plasticizer is suitable or unsuitable for use in PHB. By using *Microbulbifer* sp. SOL66 that can almost completely degrade PHB in 2–3 days, a benign plasticizer with good properties and good biodegradability could be selected very quickly, while overcoming the limitations of time and place.

2. Materials and Methods

2.1. Chemicals

All chemicals used in this study were of analytical grade. PHB pellets (Average MW: 550,000 g/mol) were obtained from Goodfellow Cambridge Ltd. (Huntingdon, UK). Chloroform and ethanol were obtained from Fisher Scientific (Hampton, NH, USA). Bis(2-ethylhexyl) adipate, bis(2-ethylhexyl) sebacate, lauric acid, triacetin, tributyl citrate, tributyl 2-acetylcitrate, l-Linalool, and geraniol were obtained from Sigma-Aldrich (St. Louis, MO, USA).

2.2. Toxicity Test of Various Plasticizers to the Growth of *Microbulbifer* sp. SOL66

To confirm the effect of various plasticizers on microbial cells, growth measurements were conducted under the effect of different concentrations of plasticizers. Bis(2-ethylhexyl) adipate, bis(2-ethylhexyl) sebacate, triacetin, tributyl citrate, tributyl 2-acetylcitrate, l-Linalool, and geraniol were used in concentrations from 0.25 mM to 16 mM. Only lauric acid was used in concentrations from 0.02 mM to 1.28 mM due to solubility. *Microbulbifer* sp. SOL66 cells were precultured in 5 mL marine broth (MB; Kisanbio, Seoul, Republic of Korea) in a shaking incubator at 37 °C, 200 rpm, for 24 h. Cells taken from the preculture were used to inoculate into the main culture process. In this process, pipetting robot (Integra, Le Locle, Switzerland) was used for the dilution of each plasticizer. Before the experiment, all kinds of plasticizers were dissolved in the HPLC grade ethanol to give a final concentration of 1.6 M of each plasticizer, except lauric acid. It cannot be dissolved significantly so a stock solution of 128 mM was prepared. All of the stock solutions were sterilized by using a filter (pore size, 0.2 μm). A 96-well cell culture plate (Thermo Fisher Scientific, MA, USA) was used for the culture process. MB liquid media 200 μL were inoculated with 2% of *Microbulbifer* sp. SOL66 in all the spotted wells, except for the first column. In the first column, 400 μL of MB liquid media containing 4% of *Microbulbifer* sp. SOL66 were spotted. After 4 μL of each plasticizer was added to the first column, a 96-well cell culture plate was set on the pipetting robot instrument. Then, according to the dilution sequence pre-arranged on this instrument, the solution was diluted two-fold while passing to the next column sequentially. In the dilution process, the mixing process through pipetting 5 times was also included for each dilution. After that, it was incubated in a shaking incubator at 37 °C, 200 rpm, for 24 h. The optical density after incubation was measured at 595 nm using a 96-well microplate reader (Tecan, Mannedorf, Switzerland) [20].

2.3. Degradation of PHB Containing Various Plasticizers under the Liquid Condition

For the liquid culture degradation test, PHB films with a thickness of 0.04 mm containing various plasticizers were prepared with 10% (*w*/*w*) and 20% (*w*/*w*) of concentrations. PHB pellets were placed in a glass bottle, into which 50 mL of chloroform was added. Then, each plasticizer was added to 10% and 20% of the total mass. The bottles were heated at 60 °C in a water bath until the contents were completely dissolved. Heated chloroform containing plasticized PHB was poured into a glass Petri dish in a fume hood, and a film was made by evaporating all of the solvents. Next, all films were cut by 20 mg each and then sterilized by immersing in 70% ethanol and drying in UV light. Then, each film was cultivated with the precultured *Microbulbifer* sp. SOL66 in 5 mL of MB liquid media [21]. After 2 days of cultivation at 37 °C, 200 rpm, optical density was measured with the same process described in the previous section. Degraded PHB films were recovered and washed with distilled water to remove impurities containing cell debris. In the case of time-dependent analysis, PHB films degraded by *Microbulbifer* sp. SOL66 after 1, 2, and 3 days of cultivation were recovered and treated with the same procedure. Recovered films were analyzed using a GC-MS instrument to measure weight loss.

2.4. GC-MS Analysis

The degree of degradation was calculated with GC-MS data. Before using the GC-MS instrument, fatty acid methyl ester (FAME) derivatization was conducted for preparing the GC-MS sample [21,22]. Then, 1 mL of methanol/sulfuric acid (85:15 *v*/*v*) and 1 mL of chloroform were mixed in each glass vial containing lyophilized PHB films. Additionally, each vial was sealed with a screw cap and Teflon tape and was heated using a heating block at 100 °C for 2 h for FAME derivatization. After the heating process, the samples were subsequently cooled at a low temperature for approximately 10 min, 1 mL of HPLC-grade water was added to each vial for making layer separation, and the samples were gently vortexed for 1 min. The bottom layer (chloroform) was transferred to a new 1.5 mL e-tube containing sodium sulfate anhydrous to remove water. Each sample was filtered (pore size, 0.2 μm) into a clean borosilicate glass tube. The prepared sample was analyzed by GC-MS (Perkin Elmer, Waltham, MA, USA) equipped with a fused silica capillary column (Elite-5 ms, 30 m × 0.25 mm i.d. × 0.25 μm) and subjected to a linear temperature gradient for analysis (40 °C for 1 min, increased at 15 °C/min to 120 °C and held for 2 min, and then increased at 10 °C/min to 280 °C and held for 10 min). The injector port temperature was 250 °C. Mass spectra were obtained by electron impact ionization at 70 eV, and scan spectra were obtained within the range of 45–450 m/z. Selected ion monitoring was used for the detection and fragmentation analysis of the major products. A standard curve was constructed using each kind of film to quantify the degraded PHB films. When a plasticizer was added to each film, the degree of degradation was calculated by considering the ratio.

2.5. Chemical and Mechanical Analysis

DSC analysis of the plasticized PHB films and non plasticized PHB films was performed using the Discovery DSC instrument (TA Instruments, New Castle, DE, USA) at temperatures ranging from −80 °C to 200 °C, while the heating and cooling rate was 10 °C/min in an N_2 atmosphere [23–25].

To analyze the crystalline/amorphous characteristics of plasticized and non-plasticized PHB films, XRD analysis was performed using an X-ray diffractometer (Rigaku Corporation, Tokyo, Japan) with a Cu Kα (λ = 1.54 Å) source, operating at 40 kV and

30 mA as the applied voltage and current, respectively. The 2θ range was from 3° to 60° at a scanning rate of 2°/min [25–28].

Thermogravimetric analysis (TGA) was performed using a TGA N-1000 (Scinco, Seoul, Korea). Samples were heated from 30 °C to 700 °C, at a heating rate of 10 °C/min under nitrogen flow (20 mL/min) [29–31].

Mechanical properties of plasticized and non-plasticized PHB films were measured using an EZ-SX universal testing machine (Shimadzu, Kyoto, Japan) [32–34]. Before measurement, PHB films were cut into a rectangular sample of size 60 mm × 10 mm (length × width). Thickness, which depends on the presence and amount of plasticizer, was measured using an 1111–100 A mini digital caliper (Insize, Loganville, GA, USA). Data were collected at a strain rate of 20 mm/min. After measurement, tensile strength, young's modulus, and elongation at break were calculated from a plot formed by TRAPEZIUM X software.

2.6. SEM Analysis

Scanning electron microscopy (SEM) analysis was performed to analyze the surface changes in the plasticized and non-plasticized PHB films after degradation by *Microbulbifer* sp. SOL66. Recovered PHB films after 2 days of degradation by *Microbulbifer* sp. SOL66 were washed and lyophilized before SEM analysis. The lyophilized films were then coated with gold dust at 5 mA for 300 s, and back-scatter electron images were obtained using a TM4000Plus SEM instrument (Hitachi, Tokyo, Japan) at a voltage of 5 kV [35].

2.7. GPC Analysis

Gel permeation chromatography (GPC) analysis was performed to detect the molecular weight changes in plasticized and non-plasticized PHB films after degradation by *Microbulbifer* sp. SOL66. The sample preparation process was carried out with the same method as our previous research [21]. A high-performance liquid chromatography (HPLC) apparatus was used for measuring molecular weight change, consisting of a loop injector (Rheodyne 7725i), an isocratic pump with dual heads (YL9112), column oven (YL9131), columns (Shodex, K-805, 8.0 mm I.D. × 300 mm; Shodex, K-804, 8.0 mm I.D. × 300 mm), and RI detector (YL9170). Sixty microliters of the sample from which bubbles were removed were injected. Only twenty microliters of injected samples were used as analytes for analysis in the injector part. The mobile phase was chloroform, and this flow rate was maintained at 1.0 mL/min. The temperature was maintained at 40 °C for analysis. The data were analyzed using YL-Clarity software for a single YL HPLC instrument (YOUNG IN Chromass, Anyang, Republic of Korea). The molecular weight was calculated with polystyrene standards ranging from 5000 to 2,000,000 g/mol [35].

3. Results

3.1. Comparison of Properties of Various Plasticizers

Eight kinds of plasticizers containing bis(2-ethylhexyl) adipate, bis(2-ethylhexyl) sebacate, lauric acid, triacetin, tributyl citrate, tributyl 2-acetylcitrate, L-Linalool, and geraniol which were often used for increasing properties of conventional plastics and bioplastics were included in the PHB film. As listed in (Table 1), all of the plasticizers that have been already reported for plasticizing PHB have very complex molecular structures and various molecular weights. Bis(2-ethylhexyl) adipate is the most commonly used plasticizer [36] because it can improve the flexibility of plastics and it has been regarded as a safe plasticizer. However, there is a precedent in which hepatic, brain, and cardiac injuries were found in rats when exposed to a large amount of this substance [37]. Although bis(2-ethylhexyl) sebacate is relatively less dangerous than bis(2-ethylhexyl) adipate, it is still not a recommended plasticizer because it is still dangerous [38,39]. Additionally, lauric acid can be used as a plasticizer because it forms a hydrogen bond when incorporated into the plastic, decreasing the force of each plastic molecule, resulting in a decrease in glass transition temperature [40,41]. Triacetin is a well-known plasticizer for PHB due to its improved biodegradability of PHB [42,43]. Tributyl citrate and tributyl 2-acetylcitrate are widely used as plasticizers in bioplastics such as PHB due to their high biocompatibility [27,44,45]. L-Linalool and geraniol are natural monoterpenoids, which are components of essential oils derived from plants, and have a unique scent. Since it can be obtained from nature rather than chemically synthesized, it has the advantage of being environmentally friendly [46]. Since all plasticizers have their strengths and weaknesses, both physical properties and biodegradability were considered to select the most appropriate plasticizer for PHB.

Table 1. Molecular properties and applied references of plasticizers used in this study.

Plasticizer	Molecular structure	Molecular formula	Molecular Weight (g/mol)	References
Bis(2-ethylhexyl) adipate (DEHA)	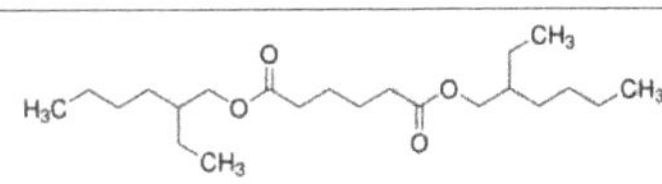	$C_{22}H_{42}O_4$	370.57	[36,37]
Bis(2-ethylhexyl) sebacate (DEHS)	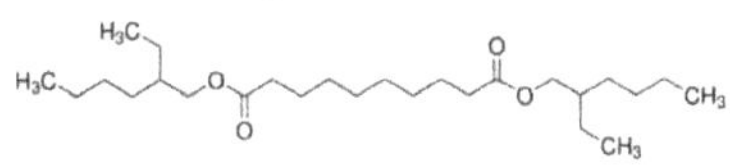	$C_{26}H_{50}O_4$	426.67	[38,39]
Lauric acid (LA)	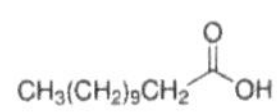	$C_{12}H_{24}O_2$	200.32	[40,41]
Triacetin (TA)	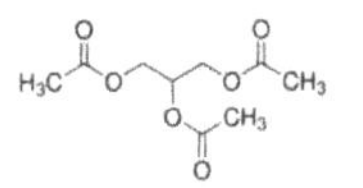	$C_9H_{14}O_6$	218.20	[42,43]
Tributyl citrate (TBC)	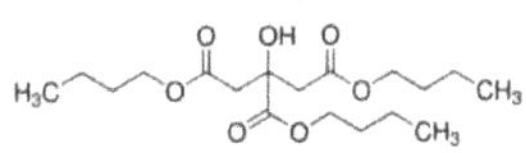	$C_{18}H_{32}O_7$	360.44	[44,45]
Tributyl 2-acetylcitrate (TAC)	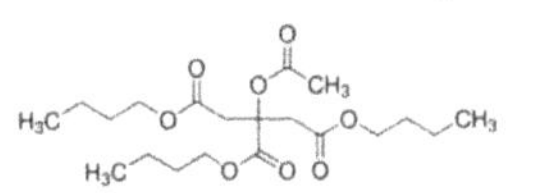	$C_{20}H_{34}O_8$	402.48	[27,45]
L-Linalool (L)	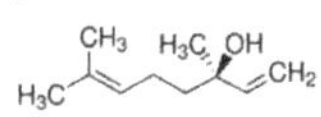	$C_{10}H_{18}O$	154.25	[46]
Geraniol (G)	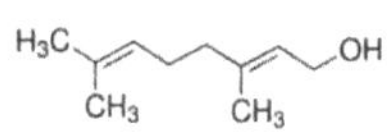	$C_{10}H_{18}O$	154.25	[46]

3.2. *Effect of Various Plasticizers on Microbulbifer* sp. *SOL66 Cell*

Before comparing biodegradability, a toxicity test on bacteria *Microbulbifer* sp. SOL66 was conducted with all 8 plasticizers. If a plasticizer that prevents cell growth is used, the biodegradation process using microorganisms is disrupted. So, compatibility between the plasticizer and *Microbulbifer* sp. SOL66 was confirmed. Although the toxicity of a plasticizer does not impede all biodegradation processes, confirmation of toxicity can provide a certain reference point for evaluating the suitability of the plasticizer for PHB, which can be easily degraded by *Microbulbifer* sp. SOL66. As a result of the toxicity test with various plasticizers, the effect of each plasticizer addition on the growth of *Microbulbifer* sp. SOL66 was varied (Figure 1).

In all cases, the growth of *Microbulbifer* sp. SOL66 was reduced when a plasticizer was added. However, in the case of triacetin, the cell growth gradually increased as the concentration increased. Except for this, the addition of bis(2-ethylhexyl) adipate, bis(2-ethylhexyl) sebacate, lauric acid, tributyl citrate, and tributyl 2-acetylcitrate showed a generally decreasing tendency as the concentration increased. When L-Linalool and geraniol, which are natural monoterpenoid plasticizers, were added, *Microbulbifer* sp. SOL66 did not grow at all when more than 8 mM and 4 mM were added, respectively. However, this does not mean that plasticizers that reduce or inhibit microbial growth are not good when added to PHB. It suggests that plasticizers may adversely affect cell growth when applied directly to cells. The impact of toxicity as a molecule might be different from the role inside of the polymer.

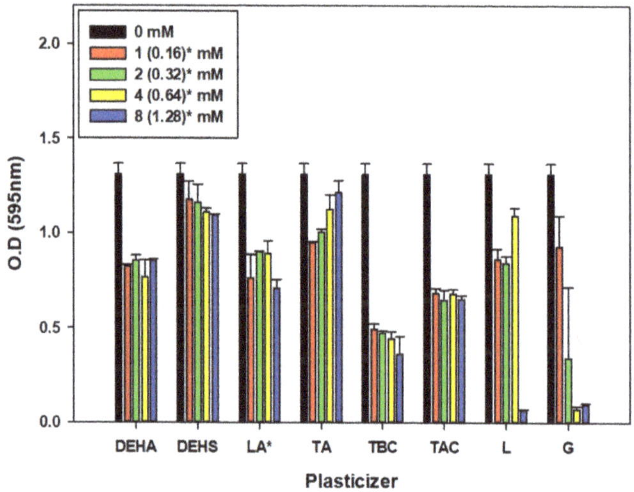

Figure 1. Growth tendency with the addition of various plasticizers. The cell growth pattern of *Microbulbifer* sp. SOL66 was expressed as optical density with the addition of plasticizers at concentrations of 1 mM, 2 mM, 4 mM, and 8 mM, lauric acid* was added at 0.16 mM, 0.32 mM, 0.64 mM, and 1.28 mM due to solubility issues.

3.3. Thermal Properties Analyzed by DSC

To confirm the change in chemical properties by the addition of plasticizer, DSC analysis was performed to measure the glass transition temperature (T_g), melting temperature (T_m), and crystallization temperature (T_c). In general, it is known that adding a plasticizer creates free space between the polymers, thereby reducing the T_g value and softening the physical properties by allowing the components of the polymer to move freely [13,47,48]. The DSC result showed that the T_g value of plasticized PHB film is low compared with non-plasticized PHB, except for PHB film containing 20% of L-Linalool (Table 2). It can be supposed that the addition of various kinds of plasticizers increased the softness of PHB. Not only the T_g value but also the T_m value and T_c value changed according to the addition of the various plasticizers, but the crystallinity value (X_c) derived from the enthalpy change at melting temperature (ΔH_m) is more important in analyzing the properties. X_c value was calculated using Equation (1) [23,29,30].

$$X_c(\%) = \frac{\Delta H_m}{\Delta H_m 100\%} \times \frac{100}{W_{PHB}}(\%) \quad (1)$$

Table 2. Thermal properties of plasticized PHB analyzed by using DSC.

Plasticizer	Concentration (%)	T_g (°C)	T_m (°C)	T_c (°C)	ΔH_m (J/g)	X_c (%)
Control	-	51.1	166.7	112.8	63.0	43.1
Bis(2-ethylhexyl) adipate (DEHA)	10	48.2	167.0	112.2	64.9	49.4
	20	46.8	163.0	110.5	56.1	48.0
Bis(2-ethylhexyl) sebacate (DEHS)	10	47.7	167.5	112.6	62.2	47.3
	20	48.5	166.2	112.1	50.3	43.1
Lauric acid (LA)	10	40.6	166.3	109.6	70.9	53.9
	20	40.2	160.5	104.0	61.3	52.5
Triacetin (TA)	10	44.0	165.0	110.9	67.7	51.5
	20	42.0	159.1	107.8	62.4	53.4
Tributyl citrate (TBC)	10	46.4	166.8	111.2	66.9	50.9
	20	42.9	164.0	108.5	51.8	44.3
Tributyl 2-acetylcitrate (TAC)	10	49.1	168.6	112.0	68.5	52.1
	20	45.1	161.0	105.9	76.0	65.0
L-Linalool (L)	10	49.7	169.0	113.5	50.5	38.4
	20	52.6	171.4	116.3	55.9	47.8
Geraniol (G)	10	44.3	166.2	112.1	61.3	46.7
	20	40.7	155.1	104.5	60.4	51.7

The melting enthalpy of a 100% crystalline 3-hydroxybutyric acid (ΔH_m 100%) was reported to be 146 J/g and W_{PHB} means the weight fraction of PHB in the plasticized PHB [49]. By using this, X_c values of plasticized PHB were calculated. The crystallinity of non-plasticized PHB was 43.1%, whereas the crystallinity of plasticized PHB was increased, except when 20% of bis(2-ethylhexyl) sebacate and 10% of L-Linalool were added to PHB. This is because adding a plasticizer improves the mobility of the polymer chain and promotes crystallization [44].

3.4. *Effect of Various Plasticizers on the Biodegradability of PHB by Microbulbifer* sp. *SOL66*

After confirming the toxicity of various plasticizers as molecules themselves, the effect of plasticizers inside of PHB on the growth of cells was compared. In addition, the biodegradability of plasticized PHB was compared to screen the optimal plasticizer for better quality PHB that can be degraded by *Microbulbifer* sp. SOL66. Eight kinds of plasticizers were added to PHB film with 10% and 20% concentrations, respectively. After the sterilization process, plasticized PHB films were cultured together with *Microbulbifer* sp. SOL66 in a 37 °C shaking incubator for two days. After cultivation, the growth expressed as optical density was first measured to confirm the effect of the plasticizer contained in the PHB film on the *Microbulbifer* sp. SOL66 cells. In the case of culturing the PHB films containing 10% of the plasticizers, growth itself was reduced compared to the case of culturing the PHB film without the plasticizer as a whole (Figure 2a). However, there was no case where growth did not occur. On the other hand, even when the PHB films containing 20% of the plasticizers were cultured, the growth pattern was similar depending on the type of added plasticizer, but in the case of culture of the PHB film containing 20% of geraniol, *Microbulbifer* sp. SOL66 did not grow (Figure 2b). The result of completely inhibiting the growth of *Microbulbifer* sp. SOL66 when 20% of geraniol was included in the film was similar to the previous toxicity test, suggesting geraniol is not a suitable plasticizer for PHB when the evaluation of PHB degradation is conducted with *Microbulbifer* sp. SOL66., but the growth pattern when the plasticizer was included in the film was different from that when the plasticizer was directly applied to microbial cells.

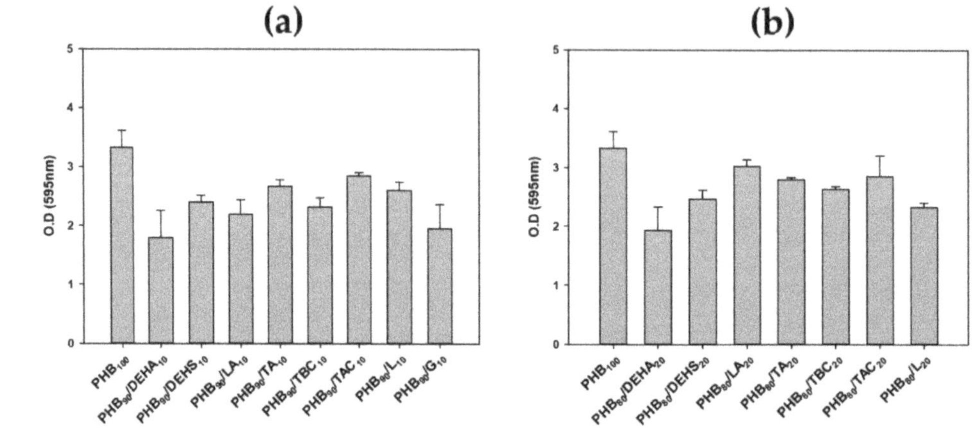

Figure 2. The growth pattern of *Microbulbifer* sp. SOL66 cultured with (**a**) 10% plasticized and (**b**) 20% plasticized PHB films.

As a result of analyzing the degree of degradation of the PHB film containing various plasticizers through GC-MS, when most plasticizers were added, the degree of degradation tended to decrease. In the case of degradation of PHB film containing 10% of plasticizers, the biodegradability of tributyl citrate and tributyl 2-acetylcitrate was similar to that of the film without plasticizers (Figure 3a). When 20% of plasticizers were included, the biodegradability of triacetin and tributyl citrate was comparable to that of the non-plasticized PHB (Figure 3b). When 20% of triacetin was added, the degree of degradation was higher than that of the non-plasticized PHB, but in the case of 10%, the degree of degradation was not very high. On the other hand, in the case of tributyl citrate, since it showed excellent robustness of biodegradability in both 10% and 20% plasticized PHB, it could be concluded that tributyl citrate was suitable as a plasticizer to be added to PHB, which is greatly degraded with *Microbulbifer* sp. SOL66, suggesting the toxic effect of plasticizer itself and degradation of plasticized PHB is different.

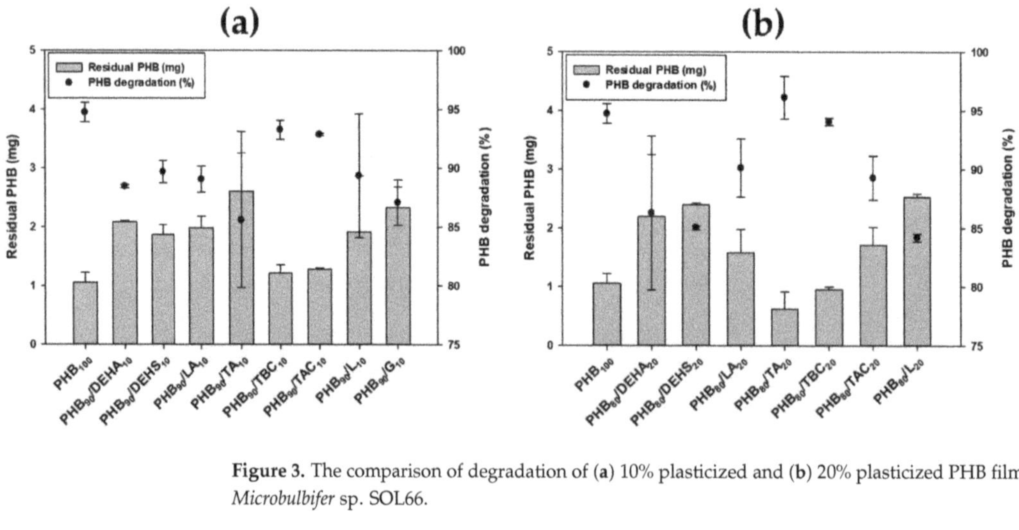

Figure 3. The comparison of degradation of (**a**) 10% plasticized and (**b**) 20% plasticized PHB films by *Microbulbifer* sp. SOL66.

3.5. Comparison of Other Properties with Using Various Analytical Instruments

As PHB with tributyl citrate showed higher degradation than PHB without tributyl citrate, the surface change with the addition of 10% and 20% of tributyl citrate after two days of degradation by *Microbulbifer* sp. SOL66 was compared using SEM. Since the surfaces of the non-plasticized PHB film and the plasticized PHB film before degradation were similar, the surface area characteristics of the film could be confirmed by checking the surface after degradation. The SEM result showed that the non-plasticized PHB film looked as if degradation had progressed as the cracked cross-section was split at an angle, and the cross-sectional portion where the tributyl citrate-added film was degraded was curved rather than angled (Figure 4). Through the comparison of the parts where the degradation has progressed in this way, it was possible to visually compare the difference in the physical properties of the surface.

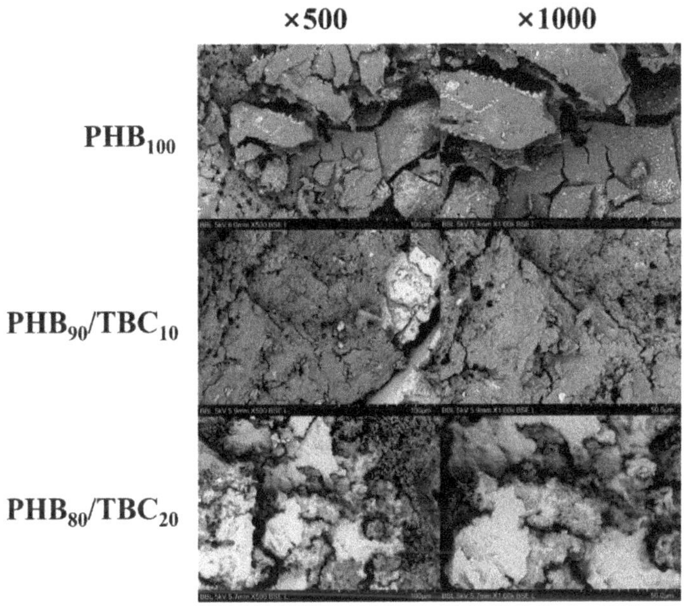

Figure 4. Representative images of surface changes observed by scanning electron microscopy. Comparison of differences in surface changes after two days of degradation by *Microbulbifer* sp. SOL66 through observation of 500 and 1000 magnifications.

Next, the change in molecular weight according to the addition of tributyl citrate before and after degradation by *Microbulbifer* sp. SOL66 was compared. As indicated in (Table 3), there was little difference between the number average molecular weight (M_n) values of plasticized and non-plasticized PHB films before degradation by *Microbulbifer* sp. SOL66. However, there was a slight difference after 2 days of degradation. When tributyl citrate was added, the number average molecular weight decreased more than when it was not added. Conversely, polydispersity index (PDI) values after degradation were higher when tributyl citrate was added. Through the difference in molecular weight change, it was found that adding tributyl citrate to the PHB film could cause a change in degradation by *Microbulbifer* sp. SOL66.

Table 3. Comparison of molecular weight change before and after degradation by *Microbulbifer* sp. SOL66 according to the addition of 0%, 10%, and 20% of tributyl citrate.

Day	TBC Concentration (%)	$M_n \times 10^5$	PDI
0	0	5.33	1.46
	10	5.39	1.44
	20	5.33	1.44
2	0	0.29	1.56
	10	0.27	1.74
	20	0.26	1.67

To find out the reason for high microbial degradation under the influence of plasticizers, XRD spectra of non-plasticized PHB film and PHB film plasticized with tributyl citrate were presented and compared (Figure 5a). The shape of the spectra was similar with or without plasticizer addition. In all three spectra, two strong peaks were commonly observed at $2\theta = 13.5$ and $2\theta = 16.9$, which are assigned with (020) and (110) diffraction planes. These peaks correspond to orthorhombic α-form crystals of PHB [50]. However, when a plasticizer was added, the intensities of these two peaks were slightly changed, meaning that crystal structures were slightly changed.

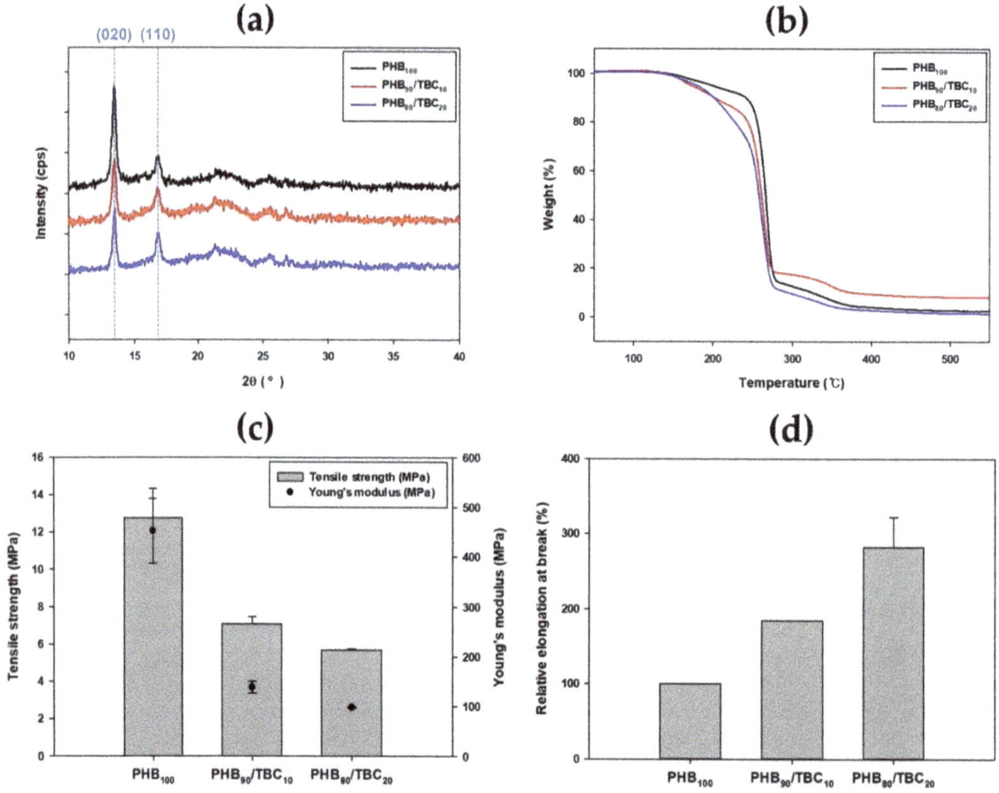

Figure 5. Changes in various properties according to the addition of tributyl citrate. Comparison of (**a**) XRD spectra and (**b**) thermo gravimetric analysis (TGA) result. Mechanical properties change expressed as (**c**) tensile strength (MPa), Young's modulus (MPa), and (**d**) relative elongation at break (%) measured using a universal testing machine (UTM).

Next, the pattern of thermal degradation according to the presence or absence of the plasticizer was compared through the TGA results (Figure 5b). The pattern of abruptly decreasing weight near 250 °C was similar, but when tributyl citrate as a plasticizer was added, the weight was already greatly reduced before entering the rapidly decreasing section [29]. It was confirmed that the stability to heat decreased according to the concentration of the plasticizer added. Additionally, the higher the concentration of tributyl citrate, the lower the stability was. This is because plasticizer loss occurs when a high temperature is applied [27].

Lastly, the change in mechanical properties expressed as tensile strength, Young's modulus, and relative elongation at break were compared through the universal testing machine results. When tributyl citrate as a plasticizer was added, tensile strength and Young's modulus were decreased by more than half (Figure 5c). However, the decrease in both values according to the increase in the concentration of the plasticizer was not very large. On the other hand, there was a large difference in the length of the film stretched (Figure 5d). When the elongation of the film without the addition of tributyl citrate was set to 100, and a relative comparison was conducted, the elongation was nearly doubled when 10% of tributyl citrate was included in PHB film, and it was almost tripled when 20% of tributyl citrate was included. This is because the addition of tributyl citrate as a plasticizer reduced the intermolecular interactions of PHB [51].

3.6. Time-Dependent Analysis of PHB Degradation

Finally, the time-dependent degradation rate of non-plasticized PHB film, PHB film including 10% tributyl citrate, and PHB film including 20% tributyl citrate was compared (Figure 6). These films are degraded by *Microbulbifer* sp. SOL66 for 1, 2, and 3 days were recovered and analyzed. After 3 days, the final degree of degradation was similar to more than 90%. However, non-plasticized PHB film was degraded by about 70% after 1 day of degradation and was almost completely degraded as the degree of degradation increased gradually, whereas the plasticized PHB films showed a very high degree of degradation of more than 80% even after one day of degradation. It suggests that tributyl citrate not only changes physical properties and increases the elongation properties of the PHB film, but is also very excellent in terms of biodegradability by *Microbulbifer* sp. SOL66.

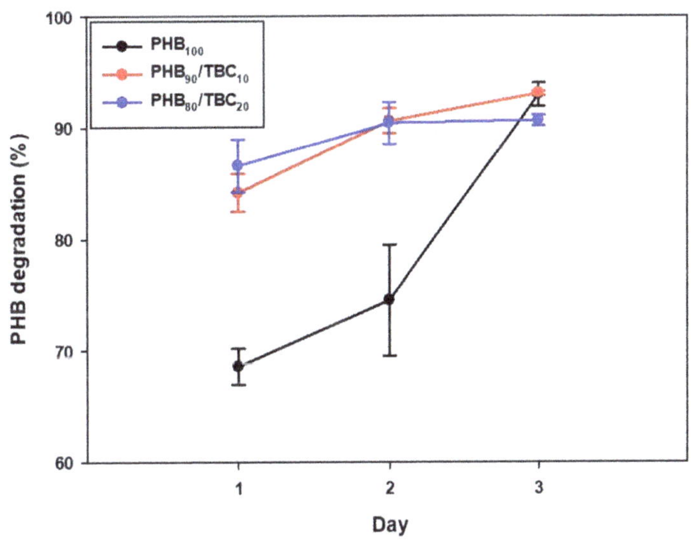

Figure 6. Time-dependent analysis of PHB biodegradation by *Microbulbifer* sp. SOL66. Comparison of the degradation rate of non-plasticized PHB film, and tributyl citrate plasticized PHB films.

4. Conclusions

Interest in biodegradable bioplastics has increased significantly as a way to reduce environmental pollution caused by non-degradable conventional plastics. However, unlike PBAT, which stretches well, PHB is brittle and is easily broken when a force is applied. In the case of PHB, its main advantage is its eco-friendly production process, as it can be produced and degraded by microorganisms. To overcome its property-related shortcomings, various types of plasticizers can be added to PHB.

Our study concluded that in the presence of plasticizer tributyl citrate, the degree of degradation was excellent compared to other plasticizers, and the robustness of the degradation rate was excellent even when the concentration of the plasticizer was increased. In addition, this suggests that there is a speedy screening process of plasticizers for PHB using *Microbulbifer* sp. SOL66, which has superior PHB degradation ability. Tributyl citrate plasticizer can be used to prepare bioplastic to have better qualities with a higher biodegradation rate, which can help to manage emerging plastic-related issues.

Author Contributions: J.Y.C., J.A., J.-M.J., J.-J.Y. and Y.-H.Y. conceived and designed the study; J.Y.C., S.H.K., H.J.J., D.H.C. and B.C.K. performed the experiments and drafted the manuscript; J.Y.C. and Y.-H.Y. interpreted the experimental results; J.Y.C., S.K.B., J.L. and Y.-H.Y. revised the manuscript. All authors have read and agreed to the published version of the manuscript.

Funding: This study was supported by the Research Program to solve social issues with the National Research Foundation of Korea (NRF), funded by the Ministry of Science and ICT [grant number 2017M3A9E4077234], National Research Foundation of Korea (NRF) [grant numbers NRF-2022R1A2C2003138 and NRF-2019M3E6A1103979]. This study was also supported by the R&D Program of MOTIE/KEIT [grant number 20009508 and 20018072].

Institutional Review Board Statement: Not applicable.

Informed Consent Statement: Not applicable.

Data Availability Statement: Not applicable.

Conflicts of Interest: The authors declare no conflict of interest.

References

1. Jõgi, K.; Bhat, R. Valorization of food processing wastes and by-products for bioplastic production. *Sustain. Chem. Pharm.* **2020**, *18*, 100326. [CrossRef]
2. Mangaraj, S.; Yadav, A.; Bal, L.M.; Dash, S.K.; Mahanti, N.K. Application of Biodegradable Polymers in Food Packaging Industry: A Comprehensive Review. *J. Packag. Technol. Res.* **2019**, *3*, 77–96. [CrossRef]
3. Bharti, S.N.; Swetha, G. Need for Bioplastics and Role of Biopolymer PHB: A Short Review. *J. Pet. Environ. Biotechnol.* **2016**, *7*, 5–8. [CrossRef]
4. Mohd Ishak, Z.A.; Ahmad Thirmizir, M.Z. Editorial corner-a personal view producing green composites via polymer blending. *Express Polym. Lett.* **2021**, *15*, 910–911. [CrossRef]
5. Popa, M.S.; Frone, A.N.; Panaitescu, D.M. Polyhydroxybutyrate blends: A solution for biodegradable packaging? *Int. J. Biol. Macromol.* **2022**, *207*, 263–277. [CrossRef]
6. Choi, B.; Yoo, S.; Park, S. Il Carbon footprint of packaging films made from LDPE, PLA, and PLA/PBAT blends in South Korea. *Sustainability* **2018**, *10*, 2369. [CrossRef]
7. Jian, J.; Xiangbin, Z.; Xianbo, H. An overview on synthesis, properties and applications of poly(butylene-adipate-co-terephthalate)–PBAT. *Adv. Ind. Eng. Polym. Res.* **2020**, *3*, 19–26. [CrossRef]
8. Park, Y.L.; Bhatia, S.K.; Gurav, R.; Choi, T.R.; Kim, H.J.; Song, H.S.; Park, J.Y.; Han, Y.H.; Lee, S.M.; Park, S.L.; et al. Fructose based hyper production of poly-3-hydroxybutyrate from *Halomonas* sp. YLGW01 and impact of carbon sources on bacteria morphologies. *Int. J. Biol. Macromol.* **2020**, *154*, 929–936. [CrossRef]
9. Turco, R.; Santagata, G.; Corrado, I.; Pezzella, C.; Di Serio, M. In vivo and Post-synthesis Strategies to Enhance the Properties of PHB-Based Materials: A Review. *Front. Bioeng. Biotechnol.* **2021**, *8*, 619266. [CrossRef]
10. Raza, Z.A.; Khalil, S.; Abid, S. Recent progress in development and chemical modification of poly(hydroxybutyrate)-based blends for potential medical applications. *Int. J. Biol. Macromol.* **2020**, *160*, 77–100. [CrossRef]
11. El-Hadi, A.; Schnabel, R.; Straube, E.; Müller, G.; Henning, S. Correlation between degree of crystallinity, morphology, glass temperature, mechanical properties and biodegradation of poly (3-hydroxyalkanoate) PHAs and their blends. *Polym. Test.* **2002**, *21*, 665–674. [CrossRef]

12. Soleymani Eil Bakhtiari, S.; Karbasi, S.; Toloue, E.B. Modified poly(3-hydroxybutyrate)-based scaffolds in tissue engineering applications: A review. *Int. J. Biol. Macromol.* **2021**, *166*, 986–998. [CrossRef]
13. Vieira, M.G.A.; Da Silva, M.A.; Dos Santos, L.O.; Beppu, M.M. Natural-based plasticizers and biopolymer films: A review. *Eur. Polym. J.* **2011**, *47*, 254–263. [CrossRef]
14. Güney, A.; Özdilek, C.; Kangal, M.O.; Burat, F. Flotation characterization of PET and PVC in the presence of different plasticizers. *Sep. Purif. Technol.* **2015**, *151*, 47–56. [CrossRef]
15. Aguilar, J.M.; Bengoechea, C.; Pérez, E.; Guerrero, A. Effect of different polyols as plasticizers in soy based bioplastics. *Ind. Crops Prod.* **2020**, *153*, 112522. [CrossRef]
16. Alonso-González, M.; Felix, M.; Romero, A. Influence of the plasticizer on rice bran-based eco-friendly bioplastics obtained by injection moulding. *Ind. Crops Prod.* **2022**, *180*, 114767. [CrossRef]
17. Andrade, R.C. *Biodegradation of Commonly Used Plasticizers by Pure Bacterial Cultures*; McGill University: Montreal, QC, Canada, 2011; ISBN 9780494841518.
18. Park, S.L.; Cho, J.Y.; Kim, S.H.; Lee, H.-J.; Kim, S.H.; Suh, M.J.; Ham, S.; Bhatia, S.K.; Gurav, R.; Park, S.-H.; et al. Novel Polyhydroxybutyrate-Degrading Activity of the Microbulbifer Genus as Confirmed by *Microbulbifer* sp. SOL03 from the Marine Environment. *J. Microbiol. Biotechnol.* **2022**, *32*, 27–36. [CrossRef]
19. Park, S.L.; Cho, J.Y.; Kim, S.H.; Bhatia, S.K.; Gurav, R.; Park, S.H.; Park, K.; Yang, Y.H. Isolation of microbulbifer sp. Sol66 with high polyhydroxyalkanoate-degrading activity from the marine environment. *Polymers* **2021**, *13*, 4257. [CrossRef]
20. Song, H.S.; Bhatia, S.K.; Choi, T.R.; Gurav, R.; Kim, H.J.; Lee, S.M.; Park, S.L.; Lee, H.S.; Joo, H.S.; Kim, W.; et al. Increased antibiotic resistance of methicillin-resistant staphylococcus aureus USA300 Δpsm mutants and a complementation study of Δpsm mutants using synthetic phenol-soluble modulins. *J. Microbiol. Biotechnol.* **2021**, *31*, 115–122. [CrossRef]
21. Cho, J.Y.; Park, S.L.; Kim, S.H.; Jung, H.J.; Cho, D.H.; Kim, B.C.; Bhatia, S.K.; Gurav, R.; Park, S.H.; Park, K.; et al. Novel Poly(butylene adipate-co-terephthalate)-degrading *Bacillus* sp. JY35 from wastewater sludge and its broad degradation of various bioplastics. *Waste Manag.* **2022**, *144*, 1–10. [CrossRef]
22. Choi, T.R.; Park, Y.L.; Song, H.S.; Lee, S.M.; Park, S.L.; Lee, H.S.; Kim, H.J.; Bhatia, S.K.; Gurav, R.; Choi, K.Y.; et al. Fructose-based production of short-chain-length and medium-chain-length polyhydroxyalkanoate copolymer by arctic pseudomonas sp. B14-6. *Polymers* **2021**, *13*, 1398. [CrossRef] [PubMed]
23. Xiao, H.; Lu, W.; Yeh, J.-T. Effect of Plasticizer on the Crystallization Behavior of Poly(lactic acid). *J. Appl. Polym. Sci.* **2009**, *113*, 112–121. [CrossRef]
24. Anbukarasu, P.; Sauvageau, D.; Elias, A. Tuning the properties of polyhydroxybutyrate films using acetic acid via solvent casting. *Sci. Rep.* **2015**, *5*, 17884. [CrossRef]
25. Armentano, I.; Fortunati, E.; Burgos, N.; Dominici, F.; Luzi, F.; Fiori, S.; Jiménez, A.; Yoon, K.; Ahn, J.; Kang, S.; et al. Processing and characterization of plasticized PLA/PHB blends for biodegradable multiphase systems. *Express Polym. Lett.* **2015**, *9*, 583–596. [CrossRef]
26. Erceg, M.; Jozic, D. Preparation and characterization of poly(3-hydroxy-butyrate)/Cloisite25A nanocomposites. *E-Polymers* **2013**, *13*, 25. [CrossRef]
27. Frone, A.N.; Nicolae, C.A.; Eremia, M.C.; Tofan, V.; Ghiurea, M.; Chiulan, I.; Radu, E.; Damian, C.M.; Panaitescu, D.M. Low molecular weight and polymeric modifiers as toughening agents in poly(3-hydroxybutyrate) films. *Polymers* **2020**, *12*, 2446. [CrossRef] [PubMed]
28. Panaitescu, D.M.; Frone, A.N.; Chiulan, I.; Nicolae, C.A.; Trusca, R.; Ghiurea, M.; Gabor, A.R.; Mihailescu, M.; Casarica, A.; Lupescu, I. Role of bacterial cellulose and poly (3-hydroxyhexanoate-co-3-hydroxyoctanoate) in poly (3-hydroxybutyrate) blends and composites. *Cellulose* **2018**, *25*, 5569–5591. [CrossRef]
29. Erceg, M.; Kovačić, T.; Klarić, I. Thermal degradation of poly(3-hydroxybutyrate) plasticized with acetyl tributyl citrate. *Polym. Degrad. Stab.* **2005**, *90*, 313–318. [CrossRef]
30. Araque, L.M.; De Morais, A.C.L.; Alves, T.S.; Azevedo, J.B.; Carvalho, L.H.; Barbosa, R. Preparation and characterization of poly(hydroxybutyrate) and hollow glass microspheres composite films: Morphological, thermal, and mechanical properties. *J. Mater. Res. Technol.* **2019**, *8*, 935–943. [CrossRef]
31. Seoane, I.T.; Cerrutti, P.; Vazquez, A.; Cyras, V.P.; Manfredi, L.B. Ternary nanocomposites based on plasticized poly(3-hydroxybutyrate) and nanocellulose. *Polym. Bull.* **2018**, *76*, 967–988. [CrossRef]
32. Inoue, S.; Yamaguchi, S.; Uyama, H.; Yamashiro, T.; Imazato, S. Influence of constant strain on the elasticity of thermoplastic orthodontic materials. *Dent. Mater. J.* **2020**, *39*, 415–421. [CrossRef]
33. Chen, S.; Hori, N.; Kajiyama, M.; Takemura, A. Compatibilities and properties of poly lactide/poly (methyl acrylate) grafted chicken feather composite: Effects of graft chain length. *J. Appl. Polym. Sci.* **2020**, *137*, 48981. [CrossRef]
34. Duong, H.C.; Chuai, D.; Woo, Y.C.; Shon, H.K.; Nghiem, L.D.; Sencadas, V. A novel electrospun, hydrophobic, and elastomeric styrene-butadiene-styrene membrane for membrane distillation applications. *J. Membr. Sci.* **2018**, *549*, 420–427. [CrossRef]
35. Cho, J.Y.; Lee Park, S.; Lee, H.J.; Kim, S.H.; Suh, M.J.; Ham, S.; Bhatia, S.K.; Gurav, R.; Park, S.H.; Park, K.; et al. Polyhydroxyalkanoates (PHAs) degradation by the newly isolated marine *Bacillus* sp. JY14. *Chemosphere* **2021**, *283*, 131172. [CrossRef]
36. Sivaprakash, G.; Raja, R.K.; Mohanrasu, K.; Dinesh, G.H.; Arun, A. Microbial nanotechnology in food industry: Antimicrobial packaging. In *Handbook of Microbial Nanotechnology*; Academic Press: Cambridge, MA, USA, 2022; pp. 311–329. [CrossRef]

37. Behairy, A.; Abd El-Rahman, G.I.; Aly, S.S.H.; Fahmy, E.M.; Abd-Elhakim, Y.M. Di(2-ethylhexyl) adipate plasticizer triggers hepatic, brain, and cardiac injury in rats: Mitigating effect of Peganum harmala oil. *Ecotoxicol. Environ. Saf.* **2021**, *208*, 111620. [CrossRef]
38. Silverstein, B.D.; White, O.J.; Brower, J.E.; Bernholc, N.M. *Biological Effects Summary Report di(2-ethylhexyl) Sebacate*; Brookhaven National Laboratory: Upton, NY, USA, 1983.
39. Chodak, I. Polyhydroxyalkanoates: Origin, properties and applications. In *Monomers, Polymers and Composites from Renewable Resources*; Elsevier: Amsterdam, The Netherlands, 2008; pp. 451–477. [CrossRef]
40. Adorna, J.A.; Aleman, C.K.A.; Gonzaga, I.L.E.; Pangasinan, J.N.; Sisican, K.M.D.; Dang, V.D.; Doong, R.A.; Ventura, R.L.G.; Ventura, J.R.S. Effect of Lauric Acid on the Thermal and Mechanical Properties of Polyhydroxybutyrate (PHB)/Starch Composite Biofilms. *Int. J. Polym. Sci.* **2020**, *2020*, 7947019. [CrossRef]
41. Abdelwahab, M.A.; Flynn, A.; Chiou, B.S.; Imam, S.; Orts, W.; Chiellini, E. Thermal, mechanical and morphological characterization of plasticized PLA-PHB blends. *Polym. Degrad. Stab.* **2012**, *97*, 1822–1828. [CrossRef]
42. Nosal, H.; Moser, K.; Warzała, M.; Holzer, A.; Stańczyk, D.; Sabura, E. Selected Fatty Acids Esters as Potential PHB-V Bioplasticizers: Effect on Mechanical Properties of the Polymer. *J. Polym. Environ.* **2020**, *29*, 38–53. [CrossRef]
43. Zarski, A.; Bajer, K.; Kapuśniak, J. Review of the most important methods of improving the processing properties of starch toward non-food applications. *Polymers* **2021**, *13*, 832. [CrossRef]
44. Chaos, A.; Sangroniz, A.; Gonzalez, A.; Iriarte, M.; Sarasua, J.R.; del Río, J.; Etxeberria, A. Tributyl citrate as an effective plasticizer for biodegradable polymers: Effect of plasticizer on free volume and transport and mechanical properties. *Polym. Int.* **2019**, *68*, 125–133. [CrossRef]
45. Râpă, M.; Darie-Nită, R.N.; Grosu, E.; Tănase, E.E.; Trifoi, A.R.; Pap, T.; Vasile, C. Effect of plasticizers on melt processability and properties of PHB. *J. Optoelectron. Adv. Mater.* **2015**, *17*, 1778–1784.
46. Mangeon, C.; Michely, L.; Rios De Anda, A.; Thevenieau, F.; Renard, E.; Langlois, V. Natural Terpenes Used as Plasticizers for Poly(3-hydroxybutyrate). *ACS Sustain. Chem. Eng.* **2018**, *6*, 16160–16168. [CrossRef]
47. Immergut, E.H.; Mark, H.F. Principles of Plasticization. *Adv. Chem.* **1965**, *48*, 1–26. [CrossRef]
48. Sothornvit, R.; Krochta, J.M. Plasticizers in edible films and coatings. In *Innovations in Food Packaging*; Academic Press: Cambridge, MA, USA, 2005; pp. 403–433. [CrossRef]
49. Chiulan, I.; Mihaela Panaitescu, D.; Nicoleta Frone, A.; Teodorescu, M.; Andi Nicolae, C.; Căşărică, A.; Tofan, V.; Sălăgeanu, A. Biocompatible polyhydroxyalkanoates/bacterial cellulose composites: Preparation, characterization, and in vitro evaluation. *J. Biomed. Mater. Res. Part A* **2016**, *104*, 2576–2584. [CrossRef] [PubMed]
50. Dai, X.; Zhang, J.; Ren, Z.; Li, H.; Sun, X.; Yan, S. A grazing incident XRD study on the structure of poly(3-hydroxybutyrate) ultrathin films sandwiched between Si wafers and amorphous polymers. *Polym. Chem.* **2016**, *7*, 3705–3713. [CrossRef]
51. Sanyang, M.L.; Sapuan, S.M.; Jawaid, M.; Ishak, M.R.; Sahari, J. Effect of plasticizer type and concentration on tensile, thermal and barrier properties of biodegradable films based on sugar palm (Arenga pinnata) starch. *Polymers* **2015**, *7*, 1106–1124. [CrossRef]

polymers

MDPI

Article

Production and Characterisation of an Exopolysaccharide by *Bacillus amyloliquefaciens*: Biotechnological Applications

Enrique Sánchez-León [1], Elisa Huang-Lin [1], Ricardo Amils [1,2] and Concepción Abrusci [1,2,*]

[1] Departamento de Biología Molecular, Facultad de Ciencias, Universidad Autónoma de Madrid, UAM, Cantoblanco, 28049 Madrid, Spain
[2] Centro de Biología Molecular Severo Ochoa, CSIC-UAM, 28049 Madrid, Spain
* Correspondence: concepcion.abrusci@uam.es; Tel.: +34-91-497-82-57

Abstract: The *Bacillus amyloliquefaciens* RT7 strain was isolated from an extreme acidic environment and identified. The biodegradation capabilities of the strain using different carbon sources (glucose, oleic acid, Tween 80, PEG 200, and the combination of glucose–Tween 80) were evaluated via an indirect impedance technique. The glucose–Tween 80 combination was further studied using nuclear magnetic resonance (NMR). The exopolysaccharide (EPS_{RT7}) that had been produced with the strain when biodegrading glucose–Tween 80 was isolated and characterised using different techniques (GC–MS, HPLC/MSMS, ATR–FTIR, TGA, and DSC), and its molecular weight was estimated. The results show that the average molecular weight of EPS_{RT7} was approximately 7.0794×10^4 Da and a heteropolysaccharide composed of mannose, glucose, galactose, and xylose (molar ratio, 1:0.5:0.1:0.1) with good thermostability. EPS_{RT7} showed good emulsifying activity against different natural oils and hydrocarbons at high concentrations (2 mg/mL) and at the studied pH range (3.1–7.2). It also presented good emulsifying activity compared to that of commercial emulsifiers. Lastly, EPS_{RT7} showed antioxidant capacity for different free radicals, a lack of cytotoxicity, and antioxidant activity at the cellular level. EPS_{RT7} has promising applications in bioremediation processes and other industrial applications.

Keywords: exopolysaccharide; biodegradation; *Bacillus*; emulsifying; antioxidant

Citation: Sánchez-León, E.; Huang-Lin, E.; Amils, R.; Abrusci, C. Production and Characterisation of an Exopolysaccharide by *Bacillus amyloliquefaciens*: Biotechnological Applications. *Polymers* **2023**, *15*, 1550. https://doi.org/10.3390/polym15061550

Academic Editor: Shashi Kant Bhatia

Received: 3 March 2023
Revised: 17 March 2023
Accepted: 18 March 2023
Published: 21 March 2023

1. Introduction

The dispersion of large amounts of toxic polluting agents to the environment caused by natural or human activities leads to adverse impacts on population and ecosystem health [1]. Conventional remediation techniques such as the use of surfactants have received great attention when counteracting polluting activities [1].

Surfactants are compounds that reduce surface and interfacial tension at the interfaces of liquids, solids and gases in order to create emulsions with liquids [2]. These compounds are highly used in the industry in order to remediate contaminated sites from environmental pollutants such as hydrocarbons [3]. The most popular surfactants are synthetic chemical surfactants [4], which are generally toxic and lack biodegradability, leading to bioaccumulation [5]. Manufacturing these surfactants and their byproducts can adversely impact the environment. Surfactants can be disposed of in rivers or sewage treatment plants, which results in marine ecosystem pollution [6].

In order to replace these compounds, attention is given to compounds that are kinder to the natural environment, such as bioemulsifiers (BEs), biosurfactants (BSs) and exopolysaccharides (EPSs) [7]. EPSs produced by microorganisms are compounds with significant potential in various commercial applications such as the emulsification of various hydrophobic substrates, food, or the pharmaceutical industry [8,9]. They have significant advantages when it comes to biodegradability and effectiveness [10]. Within this category is a group of highly interesting polymers, such as exopolysaccharides from microorganisms (microbial exopolysaccharides (EPSs)). Their physicochemical characteristics are especially interesting to researchers, such as their high molecular weight [11], the presence of different groups in their composition, and their thermostability and biocompatibility [12]. Another factor to consider is the environment where contamination can occur; many of these environments are characterised by extreme conditions such as elevated or low temperatures, alkaline or acidic pH, high pressure, or high saline concentrations. The bioremediation of these sites is typically difficult [13].

The importance of looking for microorganisms in extreme environments is due to the immense biotechnological potential of their exopolysaccharides (EPSs) [14], since they would be suitable in extreme environments. One of the most versatile genera is *Bacillus,* which is found in different ecological niches [15] and can propagate under adverse conditions [16], rendering the study of its EPSs very interesting [17]. An example is the case of EPSs produced by *Bacillus vallismortis* WF4 [18] and *Bacillus tequilensis* GM [19], which showed significant emulsifying activity in essential oils. The EPS of *Bacillus megaterium* also showed significant emulsifying activity in hydrocarbons [20]. Emulsifying activity was also found for the EPS produced by *Bacillus amyloliquefaciens* [21].

Some EPSs have wide pharmaceutical application. This is the case of *Bacillus thermoantarcticus* [22] and *Bacillus velezensis* [23], which both presented effective antifungal activity, while the EPSs of *Bacillus subtilis* [24] and *Bacillus aerophilus* [25] demonstrated antioxidant activity.

EPSs produced by different strains of the same species have very diverse pharmaceutical, biotechnological, and industrial applications. For example, different strains of *Bacillus licheniformis* have immunomodulatory [26], antiviral [27], and anticytotoxic activity [28].

Another factor to take into account is the culture medium used to stimulate the production of exopolysaccharides. Among the most used synthetic surfactants is Tween 80, composed of polyoxyethylene glycol sorbitan monooleate. This is an important nonionic surfactant, as it is economical and highly efficient [29]. For medical applications, Tween 80 has been recently included in some vaccines, such as the influenza and AstraZeneca COVID-19 vaccines, or as a food additive, and was widely tolerated [30]. While Tween 80 was degraded by bacteria and stimulated the biodegradation process [29], EPSs arising from degrading Tween 80 have not been reported. The hypothesis of this work is that a potential EPS produced by an extremophilic bacterium biodegrading Tween 80 would result in a polymer that could act in a wide pH range.

This study aims to produce a novel EPS_{RT7} from the biodegradation of glucose–Tween 80 with the *Bacillus amyloliquefaciens* RT7 strain, and to understand its potential applications. The extremophilic *Bacillus amyloliquefaciens* RT7 strain was isolated from an extreme acidic environment and identified through molecular biology methods. Furthermore, the biodegradation of the strain using different independent carbon sources (glucose, oleic acid, Tween 80, and PGE 200) and the joint biodegradation of glucose–Tween 80 were evaluated with an indirect impedance technique and nuclear magnetic resonance (NMR). EPS_{RT7} was characterised with different analytical techniques (GC–MS, HPLC/MSMS, ATR–FTIR, TGA and DSC) in order to determine the compositional and structural characteristics, and molecular weight of EPS_{RT7}. Lastly, potential applications of the isolated EPS_{RT7}, such as emulsifying activity against different natural oils (olive, sunflower, sesame, and coconut) and hydrocarbons (diesel oil, hexane, toluene), the stability of the emulsion at different pH levels, times, and concentrations, and emulsion efficiency against different commercial emulsifiers (Triton X-100, Tween 20 and SDS) were compared. In addition, in the in vitro antioxidant assays for different free radicals, we studied the cytotoxicity and antioxidant activity of EPS_{RT7} at the cellular level.

2. Materials and Methods

2.1. Chemical and Standards

Olive, sunflower, sesame, and coconut oils (Mercadona, Madrid, MA, ESP). Diesel, hexane, toluene, trypticase soya agar (TSA), dextrans standard, 1,1-diphenyl-2-picryl-hydrazyl radical (DPPH), H_2O_2, salicylic acid, Dulbecco's modified Eagle's medium (DMEM), polyoxyethylene sorbitan monolaurate (Tween 20), polyoxyethylene glycol sorbitan monooleate (Tween 80), sodium dodecyl sulphate (SDS), 2-[4-(2,4,4-trimethylpentan-2-yl)phenoxy]ethanol (Triton X-100), pyrogallol, HCl, ascorbic acid (Vc), fetal bovine serum (FBS), L-glutamine, penicillin, streptomycin (Sigma-Aldrich, St. Louis, MO, USA). JetQuick kit (Genomed, Leesburg, VA, USA), Sephadex G-100 column (Aldrich Chemical Company, Inc., Milwaukee, WI, USA), trifluoroacetic acid (TFA)(Aldrich®® Schnelldorf, Germany), 3-(4,5-dimethylthiazol-2-yl)-2,5-diphenyltetrazolium bromide (MTM) (GE Healthcare, Uppsala, Sweden).

2.2. Isolation of Bacterial Strain and PCR Amplification

Extremophilic bacterial strain RT7 was isolated from the sediments of the river source in Río Tinto (Huelva), Spain (37°43′19″ N 6°33′03″ W). First, 10 mL of NaCl 0.6 M was added to 1 g of sediment, and the mixture was serially diluted (10-fold). Aliquots of 100 μL were inoculated on trypticase soya agar (TSA) plates and stored overnight at 30 °C. The isolated strain was preserved at −80 °C in 30% glycerol.

For the identification of the strain, PCR amplification was conducted as described by Abrusci et al. [31]. Genomic DNA was extracted from bacterial cells using an UltraClean microbial DNA isolation kit. The purified genomic DNA was used as a template to amplify the 16S rRNA gene with PCR using primers 27F (5′-AGA GTT TGA TC (C/A) TGG CTC AG-3′) and 1492R (5′-TAC GG(CT) TAC CTT GTTACG ACT T-3′). PCR amplifications were carried out in a Thermal Cycler 2720 (Applied Biosystems). The following were performed: an initial denaturing step at 94 °C for 5 min, the completion of 30 cycles of 1 min at 94 °C, 1 min at 56 °C, and 3 min at 72 °C; and a final extension of 72 °C for 10 min [32]. Amplicons were purified using the JetQuick kit, and sequenced using the ABI PRISM Big Dye Terminator Cycle Sequencing Ready Reaction Kit (ABI) and an Applied Biosystem ABI 310 (PE Applied Biosystems, Foster City, CA, USA) automated sequencer [33]. The obtained sequences were compared to those in the GenBank database using the BLAST program (National Center for Biotechnology Information). The selected sequences were aligned with CLUSTAL X [34].

2.3. Biodegradation, Colony-Forming Units (CFUs)/mL, pH, and EPS Production

2.3.1. Biodegradation, Colony-Forming Units (CFUs)/mL, and pH

The biodegrading bacterium (strain RT7) was studied via indirect impedance measurements, performed at 30 °C. The aerobic biodegradation was prepared as previously described by Abrusci et al. [35,36]. Minimal growth medium (MGM): g/L: K_2HPO_4 0.5, KH_2PO_4 0.04, NaCl 0.1, $CaCl_2$ $2H_2O$ 0.002, $(NH_4)_2$ SO_4 0.2, $MgSO_4$ $7H_2O$ 0.02, $FeSO_4$ 0.001. Each of the carbon sources was added separately to the medium (MGM): glucose (4 g/L), oleic acid (1 g/L) and surfactants (1 g/L) (polysorbate 80 (Tween 80), and polyethylene glycol (PEG) 200). Polyethylene glycol PEG-200 and oleic acid were used as controls or model compounds in the biodegradation studies. In addition, glucose and Tween 80 were added together (pH adjusted to 7.0).

The bioassay measurements were performed as described by Abrusci et al. [36] in 7 mL bioreactors, introducing 1.5 mL of bacterial suspension. These bioreactors were introduced into disposable cylindrical cells of 20 mL filled with 1.5 mL of 2 g/L KOH aqueous solution, and impedance was measured using four stainless-steel electrodes on a Bac-Trac 4300 apparatus (SY LAB Geräte GmbH, Neupurkerdorf, Austria). The typical measurement error was 1–2%. The relative change in the KOH solution impedance value was monitored with the device every 20 min and was converted into carbon dioxide concentration by using a calibration curve of variation in impedance against CO_2 concentration.

The biodegradation percentage of different carbon sources was computed on the basis of the ratio between the cumulative amount of CO_2 actually generated by biodegradation at time t and the theoretical carbon dioxide amount, which assumed that all the carbon in the glucose and polysorbate structures was converted into CO_2 (Formula (1)).

$$\% \text{ Biodegradation} = ([CO_2]\text{Prod}/[CO_2]\text{Theor.}) \times 100 \quad (1)$$

To continue with the experiments, the most effective carbon source was chosen. The colony-forming units (CFUs) were evaluated via dilution plating incubated at 30 °C for 72 h with a TSA agar medium. A Thermo Orion pH Meter Model 2Star (Thermo Scientific, Asheville, NC, USA) was used to measure the pH values during a 72 h fermentation period.

To evaluate structural changes on Tween 80 after biodegradation, nuclear magnetic resonance (1H-NMR) was recorded in a deuterated chloroform CDCl3 solution on a Varian INOVA-400 instrument (Varian Inc., Palo Alto, CA, USA) at 400 MHz. In the case of the biodegraded products, the compound mixture was filtered to eliminate cells using a centrifuge (0.22 μm Millipore, Merck, Darmstadt, Germany, DEU). The residues were dried and dissolved in CDCl3.

2.3.2. EPS Production and Purification

The strain was inoculated on trypticase soy agar (TSA) medium and incubated for 24 h at 30 °C. The strain was later transferred to an MGM medium in 100 mL flasks filled with 20 mL with an initial inoculum of 2.5×10^7 CFU/mL ($OD_{550\text{ nm}}$) measured with a spectrophotometer (Biowave II). The flasks were incubated at 30 °C for 24 h at 110 rpm (Orbitek LJEIL model) [27]. Subsequently, 10 mL of the broth was transferred to flasks containing 1000 mL of MGM with glucose–Tween 80. The flasks were incubated at 30 °C for 72 h at 110 rpm. The tests were independently repeated three times.

The cultures were centrifuged with a DuPont Sorvall RC-5 centrifuge for 30 min at 4 °C at 13.154× *g*. The extracted supernatant was precipitated with three times the volume of ethanol (−80 °C). The EPS was then dialysed for 48 h at 4 °C with Milli-Q water, and lyophilised with a Flexy-Dry MPTM freeze 150 dryer. Subsequently, the dry weight of the EPS was measured.

The purity of the EPS (10 mL, 10 mg/mL) was evaluated with a DEAE-52 anion exchange column (2.6 × 30 cm), and deionised water was used for elution. For this, the used eluents were different concentrations of NaCl (0.2–1.5 M) at a flow rate of 1 mL/min. The phenol–sulfuric acid method was used to monitor the eluents [37]. The collected fractions were lyophilised, resulting in an EPS that was named EPS_{RT7}.

2.4. *Characterisation of EPS_{RT7}*

2.4.1. Monosaccharide Composition

The molecular weight of EPS_{RT7} was obtained via gel filtration chromatography with a Sephadex G-100 column (1.6 × 50 cm) eluting with 0.2 mol^{-1} NaCl solution at a flow rate of 1 mL/min. Standard reference dextrans (5–80 KDa) were used [38].

The monosaccharide composition was determined with Bruker gas chromatography (EVOC GC–TQ) combined with mass spectrometry using the procedure described by Huang-Lin et al. [39]. For this, EPS_{RT7} hydrolysis was conducted with 0.5 M trifluoroacetic acid (TFA) at 120 °C for 2 h. Subsequently, the samples were treated with N_2. Galactose, glucose, arabinose, fructose, and xylose were used as the standard. To determine the presence of amino acids and glucuronic acid, an Agilent Technologies 1100 series 6410B (HPLC/MSMS) and an ACE Excel 3 C18-Amide column were used as the stationary phase with a mobile phase of 0.1% formic acid in water. This was conducted at 40 °C (TQ, Waldbronn, Germany) with a flow rate of 0.2 mL/min.

2.4.2. Attenuated Total Reflectance/FT-Infrared Spectroscopy (ATR/FTIR)

The IR spectra of the EPS_{RT7} were recorded using a BX– 187 FTIR spectrometer (Perkin Elmer, Waltham, MA, USA) with an ATR attachment (Pike Technologies, Cottonwood, AZ, USA). The spectra were carried out from accumulating 32 scans at a 4 cm^{-1} resolution over a region from 400 to 4000 cm^{-1} [40].

2.4.3. Thermogravimetric (TGA) and Differential Scanning Calorimetric (DSC) Analysis

TGA was performed with a TGA Q500 (TA Instruments, New Castle, DE, USA). The EPSRT7 (1–3 mg) was placed into a platinum crucible and subjected to temperatures ranging from 20 to 800 °C at a heating rate of 10 °C/min under atmospheric pressure. DSC measurements were conducted using a DSC Q100 (TA Instruments, New Castle, DE, USA). The EPS_{RT7} (0.5–2 mg) was placed in an aluminium pan with its lid removed. The pans were heated from 20 to 600 °C at a rate of 10 °C/min. Data were analysed using TA Universal Analysis software [41].

2.5. *Emulsifying Activity Assesment*

The emulsifying activity of the EPS_{RT7} was measured at different pH levels (7.2, 5.1, 3.1) and at various concentrations (0.5, 1, and 2 mg/mL), using the method described by Meneghine et al. [42]. The experiment involved mixing 1.5 mL of an oil phase (olive oil, sunflower oil, sesame oil, coconut oil, diesel oil, hexane, or toluene) with an aqueous phase of 1.5 mL. For the aqueous phase, the commercial emulsifiers of Tween 20, sodium dodecyl sulphate (SDS), Triton X-100, and EPS_{RT7} were compared. The tubes were stirred for 2 min at 2400 rpm using a vortex. Emulsification indices E24, E48, and E168 were measured after 24, 48, and 168 h, respectively. Formula (2) was used to calculate the emulsification indices:

$$E\,[\%] = HEL/HT \times 100 \quad (2)$$

where HEL (mm) represents the height of the emulsion layer, and HT (mm) refers to the total height.

2.6. *Antioxidant Activity Assesments*

To evaluate the antioxidant properties of EPS_{RT7}, several tests were conducted using 1,1-diphenyl-2-picryl-hydrazyl radical (DPPH•), hydroxyl radical (•OH), and superoxide anion (O_2^{-}•) as indicators. Ascorbic acid (Vc) was used as the positive control. EPS_{RT7} was prepared in concentrations ranging from 0.1 to 10 mg/mL. Absorbance measurements were performed using a FLUOstar Omega BMG LABTECH (Aylesbury, UK) spectrophotometer for DPPH ($OD_{525\ nm}$), OH ($OD_{510\ nm}$), and O_2^{-}($OD_{325\ nm}$).

2.6.1. DPPH Radical Scavenging Activity

Research on the DPPH scavenging activity of EPS_{RT7} followed the procedure described by Niknezhad et al. [43]. First, 50 μL of EPS_{RT7} at different concentrations was mixed with 100 μL of DPPH (100 μM DPPH–ethanolic solution). The mixtures were stirred and left to incubate in the dark at 25 °C. The absorbance was measured after 30 min.

Formula (3) was used to determine the percentage of radical-scavenging activity for DPPH.

$$\text{DPPH scavenging activity } [\%] = [\,1 - (A_1 - A_2)/A_0] \times 100 \quad (3)$$

where A_1 represents the reaction mixture, A_2 refers to the reaction mixture without DPPH, and A_0 denotes the reaction mixture with DPPH and without EPS_{RT7}.

2.6.2. OH Radical Scavenging Activity

The $FeSO_4$–salicylic acid method described by Sun et al. [44] was used to determine the hydroxyl radical scavenging activity of EPS_{RT7}. The mixtures contained a $FeSO_4$ solution (9 mM, 40 μL), 40 μL of a 9 mM ethanol–salicylic acid solution, EPS_{RT7} diluted at various concentrations (40 μL), and H_2O_2 (8.8 mM, 40 μL). The mixtures had then been incubated at 37 °C for 30 min before absorbance was measured. Formula (4) was used to determine the percentage of hydroxyl radical scavenging activity.

$$\text{Hydroxyl radical scavenging activity } [\%] = [\,1 - (A_1 - A_2)/A_0] \times 100 \quad (4)$$

where A_1 represents the reaction mixture, A_2 refers to the reaction mixture without salicylic acid, and A_0 denotes the reaction mixture with salicylic acid and without EPS_{RT7}.

2.6.3. O_2^- Scavenging Activity

The superoxide scavenging activity of EPS_{RT7} was assessed as per Balakrishnan et al. [45]. In this method, 0.3 mL of different EPS_{RT7} concentrations was mixed with 2.6 mL of phosphate buffer (50 mM, pH 8.2) and 90 μL of pyrogallol (3 mM), and dissolved in HCl (10 mM). The absorbance was then monitored at 0 and 10 min. Formula (5) was used to determine the percentage of superoxide scavenging activity:

$$\text{Superoxide scavenging activity [\%]} = 1 - [(A_{10}/C_{10}) - (A_0/C_0)] \times 100 \quad (5)$$

where A_0 and A_{10} represent the reaction mixture at 0 and 10 min, respectively; C_0 and C_{10} represent the reaction mixture without pyrogallol at 0 and 10 min, respectively.

2.7. *Exopolysaccharide Toxicity Evaluation*

2.7.1. Cell Culture

HeLa cells, which are human epithelial cells derived from cervical carcinoma, were obtained from CLS (Cell Line 76 Service, BW, Eppelheim, Germany) and chosen as the reference cell line to assess the toxicity of EPS_{RT7}. The cells were cultured using Dulbecco's Modified Eagle's Medium (DMEM) and added to 10% fetal bovine serum (FBS), 2 mM L-glutamine, penicillin (100 IU/mL), and streptomycin (100 μg/mL) under a 5% CO_2 atmosphere at 37 °C [46].

2.7.2. Cytotoxicity Assay

HeLa cells were cultured in a 24-well culture plate at a density of 5×10^5 cells per well. Each well was treated with 100 μL of EPS_{RT7} at various concentrations (0–400 μg/mL) for 24 h. The toxicity of the exopolysaccharide was determined by measuring the reduction in the MTT reagent (3-[4,5-dimethyl-thiazol-2-yl]-2,5-diphenyltetrazoliumbromide) to formazan [46,47]. The optical density at a wavelength of 590 nm was recorded using a microplate reader.

The cytotoxicity of EPS_{RT7} on HeLa cells was assessed using Formula (6):

$$\text{Cell viability [\%]} = (A_1/A_2) \times 100 \quad (6)$$

where A_1 refers to cells treated with EPS_{RT7} and the MTT solution, and A_2 refers to cells without any treatment with the MTT solution.

2.8. *Evaluation of Antioxidant Ability on HeLa Cells*

2.8.1. Injury Inducement Model

The methodology used to create an injury model for HeLa cells was based on the procedure outlined by Huang-Lin et al. [39]. HeLa cells were seeded at a density of 5×10^4 cells per well for 24 h. The medum was then removed and replaced with 100 μL of various concentrations of H_2O_2 (0.25–2 mM) for 1 h at 37 °C under a 5% CO_2 atmosphere. Following exposure, the H_2O_2 solution was removed, and a fresh medium was added to the wells. Cell viability was measured using the MTT method as described in Section 2.7.2.

HeLa cell viability was computed using Formula (7):

$$\text{Cell viability (\%)} = (A_1/A_2) \times 100 \quad (7)$$

where A_1 represents the absorbance of HeLa cells treated with H_2O_2 and the MTT solution, while A_2 represents the absorbance of HeLa cells that were not subjected to any treatment with the MTT solution.

2.8.2. Evaluation of Protective EPS_{RT7} Effect on HeLa Cells from Oxidative Stress

EPS_{RT7}'s ability to protect HeLa cells against oxidative stress was evaluated following the methods described by Huang-Lin et al. [39]. HeLa cells were seeded at a concentration of 5×10^4 cell/well and incubated for 24 h. After that, the DMEM solutions were withdrawn and substituted with EPS_{RT7} diluted in DMEM at different concentrations (25–400 μg/mL). After 1 h, the EPS_{RT7} solutions were removed, and a new medium containing 2 mM of H_2O_2 was added and incubated for 1 h more. The MTT method described in Section 2.7.2 was used to determine cell viability. As a positive control, ascorbic acid (20 mg/mL) was used.

HeLa cell viability was calculated with Formula (8):

$$\text{Cell viability } [\%] = (A_1/A_2) \times 100 \tag{8}$$

where A_1 refers to cells that were treated with both H_2O_2 and EPS_{RT7}, and subsequently exposed to the MTT solution; A_2 refers to cells that did not receive any treatment and were exposed to the MTT solution.

2.9. Statistical Analysis

The experiments were conducted three times, and statistical analysis was performed using the Statistical Package for the Social Sciences (SPSS), version 21. The analysis of variance (ANOVA) test was used for statistical comparison, and statistical significance was at $p < 0.05$.

3. Results and Discussion

3.1. Bacterial Identification, Biodegradation, and EPS Production

The bacterial strain was isolated as described in Section 2.2, and identified after PCR amplification and sequencing using the 16S rDNA sequence. The 16S rDNA sequences were compared with those in the GenBank database and showed that the isolated strain was *B. amyloliquefaciens* RT7 (accession number, AB300821) with a similarity of 98%.

Bacillus amyloliquefaciens RT7 is a Gram-positive, endospore-forming bacterium isolated from the sediments of Rio Tinto (Huelva, Spain), which is one of the most acidic rock drainage fluvial–estuarine systems in the world [48]. This species is ubiquitous and adapts to very different ecological environments. This gives it great versatility in the biodegradation of different compounds, such as waste from the petrochemical industry, where strain *B. amyloliquefaciens* W1 [49] was able to degrade benzene, toluene, ethylbenzene, and xylene (BTEX), also in phenol-contaminated wastewater, which *B. amyloliquefaciens* WJDB-1 was able to biodegrade [50]. In addition, it was successfully used in various technological applications. For example, the *B. amyloliquefaciens* BRRI53 strain could stimulate plant growth [51], and the *B. amyloliquefaciens* BPRGS strain had high flocculant activity [52].

The biodegradation of the different independent carbon sources (glucose, oleic acid, Tween 80, and PEG 200) and the joint biodegradation of glucose–Tween 80 are shown in Figure 1a. The results indicate that the *B. amyloliquefaciens* RT7 strain biodegraded glucose by 27–78%, oleic acid by 27–60%, Tween 80 by 23–58%, PEG-200 by 5–2%, and the combination of glucose–Tween 80 by 46–86% at intervals of 24 and 72 h. The *B. amyloliquefaciens* RT7 strain was the most effective at biodegrading glucose–Tween 80.

The growth of *B. amyloliquefaciens* RT7, pH values, medium biodegradation, and exopolymer production (EPS) at 30 °C using the combination of glucose–Tween 80 as a carbon source is shown in Figure 1b. Cell growth peaked (7.53 log CFU/mL) after 30 h. During the process, the medium was not acutely acidified (from pH 7 to pH 6.5).

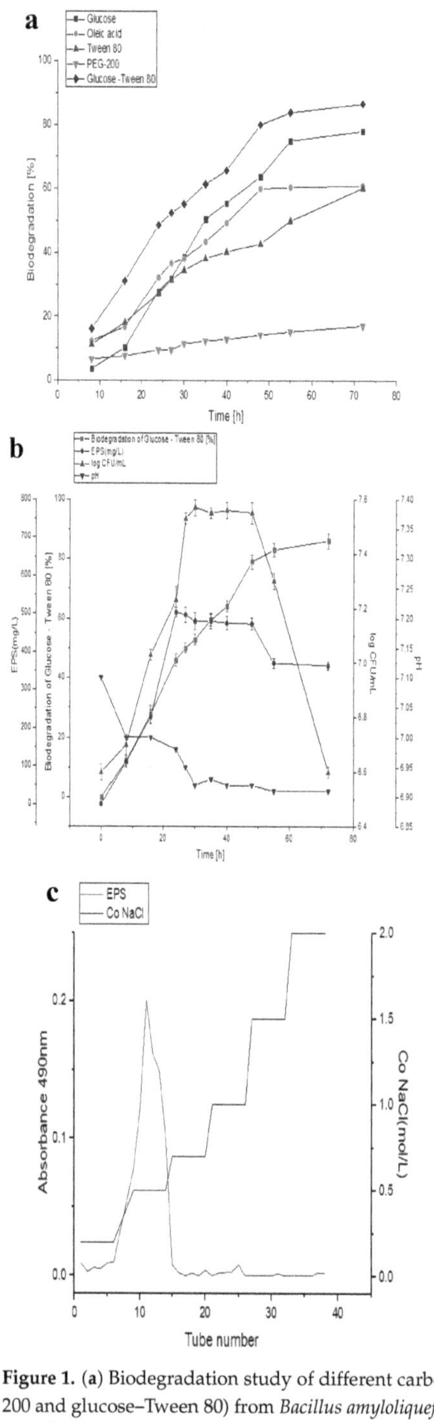

Figure 1. (**a**) Biodegradation study of different carbon sources (glucose, oleic acid, Tween 80, PEG 200 and glucose–Tween 80) from *Bacillus amyloliquefaciens*. (**b**) Optimisation of the production EPS from *B. amyloliquefaciens*. (**c**) Elution curve obtained from the purification of EPS_{RT7} under MGM with glucose–Tween 80.

To confirm that Tween 80 had undergone structural changes during biodegradation, after the bioassays, the residues were analysed with proton nuclear magnetic resonance (1H-NMR; Figure 2a). The biodegradation of oleic segments decreased the intensity of the peaks of the aliphatic protons (2.5–1.0 ppm), and caused the disappearance of the double-bond signal at 5.35 ppm and the methylene protons next to the oleic ester group at 4.23 ppm. We also found a simplification of the proton signals corresponding to the PEG fragments (HPEG–3.64, 3.77, and 3.99–4.16 ppm), indicating a decrease in the length of the initial PEG fragments after 72 h of the bioassay. This shows that the biodegradation of Tween 80 [53] was effective and likely had a positive effect on the biodegradation of glucose. Similar results were obtained with *Bacillus amyloliquefaciens* [54] when Tween 80 was combined with hydrocarbons, favouring the biodegradation of the latter. This was also confirmed in other species of the *Bacillus* genus, such as *B. subtilis* ZL09-26, where the biodegradation of phenanthrene was more efficient in the presence of Tween 80 [55]. This could have been due to the fact that Tween 80 could influence the permeability of the membrane, improving the expression of the proteins [56] and favouring the more efficient incorporation of organic nutrients such as glucose [57,58].

On the other hand, the biodegradation of this combination (glucose–Tween 80; Figure 1b) with *Bacillus amyloliquefaciens* RT7 resulted in a high production of EPS. The maximal production of the exopolysaccharide, namely, 490 mg/L, occurred at 24 h, during the exponential growth phase. However, this was previously observed as usually taking place at the beginning of the stationary phase [27,39].

The EPS production of the RT7 strain was higher than that for other strains of *B. amyloliquefaciens*. *B. amyloliquefaciens* p16 used a nutrient broth of glucose, peptone, and yeast as an energy source, and only had an EPS production of 223.87 mg/L [59]. The use of Tween 80 by *Bacillus amyloliquefaciens* RT7 as a carbon source not only allowed for greater efficiency in the use of glucose, but also effectively accelerated the synthesis of the exopolysaccharide produced by the RT7 strain. Similar processes were described in the synthesis of natural compounds, such as fengycin, accelerating their production when Tween 80 was used [60]. In addition to this, similar results were found in other genera, such as *Lactobacillus plantarum*, where Tween 80 not only facilitated the entry of nutrients, but also stimulated greater EPS production [61].

The EPS obtained from glucose and Tween 80 as an energy source was purified and showed a single characteristic peak of exopolysaccharides with high purity (Figure 2b). The result of the obtained fraction from the purified exopolymer was named EPS_{RT7}. The estimated molecular weight of EPS_{RT7} was about 7.0794×10^4 Da (Figure 2c), which fell within the typical molecular weight range for heteropolysaccharides (4×10^4 and 6×10^6 Da) [62]. EPS_{RT7} presented a high molecular weight in comparison with that of other EPS produced without the presence of Tween 80, such as strains of *B. amyloliquefaciens* GSBa-1, with an EPS composed of glucose with a molecular weight of 5.4×10^4 Da [63], and *B. amyloliquefaciens* 3MS [64], whose EPS comprised glucose, galactose, and glucuronic acid with a molecular mass of 3.76×10^4 Da.

3.2. *EPS_{RT7} Compositional Analysis and Characterisation*

3.2.1. GC–MS Analysis for EPS_{RT7}

Gas chromatography (GC–MS) analysis (Figure 3a) exhibited 4 peaks that corresponded to monosaccharides mannose (α-D-mannose and β-D-mannose), glucose (α-D-glucose and β-D-glucose), α-D-galactose, α-D-xylose, with a molar ratio of 1:0.5:0.1.0, respectively. This revealed that the polymer was a heteropolysaccharide with the absence of uronic acids and amino acids. *B. amyloliquefaciens* strains produce a wide variety of different EPS compositions. *B. amyloliquefaciens* strain JN4 [65] produced an EPS that was composed of fructose and glucose with a molar ratio of 46.1:1. *B. amyloliquefaciens* strain C-1 [66] produced two different EPSs: EPS-1 with glucose, mannose, galactose, and arabinose at a 15:4:2:1 ratio, and EPS-2, composed of glucose/mannose at a 3:1 proportion. On the other hand, heteropolysaccharide EPS-K4, isolated from the *B. amyloliquefaciens* DMBA-

K4 strain, contained rhamnose, mannose, glucose, and glucuronic acid at a molar ratio of 23.65:40.09:17.68:11.42 [67].

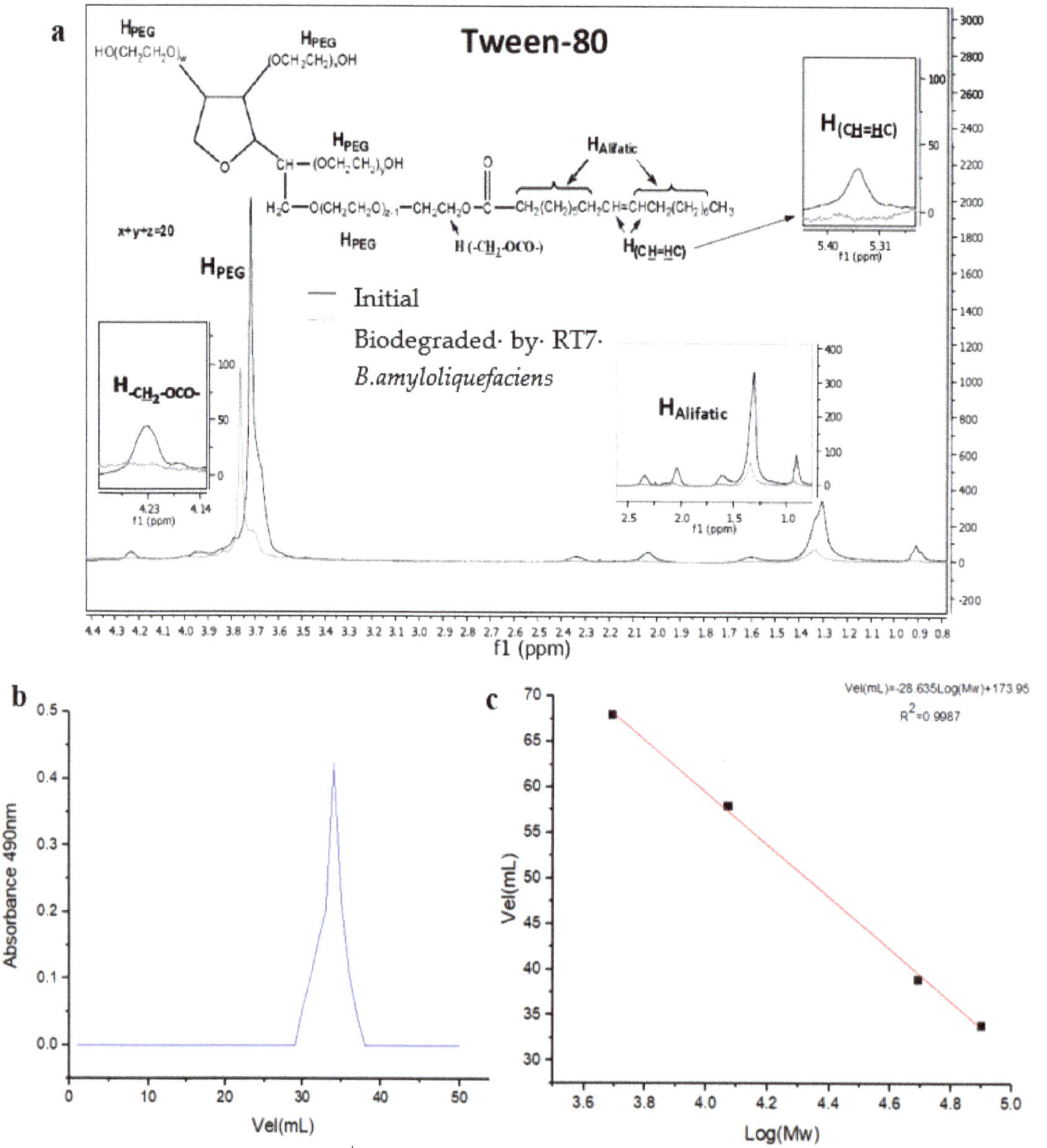

Figure 2. (**a**) The 1H-NMR spectra of Tween 80 initially and after 72 h of biodegradation. (**b**) The elution curve of EPS_{RT7} by Sephadex G-100 gel filtration. (**c**) The standard curve of relative molecular weight (Mw).

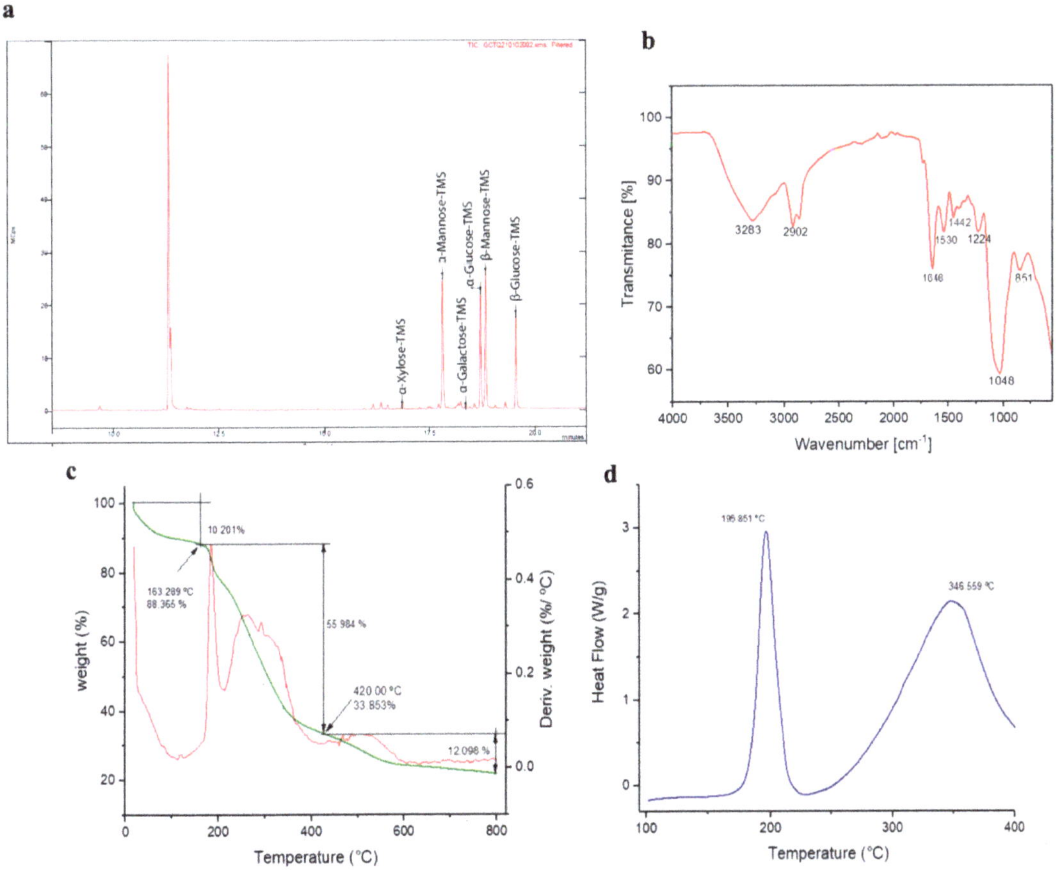

Figure 3. Characterisation analysis of EPS_{RT7}. (**a**) GC–MS; (**b**) ATR–FTIR spectra; (**c**) thermogravimetric analysis (TGA); (**d**) differential scanning calorimetry (DSC).

3.2.2. ATR–FTIR Analysis for EPS_{RT7}

In addition, ATR–FTIR spectra (Figure 3b) obtained from EPS_{RT7} produced with *B. amyloliquefaciens* RT7 showed peaks between 4000 and 400 cm^{-1}, a characteristic of carbohydrates. A band at 3283 cm^{-1} was attributed to the hydroxyl stretching vibration of the polysaccharide [68], whereas the band at 2902 cm^{-1} was ascribed to C–H stretching vibration [69]. The peaks at 1664 and 1442 cm^{-1} were attributed to the stretching vibration of C=O, which is characteristic of a carboxyl group [70,71]. The peak at 1530 cm^{-1} was ascribed to C–O vibration [72]. The region at 950–1250 cm^{-1} corresponded to CO and C–O–C stretching vibrations in carbohydrates. The absorption at 1224 cm^{-1} could have been the pyranose ring of monosaccharides in EPS_{RT7}. The strong peak at 1048 cm^{-1} was attributed to the C–O–C glycosidic bond vibration, which indicated the pyranose configuration [73]. The band at 851 cm^{-1} suggested that the EPS_{RT7} contained the α and ß configurations of the glucose unit [74].

The spectrum confirmed typical characteristics of a polysaccharide. The vibrational peaks resembled the peaks observed for the EPSs of other strains. In the case of EPS from *B. amyloliquefaciens* RK3, the polymer contained mannose and galactose [75]. These differences revealed that the exopolysaccharides of the different strains of *B. amyloliquefaciens* had a very varied composition of monosaccharides. This heterogeneity may have been due to both nutritional and environmental factors.

3.2.3. Characterisation of the Thermal Properties of EPS_{RT7}

The thermal stability of EPSs is an important characteristic for their commercial utilisation. The thermal decomposition curve (TGA) of EPS_{RT7} is shown in Figure 3c. The first step, with initial weight loss of 10.20%, was observed at around 25 and 163 °C, mainly due to the moisture loss in EPS_{RT7}. The second step had a weight loss of approximately 55.9% which reached the maximum at around 420 °C. The depolymerisation of the polysaccharide and the thermal chemical-bond scission occurred, accompanied by the dehydration of sugar units. Lastly, the EPS presented a gradual weight loss of approximately 12.09%, reaching equilibrium with only 28% of the remaining residue. Similar results associated with thermal decomposition were found in other strains of *B. amyloliquefaciens*. This was the case of the *B. amyloliquefaciens* GSBa-1 strain [63], whose EPS was formed by glucose. In the first step, 7.44% (50 to 160 °C) weight loss for water was observed. In the second step, it was 43% (around 400 °C) of mass loss. In *B. amyloliquefaciens* BPRGS [52], the first step showed 10% weight loss for water (0 to 180 °C), and the second step showed 20.21% weight loss (250 to 550 °C). On the other hand, the DSC thermogram of EPS_{RT7} (Figure 3d) showed two melting peaks at 195.8 and 346.5 °C that conferred higher thermostability. The appearance of two melting peaks was directly related to the heterogeneity in the composition of EPS_{RT7} sugars. This was in contrast with the case of *B. amyloliquefaciens* LPL061 [21]. This strain produced two distinct EPSs, both composed of mannose and glucose (EPS1 and EPS2). The two EPSs had a single melting peak of 224.09 °C (EPS1) and 301.09 °C (EPS2). This suggests that compositions with more distinct sugars may result in more than one melting points. Therefore, EPS_{RT7} with various sugars presented an important advantage over polymers with fewer sugars, since its thermostability increased, which is an important factor in various industrial applications [21,76].

3.3. *Biotechnological Applications*

3.3.1. Emulsifying Activity

The emulsification behaviour of EPS_{RT7} was investigated at different concentrations (0.5, 1, and 2 mg/mL), pH levels (7.2, 5.1, 3.1), and at different times (24 h (E24), 48 h (E48), and 168 h (E168)), with natural oils (olive, sunflower, sesame, and coconut) and hydrocarbons (diesel oil, hexane, toluene) (Figures 4 and 5). EPS_{RT7} was also tested with three commercial emulsifiers (Triton X-100, Tween 20, and SDS).

For natural oils, concentration has a statistically significant effect on emulsifying activity. Concentrations of 0.5 mg/mL resulted in nonsignificant emulsification activity across the board, with the only exception being sesame oil at pH 3.1 ((E24 86.6%), (E48 73.3%)). A concentration of 1 mg/mL generally resulted in emulsifying activity with some exceptions, but the optimal concentration was 2 mg/mL for all oils and pH and time combinations. pH had different effects on emulsifying activity depending on the oil. For olive oil (2 mg/mL), the exopolysaccharide presented significant emulsifying activity at all pH levels and studied times, with pH 3.1 being where the emulsifying activity reached its maximum (E24 (85.7%), E48 (71.4%), E168 (57.1%)). For sunflower oil (2 mg/mL), there was significant emulsifying activity for all pH levels; however, at pH 5.1, emulsification was not maintained at E168 ((E24 (85%), E48 (72%), E168 (Nd)). Sunflower oil had its highest emulsifying activity at pH 3.1 (E24 (87.5%), E48 (86.6%), E168 (62.5%)).

The EPS was initially most effective in emulsifying sesame oil at pH 3.1, although this was not maintained at 168 h (E24 (100%), E48 (82.4%), E168 (46.6%)). However, at pH 7.2, emulsification activity was maintained (E24 (86.6%), E48 (73.3%), E168 (60%)), and it was not successful at a pH 5.1. For coconut oil, emulsifying activity was similar at different pH levels for E24, but for E48, pH 5.1 was most effective, followed by pH 3.1; for E168, only pH 3.1 still presented some emulsification. Similar results were obtained with the EPS of the *B. amyloliquefacens* ZWJ strain [77], where the optimal concentration was 1.5 mg/L for two natural oils (olive oil (96.2%), sunflower oil (76%)). In the case of *B. amyloliquefaciens* LPL061 [21], the emulsifying activity was lower than that presented by EPS_{RT7}. Its EPS was only tested at a concentration of 1 mg/mL. The EPS of this strain showed emulsifying activity with the natural oils, which did not exceed 66% (olive oil (58.6%), sunflower oil (65.8%), peanut oil (60.3%), rice oil (58.5%)).

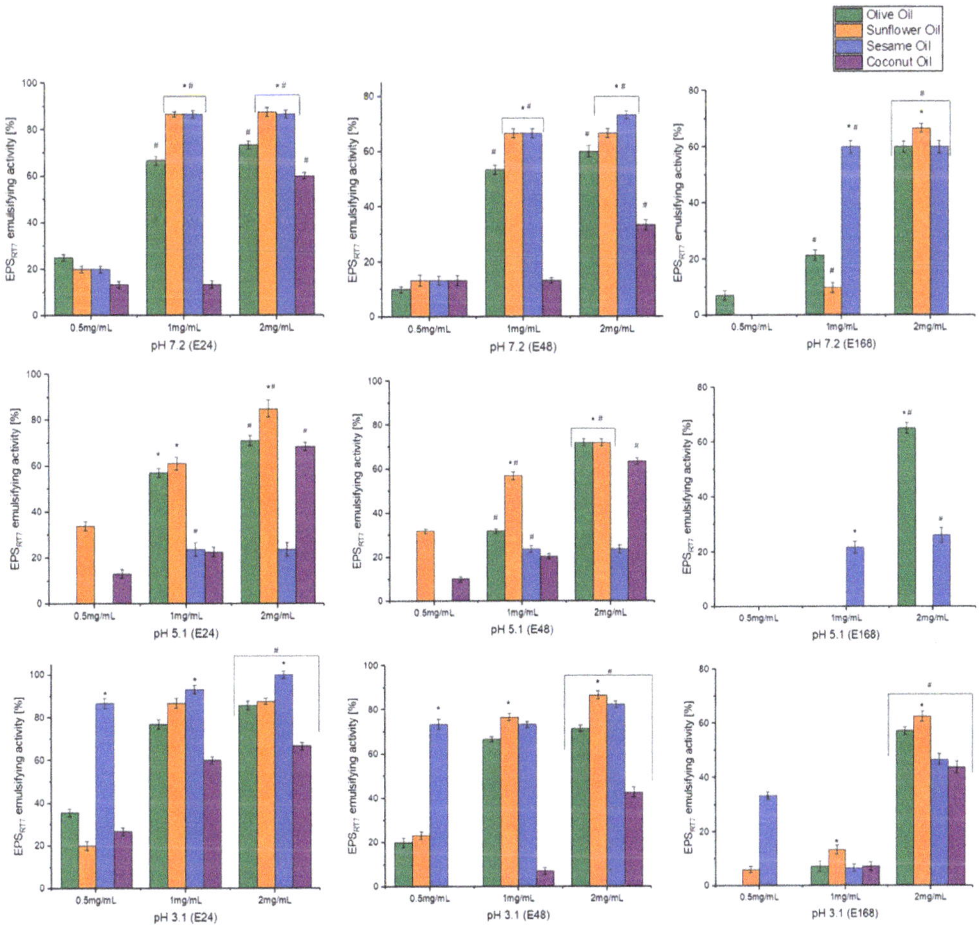

Figure 4. Emulsifying activity for natural oils with different EPS_{RT7} concentrations (0.5, 1, 2 mg/mL) and pH levels (7.2, 5.1, 3.1). Emulsion percentages of EPS_{RT7} with different oils used at 24, 48, and 168 h of study are also shown. *, statistical differences between oils for each concentration ($p < 0.05$). #, statistical differences between concentrations ($p < 0.05$).

Similarly, for hydrocarbons, a concentration of 2 mg/mL was significantly the most effective for emulsification. Diesel, (pH 7.2 (E24 (57.1%), E48 (53.3%), E168 (53.3%)) and pH 3.1, (E24 (56.6%), E48 (56.6%), E168 (57.1%)), and toluene (pH 3.1 (E24 (53.8%) E48 (50%)) presented the highest emulsification activity, which was maintained across different pH levels. Hexane only presented emulsification activity of over 50 for 2 mg/mL and pH 5.1, measured at E24. There are few studies that found EPSs produced by other strains of *B. amyloliquefaciens* with the ability to emulsify hydrocarbons. The An6 strain of *B. amyloliquefaciens* [78] produced a biosurfactant that had emulsifying activity of 80% with diesel at a pH range of 5.0–9.0, but this biosurfactant was not confirmed as an EPS, as its chemical composition was inconclusive. This ability has previously been found in other genera. *Bacillus subtilis* AF17 [79] produced an EPS that was capable of emulsifying diesel by 17%, hexane by 72%, and toluene by 84% (5 mg/mL). The stabilisation of the EPS emulsions was specific for certain hydrophobic compounds [80,81].

The EPS was compared to commercial emulsifiers (Triton X-100, Tween 20 and SDS) for both natural oils and hydrocarbons, and had similar emulsifying activity to theirs at pH 7.2 (Figure 6). At a concentration of 0.5 mg/mL, both EPS_{RT7} and commercial emulsifiers were not effective (below 50%) except for Tween 20 for olive oil (E24 87.3%), sunflower oil (E24 50.0%), and coconut oil (E24 55.8%). For the concentration of 1 mg/mL, EPS_{RT7} presented significantly higher emulsification than that of SDS, except for coconut oil (E24 47.62%) and hexane (E24 52.38%), higher than that of Triton X-100 except for olive oil (E24 86%) and coconut oil (E24 54.5%), and higher than that of Tween 20 except for olive oil (E24 95%), sunflower oil (E24 90%), and coconut oil (E24 60%). For the concentration of 2 mg/mL, EPS_{RT7} presented significantly higher emulsification than that of SDS except for diesel (E24 59.38%) and hexane (E24 57.89%), higherthan that of Triton X-100 for all natural oils and hydrocarbons, and higher than that of Tween 20 except for olive oil (E24 90%) and coconut oil (E24 66.7%).

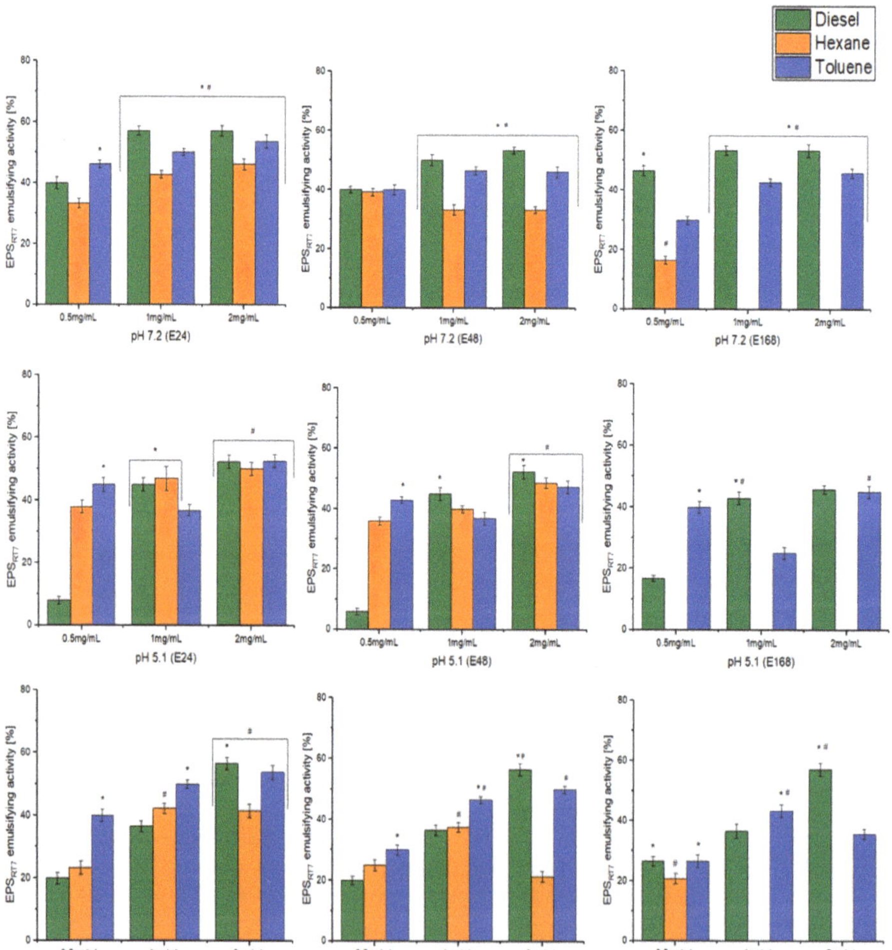

Figure 5. Emulsifying activity for hydrocarbons with different EPS_{RT7} concentrations (0.5, 1, 2 mg/mL) and pH levels (7.2, 5.1, 3.1), measured at 24, 48, and 168 h of study. *, statistical differences between different hydrocarbons for each concentration ($p < 0.05$). #, statistical differences between different concentrations ($p < 0.05$).

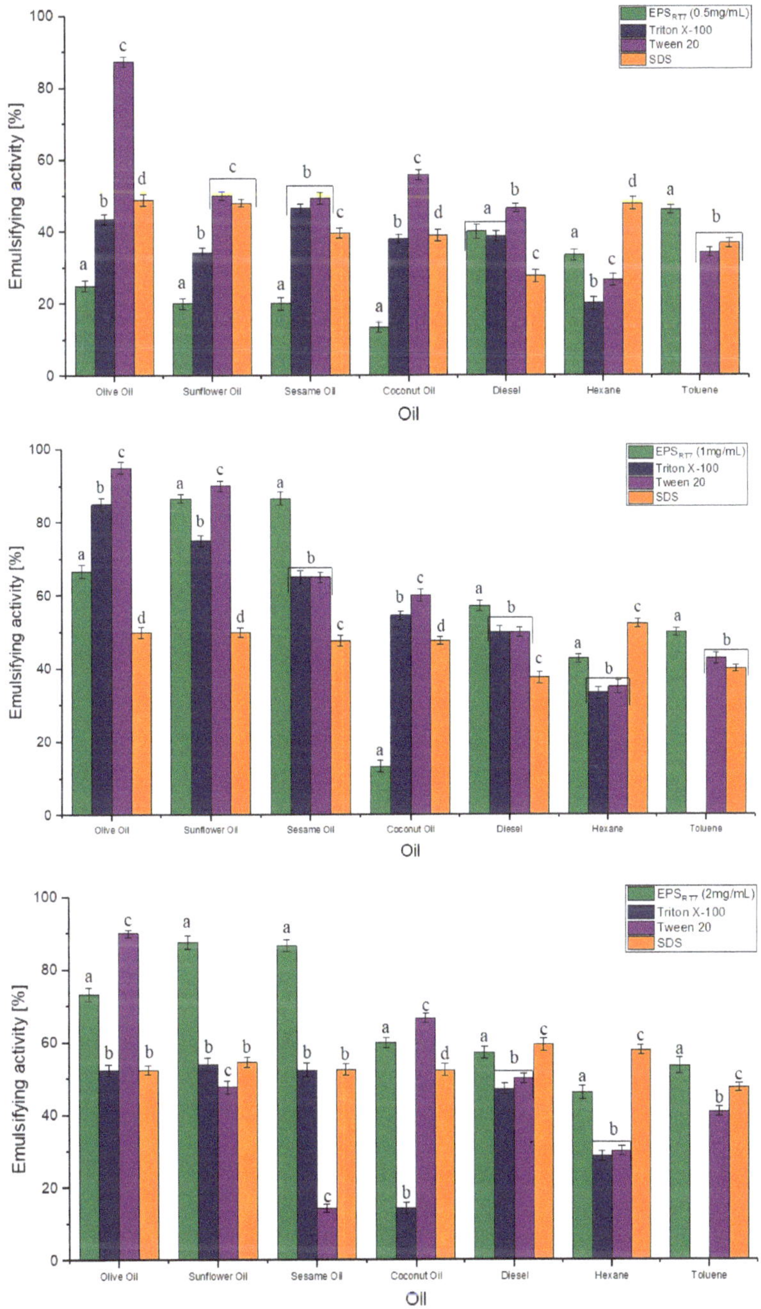

Figure 6. Comparison of emulsifying activity at different EPS$_{RT7}$ concentrations (0.5, 1, 2 mg/mL) against commercial emulsifiers (Triton X-100, Tween 20 and SDS) across different natural oils and hydrocarbons. Different letters (a–d) represent the statistical difference between different emulsifiers for each natural oils and hydrocarbons ($p < 0.05$).

The molecular compositions, molecular weight, and functional groups of EPSs have important effects on emulsification. EPS_{RT7} had good emulsifying activity at high concentrations and the studied pH range, which may be attributed to electrostatic interaction and interactions between hydrophilic groups [82]. A low pH, such as 3.1, has a negative effect on the emulsification activity of commercial polysaccharides such as xanthan and Arabic gum [43]. However, in this case, EPS_{RT7} emulsifying activity was not affected at pH 3.1. The capability of EPS_{RT7} to emulsify at different pH levels and thus its ability to bioremediate different environments, and its nontoxicity give it great advantages over other EPSs and commercial emulsifiers.

3.3.2. Antioxidant Effect

The EPS_{RT7} obtained from the biodegradation of the combination of glucose and Tween 80 with *B. amyloquefaciens* RT7 was tested in order to study its potential and benefits, particularly antioxidant and emulsifying activities. To quantify antioxidant activity, the method described in Section 2.6 was used.

DPPH evaluates the radical scavenging activity of nonenzymatic antioxidants. As shown in Figure 7a, the scavenging activity of EPS_{RT7} on DPPH radicals did not increase in a concentration-dependent manner. EPS_{RT7} presented its highest scavenging activity of 67% at 7.5 mg/mL, while for the same concentration, the scavenging capability of Vc was 82%. This was close to the one previously presented by the *B. amyloliquefaciens* GSBa-1 strain [63], where the EPS had a DPPH scavenging activity of 76.7% (5 mg/mL/Vc 90%). The potential of EPS_{RT7} for DPPH inhibition suggests that it had enough proton donors to convert free radicals into stable molecules [83].

The hydroxyl radical is one of the most reactive free radicals in a biological system [84]. This is a type of free radical with the most active chemical properties and that can cause the most harm in comparison to other free radicals, as it damages DNA base sequences [85]. The scavenging activity of the hydroxyl radical is commonly used to evaluate the ability of compounds to scavenge free radicals. Figure 7b shows the results for hydroxyl radical scavenging activity of the EPS_{RT7}. Scavenging activity increased in proportion to the concentration. At a concentration of 5 mg/mL, the scavenging capability of EPS_{RT7} was 90% and remained constant, whereas the activity of Vc was 98% at the same concentration. This efficacy is similar to that previously obtained by the *B. amyloliquefaciens* GSBa-1 strain: 89.7% (5 mg/mL/Vc 90%) [63]. However, the EPSs obtained from *B. amyloliquefaciens* C-1 [66] were much less efficient than EPS_{RT7}, since they presented hydroxyl radical activity of 60.4% for EPS-1, and less than 10% for EPS-2 at 5 mg/mL (Vc 100%). The high efficiency of EPS_{RT7} could have been due to the bond dissociation energy of EPS_{RT7} being relatively weak; therefore, it was easy to provide more electron atoms to bind to the hydroxyl radical [86].

Superoxide radicals can be harmful to cells, and their anions can increment damage to cellular components, as they generate oxidising agents and other free radicals. Figure 7c shows the superoxide anion scavenging activity of EPS_{RT7}. At the concentration of 0.25 mg/mL, the scavenging capability of EPS_{RT7} was very high, 96.5% (Vc 100%). The superoxidant anion scavenging of EPS_{RT7} was higher than the previously studied EPS of *B. amyloliquefaciens* GSBa-1 [63] of with 44.8% (5 mg/mL/Vc 99.1%), and both EPS1 and EPS2 produced by the *B. amyloliquefaciens* C1 strain (EPS1: 30.8% (5 mg/mL/Vc 99.1%), EPS2: 8.5% (5 mg/mL/Vc 99.1%)) [66]. The superoxide anion scavenging mechanism was associated with O–H bond dissociation energy [75].

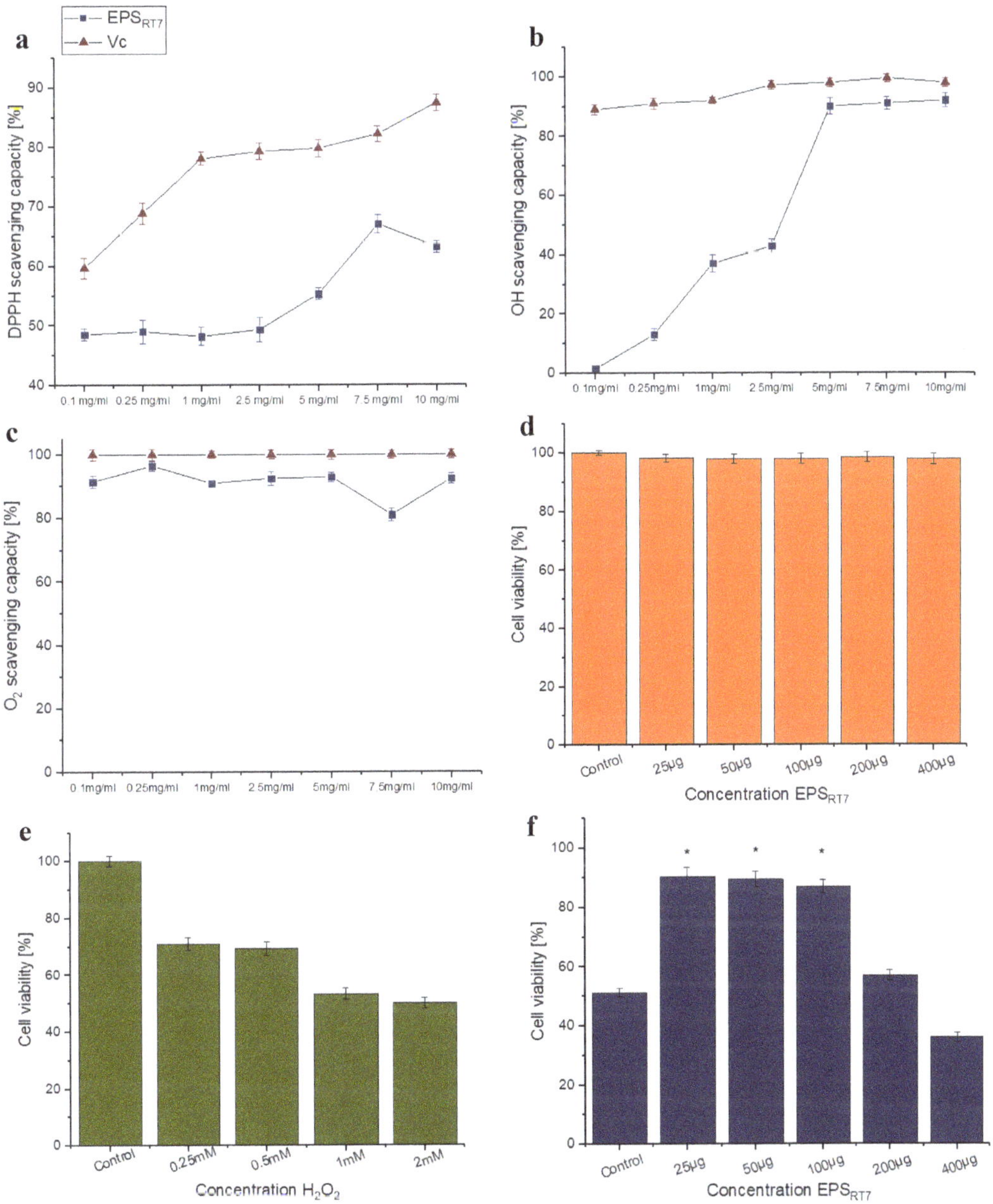

Figure 7. Antioxidant tests and toxicity evaluation with different concentrations of EPS$_{RT7}$. (**a**) DPPH free radical scavenging activity. (**b**) Hydroxyl radical scavenging activity of EPS$_{RT7}$ (**c**) Superoxide anion scavenging activity. (**d**) Hela cells viability (%) by different EPS$_{RT7}$ concentrations. (**e**) Hela cells viability (%) against oxidative stress by different H_2O_2 concentrations. (**f**) Exhibition of EPS$_{RT7}$ protection on Hela cells viability (%). (* $p < 0.05$).

3.3.3. Toxicity Evaluation and Antioxidant Ability at the Cellular Level

Figure 7 shows the biocompatibility and antioxidant capacity of EPS$_{RT7}$ at the cellular level. For all tested polymer concentrations (Figure 7d), HeLa cells did not show statistically significant damage ($p < 0.05$). This implies that EPS_{RT7} did not cause cytotoxicity in the cell line. Similar results were reported for *B. amyloliquefaciens* amy-1, where cytotoxicity assays revealed that EPSs at 50–800 μg/mL were not toxic to the human enteroendocrine cell line, NCI-H716 cells [87]. On the other hand, other biocompatibility studies for the same species revealed that, for concentrations in the range of 200–800 μg/mL of EPS, HEK293T cells moderately inhibited their growth [88].

Figure 7e shows the cell viability of HeLa cells after being treated to different concentrations of H_2O_2. Lower cell viability of below 70% was detected due to oxidative stress in the cells resulting from a build-up of reactive oxygen species (ROSs). The antioxidant effect of EPS_{RT7} on the cell line (Figure 7f) was tested at different concentrations. A statistically significant increase in cell viability was observed in concentrations between 25 and 100 μg/mL. The results demonstrate that EPS_{RT7} concentrations in the range of 25–100 μg/mL statistically significantly improved cellular viability. These results indicate that EPS_{RT7} significantly protected HeLa cells from H_2O_2-induced cytotoxicity. Cells incubated with 25, 50, and 100 μg/mL EPS_{RT7} grew better (90.2, 89.4, 86.9%, respectively) than cells of the control group did, indicating that EPS_{RT7} had a growth-promoting effect on normal cells. Similar results were reported with the pretreatment of EPS isolated from *Bacillus amyloliquefaciens* significantly and time-dependently decreasing the ROSs induced by H_2O_2 for H_2O_2-treated HepG2 cells [66]. These results suggest that EPS_{RT7} from *Bacillus amyloliquefaciens* could promote the antioxidant system by stimulating enzymes to have this capacity [89].

4. Conclusions

Bacillus amyloliquefaciens RT7 was isolated from the sediments of Rio Tinto. Biodegradation was effective for the different independent carbon sources (glucose, oleic acid, Tween 80, and PEG 200) and the joint biodegradation of glucose–Tween 80. The latter was the most effective, where high EPS production occurred—490 mg/L at 24 h. Polymer characterisation identified the extracted EPS as a heteropolysaccharide composed of mannose, glucose, galactose, and xylose (molar ratio 1:0.5:0.1:0.1). O–H, C=O, and C–O groups were detected within EPS_{RT7} using structural analysis. EPS_{RT7} had an approximate molecular weight of 7.0794×10^4 Da with good thermostability. The EPS also showed emulsifying activity against some natural oils (olive, sunflower, sesame, and coconut) and hydrocarbons (diesel oil, hexane, toluene) when used at a 2 mg/mL concentration and the studied pH range, thus demonstrating its ability to bioremediate different environments. EPS_{RT7} demonstrated its potential as an antioxidant during in vitro antioxidant assays, as it showed robust radical scavenging activity. It was also nontoxic and showed cellular biocompatibility while providing protection to cells damaged by ROSs. The strong emulsifying activity coupled with its antioxidant effect and lack of cytotoxicity suggest it could have promising applications in bioremediation processes, and offers great advantages over other EPSs and commercial emulsifiers.

Author Contributions: E.S.-L.: formal analysis, investigation, writing—original draft; E.H.-L., resources, writing—original draft; R.A.: funding acquisition, project administration, manuscript review; C.A.: conceptualisation, formal analysis, investigation, data curation, funding acquisition, methodology, validation, resources, supervision, writing—original draft, writing—review and editing. All authors have read and agreed to the published version of the manuscript.

Funding: This research was funded by the Spanish Ministry of Science and Innovation for financial support (project PID2019-104812GB-I00) and FUAM, Universidad Autónoma de Madrid, Spain (project no. 820053).

Institutional Review Board Statement: Not applicable.

Data Availability Statement: Data sharing is not available for this article.

Conflicts of Interest: The authors declare no conflict of interest.

References

1. Rasheed, T.; Shafi, S.; Bilal, M.; Hussain, T.; Sher, F.; Rizwan, K. Surfactants-Based Remediation as an Effective Approach for Removal of Environmental Pollutants—A Review. *J. Mol. Liq.* **2020**, *318*, 113960. [CrossRef]
2. Meliani, A. Enhancement of Hydrocarbons Degradation by Use of Pseudomonas Biosurfactants and Biofilms. *J. Pet. Environ. Biotechnol.* **2014**, *5*, 1. [CrossRef]
3. Ławniczak, Ł.; Woźniak-Karczewska, M.; Loibner, A.P.; Heipieper, H.J.; Chrzanowski, Ł. Microbial Degradation of Hydrocarbons—Basic Principles for Bioremediation: A Review. *Molecules* **2020**, *25*, 856. [CrossRef] [PubMed]
4. Bhadani, A.; Kafle, A.; Ogura, T.; Akamatsu, M.; Sakai, K.; Sakai, H.; Abe, M. Current Perspective of Sustainable Surfactants Based on Renewable Building Blocks. *Curr. Opin. Colloid Interface Sci.* **2020**, *45*, 124–135. [CrossRef]
5. Venhuis, S.H.; Mehrvar, M. Health Effects, Environmental Impacts, and Photochemical Degradation of Selected Surfactants in Water. *Int. J. Photoenergy* **2004**, *6*, 115–125. [CrossRef]
6. Shiu, R.F.; Lee, C.L. Effects of Anthropogenic Surfactants on the Conversion of Marine Dissolved Organic Carbon and Microgels. *Mar. Pollut. Bull.* **2017**, *117*, 156–160. [CrossRef]
7. Rehman, R.; Ali, M.I.; Ali, N.; Badshah, M.; Iqbal, M.; Jamal, A.; Huang, Z. Crude Oil Biodegradation Potential of Biosurfactant-Producing *Pseudomonas aeruginosa* and *Meyerozyma* sp. *J. Hazard. Mater.* **2021**, *418*, 126276. [CrossRef]
8. Banat, I.M.; Makkar, R.S.; Cameotra, S.S. Potential Commercial Applications of Microbial Surfactants. *Appl. Microbiol. Biotechnol.* **2000**, *53*, 495–508. [CrossRef]
9. Gudiña, E.J.; Couto, M.R.; Silva, S.P.; Coelho, E.; Coimbra, M.A.; Teixeira, J.A.; Rodrigues, L.R. Sustainable Exopolysaccharide Production by *Rhizobium viscosum* CECT908 Using Corn Steep Liquor and Sugarcane Molasses as Sole Substrates. *Polymers* **2023**, *15*, 20. [CrossRef]
10. Johnson, P.; Trybala, A.; Starov, V.; Pinfield, V.J. Effect of Synthetic Surfactants on the Environment and the Potential for Substitution by Biosurfactants. *Adv. Colloid Interface Sci.* **2021**, *288*, 102340. [CrossRef]
11. Poli, A.; Anzelmo, G.; Nicolaus, B. Bacterial Exopolysaccharides from Extreme Marine Habitats: Production, Characterization and Biological Activities. *Mar. Drugs* **2010**, *8*, 1779–1802. [CrossRef] [PubMed]
12. Freitas, F.; Alves, V.D.; Reis, M.A.M. Advances in Bacterial Exopolysaccharides: From Production to Biotechnological Applications. *Trends Biotechnol.* **2011**, *29*, 388–398. [CrossRef] [PubMed]
13. Ramezani, M.; Ghahari, S.; Ghahari, S. *Microbial Rejuvenation of Polluted Environment*; Springer: Singapore, 2021; Volume 3, ISBN 9789811574559.
14. López-Ortega, M.A.; Chavarría-Hernández, N.; del López-Cuellar, M.R.; Rodríguez-Hernández, A.I. A Review of Extracellular Polysaccharides from Extreme Niches: An Emerging Natural Source for the Biotechnology. From the Adverse to Diverse! *Int. J. Biol. Macromol.* **2021**, *177*, 559–577. [CrossRef] [PubMed]
15. Maughan, H.; Van der Auwera, G. *Bacillus* Taxonomy in the Genomic Era Finds Phenotypes to Be Essential Though Often Misleading. *Infect. Genet. Evol.* **2011**, *11*, 789–797. [CrossRef]
16. Rampelotto, P.H. Extremophiles and Extreme Environments. *Life* **2013**, *3*, 482–485. [CrossRef]
17. Satpute, S.K.; Banat, I.M.; Dhakephalkar, P.K.; Banpurkar, A.G.; Chopade, B.A. Biosurfactants, Bioemulsifiers and Exopolysaccharides from Marine Microorganisms. *Biotechnol. Adv.* **2010**, *28*, 436–450. [CrossRef]
18. Song, B.; Zhu, W.; Song, R.; Yan, F.; Wang, Y. Exopolysaccharide from *Bacillus vallismortis* WF4 as an Emulsifier for Antifungal and Antipruritic Peppermint Oil Emulsion. *Int. J. Biol. Macromol.* **2019**, *125*, 436–444. [CrossRef]
19. Abid, Y.; Azabou, S.; Joulak, I.; Casillo, A.; Lanzetta, R.; Corsaro, M.M.; Gharsallaoui, A.; Attia, H. Potential Biotechnological Properties of an Exopolysaccharide Produced by Newly Isolated *Bacillus tequilensis*-GM from Spontaneously Fermented Goat Milk. *LWT* **2019**, *105*, 135–141. [CrossRef]
20. Thavasi, R.; Jayalakshmi, S.; Balasubramanian, T.; Banat, I.M. Production and Characterization of a Glycolipid Biosurfactant from *Bacillus megaterium* Using Economically Cheaper Sources. *World J. Microbiol. Biotechnol.* **2008**, *24*, 917–925. [CrossRef]
21. Han, Y.; Liu, E.; Liu, L.; Zhang, B.; Wang, Y.; Gui, M.; Wu, R.; Li, P. Rheological, Emulsifying and Thermostability Properties of Two Exopolysaccharides Produced by *Bacillus amyloliquefaciens* LPL061. *Carbohydr. Polym.* **2015**, *115*, 230–237. [CrossRef]
22. Nicolaus, B.; Panico, A.; Manca, M.C.; Lama, L.; Gambacorta, A.; Maugeri, T.; Gugliandolo, C.; Caccamo, D. A Thermophilic *Bacillus* Isolated from an Eolian Shallow Hydrothermal Vent, Able to Produce Exopolysaccharides. *Syst. Appl. Microbiol.* **2000**, *23*, 426–432. [CrossRef] [PubMed]
23. Binmad, S.; Numnuam, A.; Kaewtatip, K.; Kantachote, D.; Tantirungkij, M. Characterization of Novel Extracellular Polymeric Substances Produced by *Bacillus velezensis* P1 for Potential Biotechnological Applications. *Polym. Adv. Technol.* **2022**, *33*, 2470–2479. [CrossRef]
24. Zhao, X.; Chen, G.; Wang, F.; Zhao, H.; Wei, Y.; Liu, L.; Zhang, H. Extraction, Characterization, Antioxidant Activity and Rheological Behavior of a Polysaccharide Produced by the Extremely Salt Tolerant *Bacillus subtilis* LR-1. *LWT* **2022**, *162*, 113413. [CrossRef]
25. Gangalla, R.; Sampath, G.; Beduru, S.; Sarika, K.; Kaveriyappan Govindarajan, R.; Ameen, F.; Alwakeel, S.; Thampu, R.K. Optimization and Characterization of Exopolysaccharide Produced by *Bacillus aerophilus* Rk1 and Its in Vitro Antioxidant Activities. *J. King Saud Univ.—Sci.* **2021**, *33*, 101470. [CrossRef]

26. Arena, A.; Maugeri, T.L.; Pavone, B.; Iannello, D.; Gugliandolo, C.; Bisignano, G. Antiviral and Immunoregulatory Effect of a Novel Exopolysaccharide from a Marine Thermotolerant *Bacillus licheniformis*. *Int. Immunopharmacol.* **2006**, *6*, 8–13. [CrossRef] [PubMed]
27. Sánchez-León, E.; Bello-Morales, R.; López-Guerrero, J.A.; Poveda, A.; Jiménez-Barbero, J.; Gironès, N.; Abrusci, C. Isolation and Characterization of an Exopolymer Produced by *Bacillus licheniformis*: In Vitro Antiviral Activity against Enveloped Viruses. *Carbohydr. Polym.* **2020**, *248*, 116737. [CrossRef]
28. Spanò, A.; Gugliandolo, C.; Lentini, V.; Maugeri, T.L.; Anzelmo, G.; Poli, A.; Nicolaus, B. A Novel EPS-Producing Strain of *Bacillus licheniformis* Isolated from a Shallow Vent Off Panarea Island (Italy). *Curr. Microbiol.* **2013**, *67*, 21–29. [CrossRef] [PubMed]
29. Cheng, M.; Zeng, G.; Huang, D.; Yang, C.; Lai, C.; Zhang, C.; Liu, Y. Tween 80 Surfactant-Enhanced Bioremediation: Toward a Solution to the Soil Contamination by Hydrophobic Organic Compounds. *Crit. Rev. Biotechnol.* **2018**, *38*, 17–30. [CrossRef] [PubMed]
30. Sellaturay, P.; Gurugama, P.; Harper, V.; Dymond, T.; Ewan, P.; Nasser, S. The Polysorbate Containing AstraZeneca COVID-19 Vaccine Is Tolerated by Polyethylene Glycol (PEG) Allergic Patients. *Clin. Exp. Allergy* **2022**, *52*, 12–17. [CrossRef] [PubMed]
31. Abrusci, C.; Martín-González, A.; Del Amo, A.; Catalina, F.; Collado, J.; Platas, G. Isolation and Identification of Bacteria and Fungi from Cinematographic Films. *Int. Biodeterior. Biodegrad.* **2005**, *56*, 58–68. [CrossRef]
32. Orphan, V.J.; Hinrichs, K.U.; Ussler, W.; Paull, C.K.; Taylor, L.T.; Sylva, S.P.; Hayes, J.M.; Delong, E.F. Comparative Analysis of Methane-Oxidizing Archaea and Sulfate-Reducing Bacteria in Anoxic Marine Sediments. *Appl. Environ. Microbiol.* **2001**, *67*, 1922–1934. [CrossRef] [PubMed]
33. Moore, S.S.; Whan, V.; Davis, G.P.; Byrne, K.; Hetzel, D.J.S.; Preston, N. The Development and Application of Genetic Markers for the Kuruma Prawn *Penaeus japonicus*. *Aquaculture* **1999**, *173*, 19–32. [CrossRef]
34. Thompson, J.D.; Gibson, T.J.; Plewniak, F.; Jeanmougin, F.; Higgins, D.G. The CLUSTAL X Windows Interface: Flexible Strategies for Multiple Sequence Alignment Aided by Quality Analysis Tools. *Nucleic Acids Res.* **1997**, *25*, 4876–4882. [CrossRef] [PubMed]
35. Abrusci, C.; Pablos, J.L.; Corrales, T.; López-Marín, J.; Marín, I.; Catalina, F. Biodegradation of Photo-Degraded Mulching Films Based on Polyethylenes and Stearates of Calcium and Iron as pro-Oxidant Additives. *Int. Biodeterior. Biodegrad.* **2011**, *65*, 451–459. [CrossRef]
36. Abrusci, C.; Marquina, D.; Del Amo, A.; Catalina, F. Biodegradation of Cinematographic Gelatin Emulsion by Bacteria and Filamentous Fungi Using Indirect Impedance Technique. *Int. Biodeterior. Biodegrad.* **2007**, *60*, 137–143. [CrossRef]
37. Tada, H.; Shiho, O.; Kuroshima, K.; Koyama, M.; Tsukamoto, K. An Improved Colorimetric Assay for Interleukin 2. *J. Immunol. Methods* **1986**, *93*, 157–165. [CrossRef]
38. Aullybux, A.A.; Puchooa, D.; Bahorun, T.; Jeewon, R.; Wen, X.; Matin, P. Antioxidant and Cytotoxic Activities of Exopolysaccharides from *Alcaligenes faecalis* Species Isolated from the Marine Environment of Mauritius. *J. Polym. Environ.* **2022**, *30*, 1462–1477. [CrossRef]
39. Huang-Lin, E.; Sánchez-León, E.; Amils, R.; Abrusci, C. Potential Applications of an Exopolysaccharide Produced by *Bacillus xiamenensis* RT6 Isolated from an Acidic Environment. *Polymers* **2022**, *14*, 3918. [CrossRef]
40. Morro, A.; Catalina, F.; Corrales, T.; Pablos, J.L.; Marin, I.; Abrusci, C. New Blends of Ethylene-Butyl Acrylate Copolymers with Thermoplastic Starch. Characterization and Bacterial Biodegradation. *Carbohydr. Polym.* **2016**, *149*, 68–76. [CrossRef]
41. Morro, A.; Catalina, F.; Sanchez-León, E.; Abrusci, C. Photodegradation and Biodegradation Under Thermophile Conditions of Mulching Films Based on Poly(Butylene Adipate-Co-Terephthalate) and Its Blend with Poly(Lactic Acid). *J. Polym. Environ.* **2019**, *27*, 352–363. [CrossRef]
42. Meneghine, A.K.; Moretto, C.; Castellane, T.C.L.; Carareto Alves, L.M. Production, Characterization and Bioemulsifying Activity of an Exopolysaccharide Produced by *Sphingomonas* sp. Isolated from Freshwater. *J. Polym. Environ.* **2017**, *25*, 1080–1086. [CrossRef]
43. Niknezhad, S.V.; Najafpour-Darzi, G.; Morowvat, M.H.; Ghasemi, Y. Eexopolysaccharide Production of *Pantoea sp.* BCCS 001 GH: Physical Characterizations, Emulsification, and Antioxidant Activities. *Int. J. Biol. Macromol.* **2018**, *118*, 1103–1111. [CrossRef] [PubMed]
44. Sun, M.L.; Zhao, F.; Shi, M.; Zhang, X.Y.; Zhou, B.C.; Zhang, Y.Z.; Chen, X.L. Characterization and Biotechnological Potential Analysis of a New Exopolysaccharide from the Arctic Marine Bacterium *Polaribacter* sp. SM1127. *Sci. Rep.* **2015**, *5*, 18435. [CrossRef] [PubMed]
45. Balakrishnan, B.; Prasad, B.; Rai, A.K.; Velappan, S.P.; Subbanna, M.N.; Narayan, B. In Vitro Antioxidant and Antibacterial Properties of Hydrolysed Proteins of Delimed Tannery Fleshings: Comparison of Acid Hydrolysis and Fermentation Methods. *Biodegradation* **2011**, *22*, 287–295. [CrossRef]
46. Morro, A.; Abrusci, C.; Pablos, J.L.; Marín, I.; García, F.C.; García, J.M. Inherent Antibacterial Activity and in Vitro Biocompatibility of Hydrophilic Polymer Film Containing Chemically Anchored Sulfadiazine Moieties. *Eur. Polym. J.* **2017**, *91*, 274–282. [CrossRef]
47. Pérez-Blanco, C.; Huang-Lin, E.; Abrusci, C. Characterization, Biodegradation and Cytotoxicity of Thermoplastic Starch and Ethylene-Vinyl Alcohol Copolymer Blends. *Carbohydr. Polym.* **2022**, *298*, 120085. [CrossRef]
48. Amils, R.; Fernández-Remolar, D. The IPBSL Team Río Tinto: A Geochemical and Mineralogical Terrestrial Analogue of Mars. *Life* **2014**, *4*, 511–534. [CrossRef]
49. Wongbunmak, A.; Khiawjan, S.; Suphantharika, M.; Pongtharangkul, T. BTEX Biodegradation by *Bacillus amyloliquefaciens* Subsp. *plantarum* W1 and Its Proposed BTEX Biodegradation Pathways. *Sci. Rep.* **2020**, *10*, 17408. [CrossRef]

50. Lu, D.; Zhang, Y.; Niu, S.; Wang, L.; Lin, S.; Wang, C.; Ye, W.; Yan, C. Study of Phenol Biodegradation Using *Bacillus amyloliquefaciens* Strain WJDB-1 Immobilized in Alginate-Chitosan-Alginate (ACA) Microcapsules by Electrochemical Method. *Biodegradation* **2012**, *23*, 209–219. [CrossRef]
51. Rehman, K.; Imran, A.; Amin, I.; Afzal, M. Inoculation with Bacteria in Floating Treatment Wetlands Positively Modulates the Phytoremediation of Oil Field Wastewater. *J. Hazard. Mater.* **2018**, *349*, 242–251. [CrossRef]
52. Rao, B.P.; Kasirajan, S.; Mandal, A.B.; Sudharsan, K.; Sekaran, R.C.H.G.; Mandal, A.B. Characterization of Exopolysaccharide from *Bacillus amyloliquefaciens* BPRGS for Its Bioflocculant Activity. *Artic. Int. J. Sci. Eng. Res.* **2013**, *4*, 1696–1704.
53. Li, S.; Fei, X.; Cao, L.; Chi, Y. Insights into the Effects of Carbon Source on Sequencing Batch Reactors: Performance, Quorum Sensing and Microbial Community. *Sci. Total Environ.* **2019**, *691*, 799–809. [CrossRef] [PubMed]
54. Khandelwal, A.; Sugavanam, R.; Ramakrishnan, B.; Dutta, A.; Varghese, E.; Nain, L.; Banerjee, T.; Singh, N. Free and Immobilized Microbial Culture–Mediated Crude Oil Degradation and Microbial Diversity Changes Through Taxonomic and Functional Markers in a Sandy Loam Soil. *Front. Environ. Sci.* **2022**, *9*, 794303. [CrossRef]
55. Zhang, L.; Wang, M.; Cui, H.; Qiao, J.; Guo, D.; Wang, B.; Li, X.; Huang, H. How Humic Acid and Tween80 Improve the Phenanthrene Biodegradation Efficiency: Insight from Cellular Characteristics and Quantitative Proteomics. *J. Hazard. Mater.* **2022**, *421*, 126685. [CrossRef]
56. Jiang, R.; Wu, X.; Xiao, Y.; Kong, D.; Li, Y.; Wang, H. Tween 20 Regulate the Function and Structure of Transmembrane Proteins of *Bacillus cereus*: Promoting Transmembrane Transport of Fluoranthene. *J. Hazard. Mater.* **2021**, *403*, 123707. [CrossRef]
57. Ma, P.; Li, Y.; Miao, C.; Sun, Y.; Liu, C.; Li, H. Production of Tween 80-Inducing Esterase by *Acinetobacter sp.* B1 Using Response Surface Methodology. *Microbiol. Biotechnol. Lett.* **2019**, *47*, 87–95. [CrossRef]
58. Yang, X.; Yang, Y.; Zhang, Y.; He, J.; Xie, Y. Enhanced Exopolysaccharide Production in Submerged Fermentation of *Ganoderma lucidum* by Tween 80 Supplementation. *Bioprocess Biosyst. Eng.* **2021**, *44*, 47–56. [CrossRef]
59. Deka, P.; Goswami, G.; Das, P.; Gautom, T.; Chowdhury, N.; Boro, R.C.; Barooah, M. Bacterial Exopolysaccharide Promotes Acid Tolerance in *Bacillus amyloliquefaciens* and Improves Soil Aggregation. *Mol. Biol. Rep.* **2019**, *46*, 1079–1091. [CrossRef]
60. Kaspar, F.; Neubauer, P.; Gimpel, M. Bioactive Secondary Metabolites from *Bacillus subtilis*: A Comprehensive Review. *J. Nat. Prod.* **2019**, *82*, 2038–2053. [CrossRef] [PubMed]
61. Soumya, M.P.; Sasikumar, K.; Pandey, A.; Nampoothiri, K.M. Cassava Starch Hydrolysate as Sustainable Carbon Source for Exopolysaccharide Production by *Lactobacillus plantarum*. *Bioresour. Technol. Rep.* **2019**, *6*, 85–88. [CrossRef]
62. Ruas-Madiedo, P.; Hugenholtz, J.; Zoon, P. An Overview of the Functionality of Exopolysaccharides Produced by Lactic Acid Bacteria. *Int. Dairy J.* **2002**, *12*, 163–171. [CrossRef]
63. Zhao, W.; Zhang, J.; Jiang, Y.Y.; Zhao, X.; Hao, X.N.; Li, L.; Yang, Z.N. Characterization and Antioxidant Activity of the Exopolysaccharide Produced by *Bacillus amyloliquefaciens* GSBa-1. *J. Microbiol. Biotechnol.* **2018**, *28*, 1282–1292. [CrossRef] [PubMed]
64. El-Newary, S.A.; Ibrahim, A.Y.; Asker, M.S.; Mahmoud, M.G.; El Awady, M.E. Production, Characterization and Biological Activities of Acidic Exopolysaccharide from Marine *Bacillus amyloliquefaciens* 3MS 2017. *Asian Pac. J. Trop. Med.* **2017**, *10*, 652–662. [CrossRef]
65. Cai, G.; Liu, Y.; Li, X.; Lu, J. New Levan-Type Exopolysaccharide from *Bacillus amyloliquefaciens* as an Antiadhesive Agent against Enterotoxigenic *Escherichia coli*. *J. Agric. Food Chem.* **2019**, *67*, 8029–8034. [CrossRef] [PubMed]
66. Yang, H.; Deng, J.; Yuan, Y.; Fan, D.; Zhang, Y.; Zhang, R.; Han, B. Two Novel Exopolysaccharides from *Bacillus amyloliquefaciens* C-1: Antioxidation and Effect on Oxidative Stress. *Curr. Microbiol.* **2015**, *70*, 298–306. [CrossRef] [PubMed]
67. Kuang, J.H.; Huang, Y.Y.; Hu, J.S.; Yu, J.J.; Zhou, Q.Y.; Liu, D.M. Exopolysaccharides from *Bacillus amyloliquefaciens* DMBA-K4 Ameliorate Dextran Sodium Sulfate-Induced Colitis via Gut Microbiota Modulation. *J. Funct. Foods* **2020**, *75*, 104212. [CrossRef]
68. Zhou, K.; Zeng, Y.; Yang, M.; Chen, S.; He, L.; Ao, X.; Zou, L.; Liu, S. Production, Purification and Structural Study of an Exopolysaccharide from *Lactobacillus plantarum* BC-25. *Carbohydr. Polym.* **2016**, *144*, 205–214. [CrossRef] [PubMed]
69. Thakham, N.; Thaweesak, S.; Teerakulkittipong, N.; Traiosot, N.; Kaikaew, A.; Lirio, G.A.; Jangiam, W. Structural Characterization of Functional Ingredient Levan Synthesized by *Bacillus siamensis* Isolated from Traditional Fermented Food in Thailand. *Int. J. Food Sci.* **2020**, *2020*, 7352484. [CrossRef]
70. Krishnamurthy, M.; Jayaraman Uthaya, C.; Thangavel, M.; Annadurai, V.; Rajendran, R.; Gurusamy, A. Optimization, Compositional Analysis, and Characterization of Exopolysaccharides Produced by Multi-Metal Resistant *Bacillus cereus* KMS3-1. *Carbohydr. Polym.* **2020**, *227*, 115369. [CrossRef]
71. Sahana, T.G.; Rekha, P.D. A Novel Exopolysaccharide from Marine Bacterium *Pantoea sp.* YU16-S3 Accelerates Cutaneous Wound Healing through Wnt/β-Catenin Pathway. *Carbohydr. Polym.* **2020**, *238*, 116191. [CrossRef]
72. Ayyash, M.; Abu-Jdayil, B.; Itsaranuwat, P.; Galiwango, E.; Tamiello-Rosa, C.; Abdullah, H.; Esposito, G.; Hunashal, Y.; Obaid, R.S.; Hamed, F. Characterization, Bioactivities, and Rheological Properties of Exopolysaccharide Produced by Novel Probiotic *Lactobacillus plantarum* C70 Isolated from Camel Milk. *Int. J. Biol. Macromol.* **2020**, *144*, 938–946. [CrossRef] [PubMed]
73. Taylan, O.; Yilmaz, M.T.; Dertli, E. Partial Characterization of a Levan Type Exopolysaccharide (EPS) Produced by *Leuconostoc mesenteroides* Showing Immunostimulatory and Antioxidant Activities. *Int. J. Biol. Macromol.* **2019**, *136*, 436–444. [CrossRef] [PubMed]
74. Ismail, B.; Nampoothiri, K.M. Production, Purification and Structural Characterization of an Exopolysaccharide Produced by a Probiotic *Lactobacillus plantarum* MTCC 9510. *Arch. Microbiol.* **2010**, *192*, 1049–1057. [CrossRef]

75. Gangalla, R.; Gattu, S.; Palaniappan, S.; Ahamed, M.; Macha, B.; Thampu, R.K.; Fais, A.; Cincotti, A.; Gatto, G.; Dama, M.; et al. Structural Characterisation and Assessment of the Novel *Bacillus amyloliquefaciens* RK3 Exopolysaccharide on the Improvement of Cognitive Function in Alzheimer's Disease Mice. *Polymers* **2021**, *13*, 2842. [CrossRef] [PubMed]
76. Jenny Angel, S.; Vidyadharani, G.; Santhosh, S.; Dhandapani, R. Optimization and Characterisation of Thermo Stable Exopolysaccharide Produced from *Bacillus licheniformis* WSF-1 Strain. *J. Polym. Environ.* **2018**, *26*, 3824–3833. [CrossRef]
77. Song, B.; Song, R.; Cheng, M.; Chu, H.; Yan, F.; Wang, Y. Preparation of Calcipotriol Emulsion Using Bacterial Exopolysaccharides as Emulsifier for Percutaneous Treatment of Psoriasis Vulgaris. *Int. J. Mol. Sci.* **2020**, *21*, 77. [CrossRef]
78. Ben Ayed, H.; Jemil, N.; Maalej, H.; Bayoudh, A.; Hmidet, N.; Nasri, M. Enhancement of Solubilization and Biodegradation of Diesel Oil by Biosurfactant from *Bacillus amyloliquefaciens* An6. *Int. Biodeterior. Biodegrad.* **2015**, *99*, 8–14. [CrossRef]
79. Bouallegue, A.; Chaari, F.; Casillo, A.; Corsaro, M.M.; Bachoual, R.; Ellouz-Chaabouni, S. Levan Produced by *Bacillus subtilis* AF17: Thermal, Functional and Rheological Properties. *J. Food Meas. Charact.* **2022**, *16*, 440–447. [CrossRef]
80. Kanmani, P.; Satish kumar, R.; Yuvaraj, N.; Paari, K.A.; Pattukumar, V.; Arul, V. Production and Purification of a Novel Exopolysaccharide from Lactic Acid Bacterium *Streptococcus phocae* PI80 and Its Functional Characteristics Activity in Vitro. *Bioresour. Technol.* **2011**, *102*, 4827–4833. [CrossRef]
81. Kodali, V.P.; Das, S.; Sen, R. An Exopolysaccharide from a Probiotic: Biosynthesis Dynamics, Composition and Emulsifying Activity. *Food Res. Int.* **2009**, *42*, 695–699. [CrossRef]
82. Hu, X.; Li, F.; Zhang, X.; Pan, Y. The Structure, Characterization and Dual-Activity of Exopolysaccharide Produced by *Bacillus enclensis* AP-4 from Deep-Sea Sediments. *Front. Mar. Sci.* **2022**, *9*, 976543. [CrossRef]
83. Rahnama Vosough, P.; Habibi Najafi, M.B.; Edalatian Dovom, M.R.; Javadmanesh, A.; Mayo, B. Evaluation of Antioxidant, Antibacterial and Cytotoxicity Activities of Exopolysaccharide from *Enterococcus* Strains Isolated from Traditional Iranian Kishk. *J. Food Meas. Charact.* **2021**, *15*, 5221–5230. [CrossRef]
84. Liang, N.; Kitts, D.D. Antioxidant Property of Coffee Components: Assessment of Methods That Define Mechanism of Action. *Molecules* **2014**, *19*, 19180–19208. [CrossRef] [PubMed]
85. Nguyen Vu, T.H.; Quach, N.T.; Nguyen, N.A.; Nguyen, H.T.; Ngo, C.C.; Nguyen, T.D.; Ho, P.H.; Hoang, H.; Chu, H.H.; Phi, Q.T. Genome Mining Associated with Analysis of Structure, Antioxidant Activity Reveals the Potential Production of Levan-Rich Exopolysaccharides by Food-Derived *Bacillus velezensis* Vtx20. *Appl. Sci.* **2021**, *11*, 7055. [CrossRef]
86. Yang, X.; Feng, J.; Zhu, Q.; Hong, R.; Li, L. A Relation between Exopolysaccharide from Lactic Acid Bacteria and Properties of Fermentation Induced Soybean Protein Gels. *Polymers* **2022**, *14*, 90. [CrossRef]
87. Chen, Y.C.; Huang, S.; Tu, J.H.; Yu, J.S.; Nurlatifah, A.O.; Chiu, W.C.; Su, Y.H.; Chang, H.L.; Putri, D.A.; Cheng, H.L. Exopolysaccharides of *Bacillus amyloliquefaciens* Modulate Glycemic Level in Mice and Promote Glucose Uptake of Cells through the Activation of Akt. *Int. J. Biol. Macromol.* **2020**, *146*, 202–211. [CrossRef] [PubMed]
88. Sung, W.W.; Tu, J.H.; Yu, J.S.; Ulfa, M.Z.; Chang, J.H.; Cheng, H.L. *Bacillus amyloliquefaciens* Exopolysaccharide Preparation Induces Glucagon-like Peptide 1 Secretion through the Activation of Bitter Taste Receptors. *Int. J. Biol. Macromol.* **2021**, *185*, 562–571. [CrossRef]
89. Zhou, L.; Luo, S.; Li, J.; Zhou, Y.; Chen, T.; Feng, S.; Ding, C. Simultaneous Optimization of Extraction and Antioxidant Activity from *Blumea laciniata* and the Protective Effect on Hela Cells against Oxidative Damage. *Arab. J. Chem.* **2020**, *13*, 9231–9242. [CrossRef]

polymers MDPI

Article

Identification of a Thermostable Levansucrase from *Pseudomonas orientalis* That Allows Unique Product Specificity at Different Temperatures

Cuie Guang [1], Xiaoqi Zhang [1], Dawei Ni [1], Wenli Zhang [1], Wei Xu [1,*] and Wanmeng Mu [1,2]

1 State Key Laboratory of Food Science and Technology, Jiangnan University, Wuxi 214122, China
2 International Joint Laboratory on Food Safety, Jiangnan University, Wuxi 214122, China
* Correspondence: weixu@jiangnan.edu.cn; Tel./Fax: +86-510-8591-9161

Abstract: The biological production of levan by levansucrase (LS, EC 2.4.1.10) has aroused great interest in the past few years. Previously, we identified a thermostable levansucrase from *Celerinatantimonas diazotrophica* (Cedi-LS). A novel thermostable LS from *Pseudomonas orientalis* (Psor-LS) was successfully screened using the Cedi-LS template. The Psor-LS showed maximum activity at 65 °C, much higher than the other LSs. However, these two thermostable LSs showed significantly different product specificity. When the temperature was decreased from 65 to 35 °C, Cedi-LS tended to produce high-molecular-weight (HMW) levan. By contrast, Psor-LS prefers to generate fructooligosaccharides (FOSs, DP $\leq$ 16) rather than HMW levan under the same conditions. Notably, at 65 °C, Psor-LS would produce HMW levan with an average M_W of 1.4×10^6 Da, indicating that a high temperature might favor the accumulation of HMW levan. In summary, this study allows a thermostable LS suitable for HMW levan and levan-type FOSs production simultaneously.

Keywords: levansucrase; thermostability; levan; product specificity; application

Citation: Guang, C.; Zhang, X.; Ni, D.; Zhang, W.; Xu, W.; Mu, W. Identification of a Thermostable Levansucrase from *Pseudomonas orientalis* That Allows Unique Product Specificity at Different Temperatures. *Polymers* **2023**, *15*, 1435. https://doi.org/10.3390/polym15061435

Academic Editor: Shashi Kant Bhatia

Received: 18 January 2023
Revised: 9 March 2023
Accepted: 9 March 2023
Published: 14 March 2023

1. Introduction

Levansucrase (LS, EC 2.4.1.10), inulosucrase (IS, EC 2.1.4.9), and β-fructofuranosidase (Ffase, EC 3.2.1.26) are three fructansucrases (FSs) that could use sucrose as the substrate to produce fructooligosaccharides (FOSs) and fructans (homopolymers of fructose) [1]. LS and IS belong to the glycoside hydrolase 68 (GH 68) family of enzymes, while Ffase may be categorized as GH68 or GH32 family enzymes. These enzymes can all hydrolyze sucrose and subsequently synthesize fructan, which are defined as hydrolysis reaction (H) and transfructosylation (T), respectively [2]. Both reactions start from sucrose splitting into glucose and a fructosyl moiety. The "T" reaction occurs when the fructosyl moiety is transferred to an acceptor such as the sucrose or the elongating fructan chain. The "H" reaction will happen when the acceptor is water, releasing glucose and fructose [3].

LS, IS, and Ffase have distinguished product specificities. Ffase exclusively synthesizes FOS as its main product, whereas LS and IS could produce FOS and fructan. In addition, the product generated by IS, inulin, primarily consists of β-(2,1) linkages on the polymer backbone, while the fructan generated by LS, levan, harbors β-(2,6) linkages on the main chain [4]. Meanwhile, the identified LSs from more than 40 kinds of microorganisms all alternatively produce β-(2,6) type FOSs, β-(2,6)-type low-molecular-weight (LMW, FOS < M_W < 5×10^4 Da) levan, and β-(2,6)-type high-molecular-weight (HMW, M_W > 5×10^4 Da) levan in the reaction mixture. For instance, the LSs from *Erwinia amylovora* [5] and *Zymomonas* species [6] produced FOSs with a degree of polymerization (DP) below 10 as the main product. However, they only produced a small amount of HMW or LMW levan. The LS from *Bacillus methylotrophicus* SK 21.002 is the only one that produces LMW levan with an M_W of 4–5 $\times 10^3$ Da [7]. By contrast, the LSs from *Acetobacter nitrogenifigens* RG1 [8] and *Lactobacillus reuteri* LTH5448 [9] could synthesize HMW levan as the main product, with an M_W of 7.1 × 10^6 and 3.9 × 10^7 Da, respectively. In particular, the LS from *Clostridium acetobutylicum* could exclusively synthesize levan rather than FOSs in the reaction [10].

The properties of levan were varied with M_W. For instance, the LMW levan was reported to have a potential role in peptic ulcer curing [11] and carcinogenesis initiation stage inhibitory [12], while HMW levan could act as an antivirus agent [13] and pancreatic anticancer agent [14]. Since the practical application of levan dramatically depends on its M_W, many attempts have been made to explore the potential reason for the product specificity of LSs. Enzyme concentration was regarded as a critical factor. For instance, the LS from *Bacillus subtilis* produced LMW levan (M_W = 7.2 × 10^3 Da) at a high enzyme concentration (10 U/mL), while synthesizing HMW levan (M_W = 2.3 × 10^6 Da) at a low enzyme concentration (0.1 U/mL) [15]. Additionally, sucrose concentration could also affect product specificity of LS. Relatively high initial sucrose concentrations usually result in the synthesis of FOSs or LMW levan, while lower initial sucrose concentrations favor HMW levan production [16]. On the contrary, the *E. amylovora* LS [5] generated FOSs (DP 2–6) at a low sucrose concentration (200 mM), while it synthesized HMW levan at a high sucrose concentration (>500 mM). In addition to enzyme and substrate concentrations, temperature could also affect the product specificity. Lowering temperature was found to favor the T reaction of LS. For instance, the production of HMW levan from *Z. mobilis* LS was increased when the temperature was decreased from 40 to 4 °C [17].

In this work, a novel LS from the mesophilic bacteria *Pseudomonas orientalis* (Psor-LS) was screened on a *C. diazotrophica* LS (Cedi-LS) template. As a result, the Psor-LS showed maximum activity at 65 °C, much higher than the other LSs. The Psor-LS retained 46% of its initial activity at 55 °C for 9 h and 50% at 45 °C for 60 h, exhibiting excellent thermostability. Notably, two thermostable LSs showed a great difference in their product specificity. The Cedi-LS could produce FOSs, LMW (M_W = 4.1 × 10^4 Da), and HMW (M_W = 1.8 × 10^6 Da) levan in the reaction mixture, while the Psor-LS would specifically produce FOS and HMW (M_W = 1.4 × 10^6 Da) levan rather than LMW levan. In particular, temperature was proposed to be significant to the product distribution of Cedi-LS and Psor-LS. When the temperature was changed from 65 to 35 °C, Cedi-LS tended to produce HMW with an increased M_W of 8.4 × 10^6 Da. By contrast, at 35 °C, Psor-LS would produce more FOSs and significantly decrease the HMW levan. This study examines the effect of temperature on the LS product specificity and proposes a thermostable LS suitable for the HMW levan polymer and levan-type FOSs production.

2. Materials and Methods

2.1. Chemicals, Reagents, and Strains

Escherichia coli DH5α and BL21 (DE3) cells, Isopropyl β-D-1-thiogalactopyranoside (IPTG), ampicillin sodium, Luria–Bertani (LB) medium, and other chemicals of analytical grade were purchased from Sangon Biotech Co., Ltd. (Shanghai, China). Standards, including sucrose, glucose, and fructose, were purchased from Sigma (St. Louis, MO, USA) for high-performance liquid chromatography (HPLC) analysis.

2.2. Expression and Purification of Psor-LS

The genomic DNA of *P. orientalis* is available on the NCBI database with the GenBank accession number ASM385204v1, which revealed a putative gene encoding levansucrase. The gene was fused with the pET-22(+) vector using two restriction sites *Nde* I and *Xho* I, at the 5′- and 3′- terminus, which were commercially synthesized by Generay Biotech Co., Ltd. (Shanghai, China). A 6 × histidine-tag was designed at the 3′-terminus for purification via Ni^{2+} affinity chromatography.

For levansucrase expression, the recombinant plasmid was transformed into competent *E. coli* BL21 (DE3) cells, which were then inoculated into 200 mL of LB broth containing ampicillin (100 μg/mL broth) for the selection of transformants. The cells were cultivated at 37 °C with shaking at 200 rpm until the optical density at 600 nm (OD_{600}) reached 0.6–0.8.

Then, isopropyl β-D-1-thiogalactopyranoside (IPTG) was added into the broth with the final concentration of 1 mM, and the target enzyme expression was performed at 28 °C with shaking at 200 rpm for 6~7 h.

The bacteria were collected by centrifugation at 6000× *g* for 5 min, then disrupted in 15 mL lysis buffer (50 mM sodium phosphate buffer with 100 mM NaCl, pH 7). After being disrupted by ultrasonication, the supernatant was centrifuged at 8000× *g* and 4 °C for 10 min and filtered through a 0.45 μm Millipore filter. The enzyme was purified via Ni^{2+}-affinity chromatography using Ni-IDA-Sefinose resin (Sangon Biotech Co. Ltd., Shanghai, China). A binding buffer (50 mM sodium phosphate buffer with 100 mM NaCl, pH 7.0), washing buffer (50 mM sodium phosphate buffer with 100 mM NaCl and 50 mM imidazole, pH 7.0), and elution buffer (50 mM sodium phosphate buffer with 100 mM NaCl and 500 mM imidazole, pH 7.0) were used to equilibrate the chromatography column, wash out non-specific proteins and elute target proteins, respectively. Then, the collected enzyme solution was dialyzed against 50 mM sodium phosphate buffer (pH 7.0) for more than 18 h to remove imidazole. SDS-PAGE examined the subunit molecular mass of Psor-LS with 12% (*w*/*v*) separation gel and 5% stacking gel, and the protein bands were stained with Coomassie Brilliant Blue R250. The Bradford assay [18] estimated the protein concentration using bovine serum albumin as the standard.

2.3. *Enzyme Activity and the Ratio of Transfructosylation Activity to Hydrolytic Activity (T/H) Assay*

The reaction systems of Psor-LS were set as 1 mL containing 30% (*w*/*v*) sucrose and 10 μg enzyme. The activity of Psor-LS was determined at 65 °C for 15 min. The reaction was stopped in boiling water for 10 min. One unit of total and hydrolytic activity was designated as the amount of enzyme catalyzing the release of 1 μmol glucose and fructose per minute, respectively. The transfructosylation activity was the subtraction between total and hydrolysis activity.

The ratio of transfructosylation activity to hydrolytic activity (T/H) of Psor-LS was determined at different concentrations of substrate (from 10% to 60%), different pH (4.0–9.0), and different temperatures (30–80 °C). The glucose and fructose contents of the reaction mixture were analyzed by high-performance liquid chromatography (HPLC). The Waters e2695 system (Waters Corporation, MA, USA) has a Waters 2414 RI detector and a Sugar-Pak I column (6.5 mm × 300 mm, Waters, USA). The column temperature and the mobile phase velocity were set as 85 °C and 0.4 mL/min.

2.4. *Biochemical Characterization*

Three types of buffers with different pH ranges were used to determine the influence of pH on the activity of Psor-LS: acetate buffer (pH 4.0–6.0, 50 mM), sodium phosphate buffer (pH 6.0–7.5, 50 mM), and Tris-HCl (pH 7.5–9.0, 50 mM). The interval of each pH was set as 0.5. The reaction conditions were the same as those for the enzymatic activity assay.

The reactions were performed at various temperatures within 35–80 °C at intervals of 5 °C to measure the optimal temperature of Psor-LS. The other reaction conditions were the same as the enzymatic activity determination except for the temperature. The enzyme was incubated at 45, 55, and 65 °C for different time intervals to determine the thermostability of Psor-LS. The initial activity without pre-incubation treatment was taken as 100%. The formula calculated the half-life ($t_{1/2}$) value, $t_{1/2} = \ln 2/k_d$, and the inactivation rate constant (k_d) was determined by the linear regression with the equation: $\ln (A_t/A_0) = k_d \times t$ (A_t: residual activity; A_0: initial activity; t: incubation time).

Nano DSC III (TA Instrument, New Castle, DE, USA) was introduced to detect the denaturation temperature of this recombinant Psor-LS. The Nano-DSC was equipped with flow-through capillary cells loaded with a pipette mounted on both the inlet and outlet. The sodium phosphate (pH 7.0) buffer was used as the dialysis fluid overnight for the sample, and it was injected into the reference cell for buffer baseline determination. All samples were placed horizontally in a rack on an orbital shaker in a temperature-controlled incubator with gentle agitation. All samples must be degassed under a vacuum and loaded into the DSC. Scans were conducted by elevating temperatures from 20 °C to 100 °C, with a scan rate of 1 °C min^{-1}. To measure molar heat capacities (Cp), sample scans were obtained from the subtraction of the buffer scans to measure molar heat capacities (Cp), and the experimental thermograms with protein concentration and the volume of the calorimeter cell were also normalized. The apparent denaturation temperature Tm values of Psor-LS were determined, and experiment data from Nano DSC III were analyzed using the Nano Analyze software package.

The kinetic parameters for Psor-LS against sucrose were determined at 65 °C. The reactions were prepared in potassium phosphate buffer (pH 6.5) containing sucrose 10–900 mM. The apparent Michaelis–Menten constant (Km) and the turnover number (kcat) were obtained through the nonlinear least square regression method (nonlinear regression Michaelis and Menten (http://biomodel.uah.es/en/metab/enzimas/MM-regresion.htm).

2.5. *Optimization of Levan Production*

A reaction containing 10–60% sucrose (*w*/*v*) was carried out to determine the optimal sucrose concentration. Under optimized conditions, different enzyme dosages, ranging from 5 to 100 µg/mL, were added to the reaction mixture to investigate the optimal enzyme dosages. The highest enzyme activity determined the optimal sucrose concentration. Differently, the optimal enzyme dosage was decided upon the benefit ratio, which is the ratio of the product formed in 15 min (including FOS and levan) to enzyme dosage. The biological production of Psor-LS was studied by measuring different saccharides at specific time intervals (0.5, 1, 2, 3, and 6 h).

2.6. *Purification of Polysaccharide and FOSs*

2.6.1. Purification of Polysaccharides

Sevag reagent (N-butanol: chloroform = 1:4, *v*/*v*) was used to remove protein from the reaction mixture to purify fructans. The polysaccharides in the system were separated by ethanol (final concentration 60%, *v*/*v*) precipitation. Subsequent alcohol precipitation was used to increase the purity of polysaccharides (until HPLC could detect no monosaccharide and sucrose). The samples obtained were then dried via lyophilization using a LABCONCO FreeZone (LABCONCO Co., Kansas City, MO, USA).

2.6.2. Purification of FOS

Ethanol present in the supernatant phases of alcohol precipitation in the Section 2.6.1 was removed by rotary evaporation. The FOS mixture was then purified by active carbon adsorption chromatography to remove glucose, fructose, and sucrose [19]. The chromatographic column was filled with treated activated carbon, and the column volume was 200mL. After loading the sample of 20 mL (flow rate of 1 mL/min), we let it stand for 1 h and balanced the column with purified water. The amount used was 2 column volume. Elution with 5%, 10%, 15%, 20%, and 30% concentrations and pure ethanol at a 1 mL/min flow rate was carried out. We collected one tube every 10 mL and performed determination of FOS content. The purified FOSs with high purity were then freeze-dried for 48 h.

2.7. *Nuclear Magnetic Resonance (NMR) Analysis*

The linkage between the fructosyl moieties of the biosynthesized fructans was characterized by ^{1}H NMR and ^{1}C NMR. About 35 mg of purified sample was dissolved in 500 µL deuterium oxide (D_2O) by bathing at 60 °C. The ^{1}H NMR and ^{1}C NMR spectra were recorded using an AVANCE III 400 MHz NMR spectrometer (Brucker Co., Billerica, MA, USA) at 60 °C, which used acetone (^{1}H = 2.225 ppm) and 1, 4-dioxan (^{13}C = 66.50 ppm) as internal reference standards.

2.8. M_W and Distribution Analysis

The M_W and distribution of levan were detected by high-performance gel filtration chromatography (HPGFC). The system was supplemented with a refractive-index detector and an Ultrahydrogel^{TM} Linear column (7.8 mm × 300 mm). The mobile phase is 0.1 N $NaNO_3$ with a 0.5 mL/min flow rate. The M_W reference standards are Dextran T-2000 (M_W = 2 × 10^6 Da, retention time is 14 min), Dextran T-300 (M_W = 3.0 × 10^5 Da, retention time is 15.8 min), Dextran T-150 (M_W = 1.4 × 10^5 Da, retention time is 16.4 min), Dextran T-10 (M_W = 9.7 × 10^3 Da, retention time is 19 min) and Dextran T-5 (M_W = 2.7 × 10^3 Da, retention time is 21 min). The detected temperature was 40 °C.

The degree of polymerization (DP) of FOS was determined by high-performance anion-exchange chromatography with pulsed amperometric detection (HPAEC-PAD). The column used in this system was a CarboPac PA200 column (3 mm × 250 mm) with the guard column CarboPac PA200 (3 mm × 50 mm). The eluent was 100 mM NaOH and 40 mM NaAc at the first 40 min, 100 mM NaOH and 400 mM NaAc at 40.1 min, and 100 mM NaOH and 40 mM NaAc between 40.1 and 60 min. The flow rate was 0.5 mL/min at 30 °C, and pH-Ag/AgCl as a reference electrode.

2.9. Molecular Dynamics Simulation

High-temperature molecular dynamics were performed using GROMACS (Version 2020.6) with an AMBER ff14SB force field [20,21]. The enzymes were solvated in TIP3P water and were relaxed through energy minimization to eliminate the error water insert. The system was heated to a pressure and temperature of 1 bar and 500 K using NVT and NPT ensemble balance with position constraint. The LINCS algorithm was used to constrain the hydrogen bonds in the system. The heavy atom of the protein is subjected to a position inhibition force with a constant of 1000 kJ mol^{-1} nm^{-2}. After equilibrium, the final output of the NPT simulation was subjected to without-position limitation. The root-mean-square deviation (RMSD) of backbone atom positions per 5 ps were calculated and analyzed using GROMACS analysis tools.

The root-mean-square fluctuation (RMSF) was determined at 280 K, and the other conditions were the same as above. The final structure was simulated by molecular dynamics of 100 ns.

All assays were performed in triplicate. Data management and analysis were performed using GraphPad Prism 5.0 (GraphPad Software, San Diego, CA, USA). All data are presented as the mean ± standard error of the mean.

3. Results and Discussion

3.1. Computer-Aided Enzyme Screening

Unlike the traditional BLAST tool, a computer-aided enzyme screening method combined with the Enzyme miner online server (https://loschmidt.chemi.muni.cz/enzymeminer/custom-sequences) [22] was employed to screen out the potential thermostable LS. High-temperature molecular dynamics simulations predicted the flexibilities of enzyme orthologs and thermostability. The RMSD value of the different microbial FSs is shown in Figure 1B, which can quantify the backbone atom movements of the protein. The Genbank accession numbers of these FSs are listed in Table S1. As a result, the LS from *C. diazotrophica* (Cedi-LS) exhibited the lowest RMSD value among all the enzymes, which means that Cedi-LS with the most rigid structure might have the best thermostability. Subsequently, a novel LS from mesophilic bacteria *P. orientalis* (Psor-LS) was screened on the template of Cedi-LS.

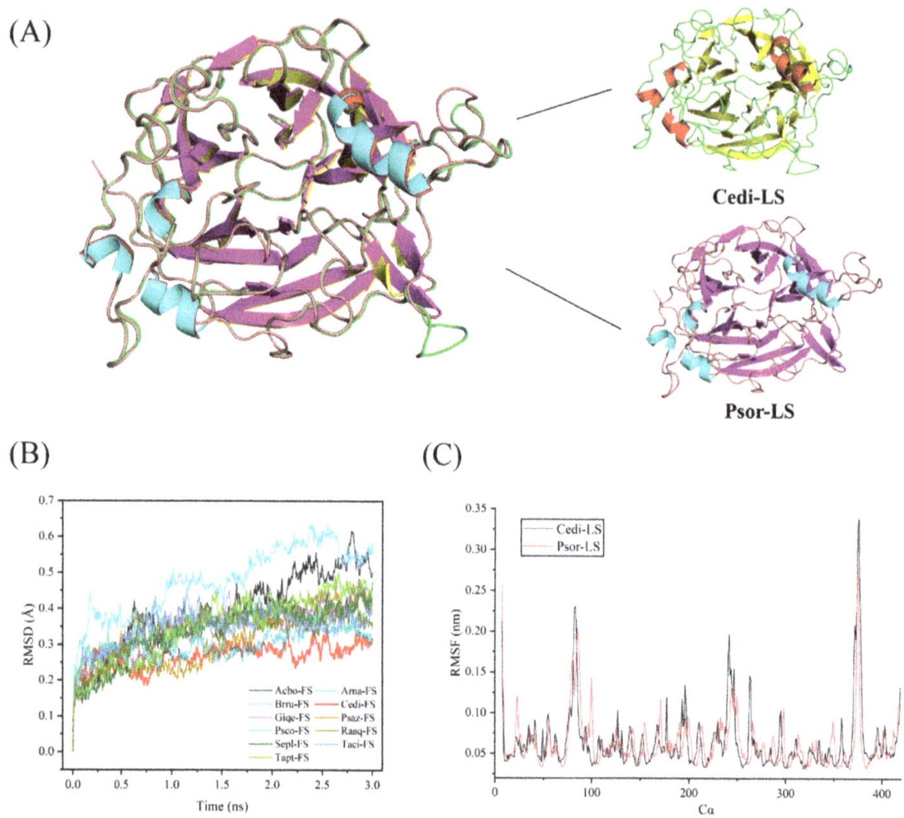

Figure 1. The structure and molecular dynamic simulation of Cedi-LS and Psor-LS. (**A**) The structure comparison of Cedi-LS and Psor-LS. The structures of Cedi-LS and Psor-LS are shown as carton. The superposition structure of Cedi-LS and Psor-LS are made by Pymol. (**B**) RMSD simulations of different microbial FSs at 500 K. (**C**) RMSF of the backbone Cα of modeled Cedi-LS and Psor-LS from the molecular dynamic simulation at 280 K.

The 3D structures of LSs were homologically modeled employing the crystal structures of *E. tasmaniensis* LS (PDB: 6FRW) as their template in the SWISSMODEL online server (https://swissmodel.expasy.org/) [23]. As shown in Figure 1A, Psor-LS showed a high coincidence degree in structure with Cedi-LS, except in loops 1 and 8. The RMSD value between two of the LSs was 0.117. The RMSF was used to study the movement of each residue in the enzyme and determine the flexibility of a particular region in the protein. As shown in Figure 1C, the RMSF results of Psor-LS and Cedi-LS showed a similar pattern, and the maximum D-value did not exceed 0.06 nm. Meanwhile, Cedi-LS and Psor-LS both showed low RMSF values in the whole structure (below 0.35 nm), indicating a very close relationship between the crystal structures of Cedi-LS and Psor-LS.

3.2. Expression and Purification of Psor-LS

The sequence of the gene encoding the Psor-LS has been deposited in the GenBank database. SDS-PAGE analysis of the recombinant protein from *P. orientalis* indicated a band around 44 kDa (Figure 2A), consistent with the calculated molecular weights of the LS (ExPASy Computer M_w tool, https://web.expasy.org/compute_pi/), suggesting a triumphant expression of the target enzyme in *E. coli*.

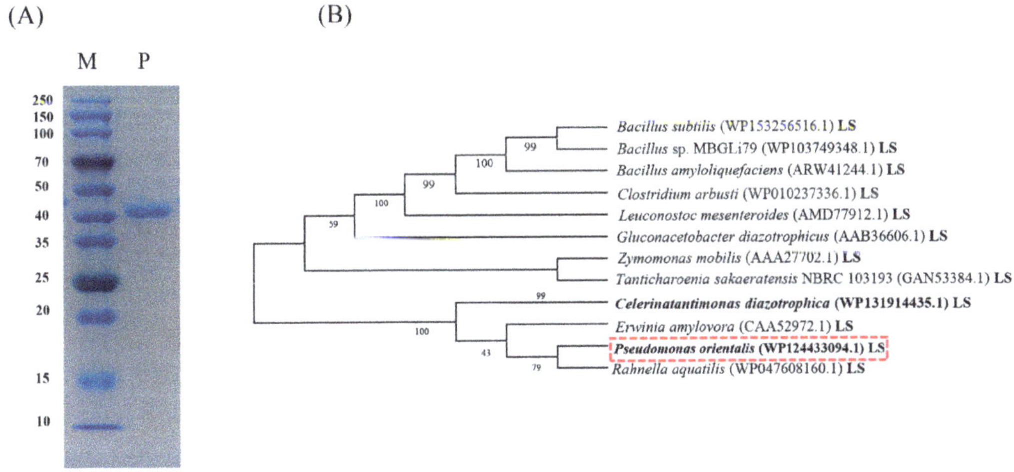

Figure 2. SDS-PAGE and phylogenetic tree. (**A**) Lane P represents the purified LS from *P. orientalis*, and lane M represents the protein marker with standard enzymes with the following molecular weights: 250, 150, 100, 70, 50, 40, 35, 25, 20, 15, and 10 kDa. (**B**) Phylogenetic tree of LSs from different sources. The phylogenetic tree was constructed by MEGA 5.1. The content in the bracket is the accession numbers in the GenBank database.

Amino acid sequence identity analysis was performed by EMBI 225 (https://www.ebi.ac.uk/Tools/services/web/tool). The Psor-LS showed the highest identity of 80% with the *R. aquatilis* LS and had more than 70% identity to LSs from *C. diazotrophica*, *Brenneria* sp. EniD312 and *E. amylovora*. By contrast, the Psor-LS showed the lowest identity of 24% with the *Leuconostoc mesenteroides* LS. They only had less than 30% identity to the LSs from *B. subtilis*, *B. amyloliquefaciens*, and *Clostridium arbusti*. The evolutional relationship to LSs from different sources is shown in Figure 2B.

3.3. Effect of pH on the Activity and T/H Ratio of Psor-LS

As shown in Figure 3A, Psor-LS showed relatively high activity (>80%) at pH values ranging from 5.0 to 7.5 but dropped when pH was below 5.0 or above 7.5. Figure 3B shows the effect of pH on the transfructosylation activity of Psor-LS. Unlike Cedi-LS, the Psor-LS was sensitive to pH since it exhibited the maximum transfructosylation activity at pH 6.0. However, less than 50% of the activity remained when pH was shifted from 6.0 to 4.5 or 8.0. Unlike Cedi-LS, the T/H of Psor-LS was lower than 1.0 in the whole pH range of 4.0 to 9.0, which means that the hydrolysis reaction was dominant for Psor-LS (Figure 3C). Most LSs exhibited optimal activity at slightly acidic (5.5) or neutral pH (7.0). For instance, the LS from *Brenneria goodwinii* [24] showed optimal activity at pH 5.5 and 6.0. The *L. mesenteroides* MTCC10508 LS had the highest activity at pH 5.5 [25].

3.4. Effect of Temperature on the Activity and T/H of Psor-LS

The effect of temperature on Psor-LS activity was measured at an optimal pH of 6.5. As a result, the optimal temperature of Psor-LS was 65 °C (Figure 4A), the same as Cedi-LS. Psor-LS could continue relatively high activity (>70%) at temperatures ranging from 45 to 70 °C, but this dropped when the temperature was above 75 °C. A slight decrease was observed below 45 °C, but Psor-LS could retain more than 50% of its relative activity at 35 °C. The variation in transfructosylation activity is shown in Figure 4B. Psor-LS could retain 80% of its transfructosylation activity at 60–70 °C, suggesting that Psor-LS exhibits excellent transfructosylation ability at high temperatures. Unlike Cedi-LS, a relatively high T/H value of Psor-LS (>1.0) was obtained below 45 °C (Figure 4C).

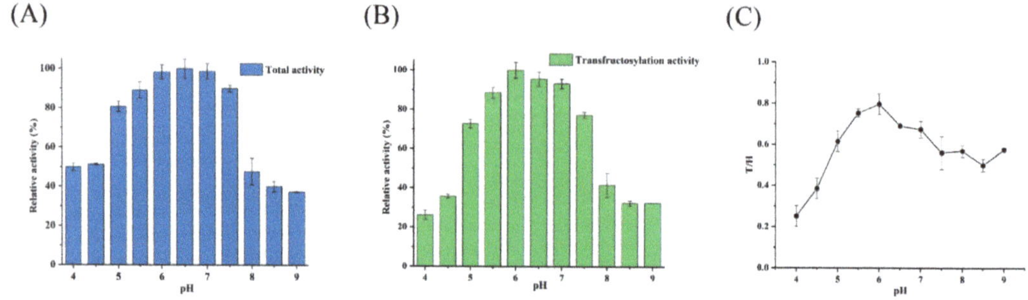

Figure 3. Effect of pH on the activity and T/H of the recombinant LS. (**A**) Effect of pH on the total activity of Psor-LS. (**B**) Effect of pH on the transfructosylation activity of Psor-LS. (**C**) Effect of pH on the T/H of Psor-LS. All of the values were the mean of triplicate experiments.

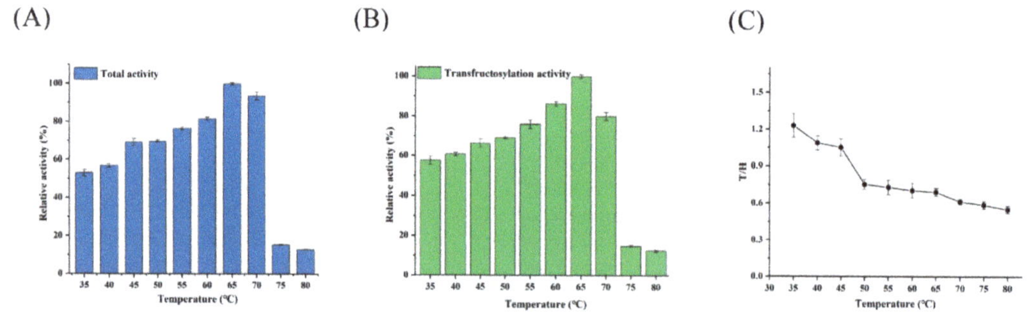

Figure 4. Effect of temperature on the activity and T/H of the recombinant LS. (**A**) Effect of temperature on the total activity of Psor-LS. (**B**) Effect of temperature on the transfructosylation activity of Psor-LS. (**C**) Effect of temperature on the T/H of Psor-LS. All of the values were the mean of triplicate experiments.

Many previous studies have shown that LS showed higher transfructosylation activities at lower temperatures, while hydrolysis activity shows the opposite. For example, the LS from *Z. mobilis* exhibited the highest transfructosylation activity at 30 °C, while its hydrolase activity optimal temperature was 50 °C [26]. The *L. reuteri* LTH5448 LS showed the highest transfructosylation and hydrolase activity at 35 °C and 45 °C, respectively [9]. This characteristic is also related to the T/H of LS. At higher temperatures (>45 °C), hydrolysis is the dominant reaction for *B. goodwinii* LS [24]. The LS from *L. reuteri* LTH5448 exhibited a higher hydrolysis ability above 50 °C, with its T/H below 1.0 [9]. By contrast, Psor-LS exhibited high transfructosylation activity at high temperatures (>55 °C), which was better than most LSs.

3.5. Thermostability Determination of Psor-LS

Melting temperature (T_m) is related to the structural stability of enzymes, and high T_m generally represents high structural stability and thermostability [27]. The T_m of Psor-LS was determined to be 65.1 °C (Figure 5A), which was significantly higher than that of FSs from *L. reuteri* 121 (50 °C) and IS from *L. gasseri* (55 °C) [27,28]. To further test the stability of Psor-LS at different temperatures, the enzyme was incubated at 45, 55, and 65 °C, respectively. Psor-LS retained 46% of initial activity at 55 °C for 9 h and half of the initial activity at 45 °C for 60 h (Figure 5B). The half-life was 69 h at 45 °C and 7.5 h at 55 °C (Figure 5C). However, Psor-LS was almost inactive when solely incubated at 65 °C for 10 min.

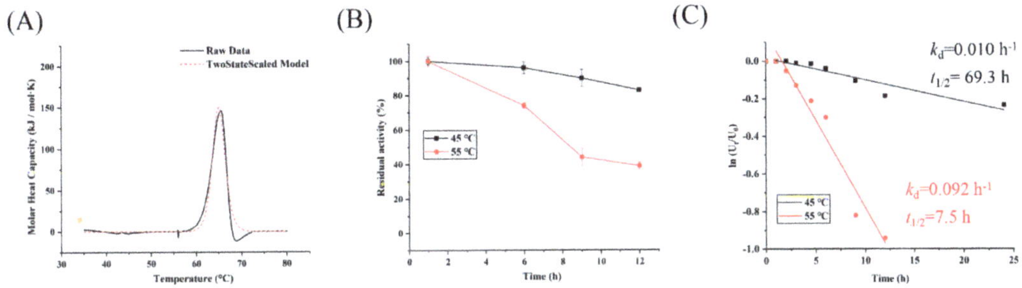

Figure 5. T_m and thermostability of recombinant LS. (**A**) T_m of Psor-LS. (**B**) Thermostability of Psor-LS at 45 and 55 °C. (**C**) The $t_{1/2}$ value of Psor-LS at 45 and 55 °C.

Except for the LS from *L. reuteri* LTH5448, most LSs showed low thermostability at temperatures above 50 °C (Table 1). For instance, the *B. licheniformis* RN-01 LS retained less than 50% of its initial activity after 1 h of incubation at 50 °C [29]. The LS from *Bacillus* sp. TH4-2 lost 50% of its initial activity at 60 °C for 30 min [30]. The LS from *B. subtilis* NRC could retain 60% of its activity after incubating at 50 °C for 2 h [31]. Before this study, the LS from *Geobacillus stearothermophilus* was reported as the most thermostable LS since it could retain more than 95% of initial activity at 4-47 °C for 6 h [32]. However, the enzyme would rapidly lose activity at higher temperatures (57 °C).

Table 1. Thermostability of LSs from different microorganisms.

Microorganisms	Optimal Temp. (°C)	Thermostability	Reference
P. orientalis	65	The half-life was 69 h at 45 °C and 7.5 h at 55 °C.	This study
B. licheniformis RN-01	50	Half of the initial activity was lost after 1 h at 50 °C.	[29]
Bacillus sp. TH4-2	60	Half of the initial activity was lost after 30 min at 60 °C.	[30]
G. stearothermophilus	57	More than 95% of the initial activity was retained at 4–47 °C for 6 h.	[32]
Z. mobilis	30	The activity lost at 50 °C for 15 min.	[33]
Brenneria sp. EniD312	45	The half-life was 2 h at 45 °C and 1.2 h at 55 °C	[34]
B. goodwinii	40	The activity lost after 0.5 h of incubation at 50 °C	[24]
L. reuteri LTH5448	35	The activity remained 63.8% at 55 °C for 12 h	[9]
Bacillus subtilis NRC	45	The activity remained 60% at 50 °C for 12 h	[31]
B. circulans	35	The half-life was 130 min at 50 °C	[35]

3.6. Kinetic Parameters Determination

Differently from previous studies, the kinetics of fructose and glucose "release" in the reaction have been calculated. The rate of glucose release is the hydrolysis rate plus the transferase reaction rate. The K_m of hydrolysis and transferase reaction is expressed as "K_m^H" and "K_m^T", respectively [36]. The kinetic parameters of Cedi-LS and Psor-LS are shown in Table 2. The determined K_m^H and K_m^T values of Cedi-LS were 57 ± 2 and 202 ± 7 mM, respectively. Cedi-LS showed higher k_{cat} values for transfructosylation than for hydrolysis, and its k_{cat}^H/k_{cat}^T was 1.35, significantly higher than that of *B. subtilis* LS (0.37) [36]. The K_m^H (117 ± 8 mM) and k_{cat}^H (620 ± 12 s^{-1}) of Psor-LS were higher than Cedi-LS, indicating that Psor-LS is more favorable to hydrolyze sucrose than Cedi-LS. Nevertheless, the kinetic parameter of the transferase reaction of Psor-LS did not conform to the nonlinear least square regression method.

Table 2. Kinetic parameters of Cedi-LS and Psor-LS.

LS	K_m (mM)		k_{cat} (s^{-1})		k_{cat}/K_m (mM^{-1} s^{-1})	
	Hydrolysis	Transfer	Hydrolysis	Transfer	Hydrolysis	Transfer
Cedi-LS	57 ± 2	202 ± 7	332 ± 22	449 ± 13	5.80	2.23
Psor-LS	117 ± 8	-	620 ± 12	-	5.27	-

3.7. The Effect of Sucrose Concentration on the Activity and T/H of Psor-LS

Sucrose concentration is an essential factor in the activity and T/H of LSs. As a result, Psor-LS exhibited the highest activity at 30% sucrose (Figure 6A). The Psor-LS could retain over 80% of its activity at substrate concentrations ranging from 10 to 60%, suggesting that Psor-LS showed a broad sucrose concentration spectrum for its activity. In this study, the total activity of Psor-LS could be saturated at 30% sucrose. A similar result was reported in the LSs from *B. methylotrophicus* SK 21.002 [7] and *Z. mobilis* [37]. The "E-F·F-G" complex accumulation and the nonproductive binding are possibilities for this inhibitory phenomenon [37]. Although the activity was decreased, the T/H of the Psor-LS showed an upward trend with the increase in sucrose concentration (Figure 6B). The Psor-LS showed relatively high T/H (>1) at 40% sucrose and exhibited the highest T/H at 60% sucrose (1.3). By comparison, the *E. amylovora* LS had a T/H of about 4 at 100 mM sucrose, and its transfructosylation reaction could reach a maximum of 97% at 1.5 M sucrose [5].

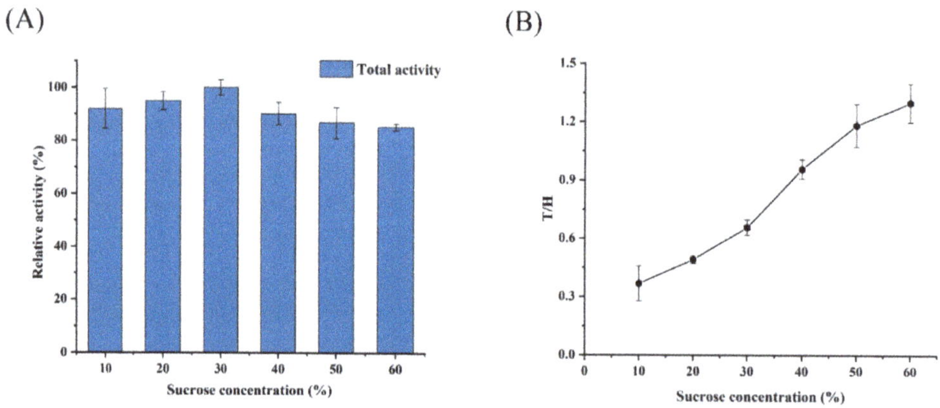

Figure 6. Effect of sucrose concentration on the activity and T/H of the recombinant LSs. (**A**) Effect of sucrose concentration on the total activity of Psor-LS. (**B**) Effect of sucrose concentration on the T/H of Psor-LS. All of the values were the mean of triplicate experiments.

3.8. Effect of Enzyme Concentration on the Levan Production and T/H of Psor-LS

Different enzyme dosages ranging from 10 to 100 μg/mL at 30% sucrose were employed to optimize the levan production. Since enzyme concentration was related to levan production, the ratio of levan production and enzyme concentration (P/E) was evaluated as "input–output" in this study. As a result, the highest P/E value of Psor-LS was exhibited at 25 μg/mL enzyme, suggesting that both enzyme and substrate were the maximum output in this enzyme dosage (input) (Figure 7A). A comparable P/E ratio was observed at 30 μg/mL enzyme concentration compared to 25 μg/mL. However, it dropped remarkably when enzyme concentration was below 25 or above 30 μg/mL. The Psor-LS showed a growing T/H with increased enzyme concentration (Figure 7B), and it exhibited an equivalent transfructosylation and hydrolysis reaction at 100 μg/mL enzyme. A similar result for IS from *Lactobacillus jensenii* was reported when the T/H was increased with increased enzyme dosage [38]. By contrast, the *B. subtilis* LS showed high T/H (2.7) at 0.1 U/mL enzyme concentration but low T/H (0.7) when enzyme concentration is increased to 10 U/mL.

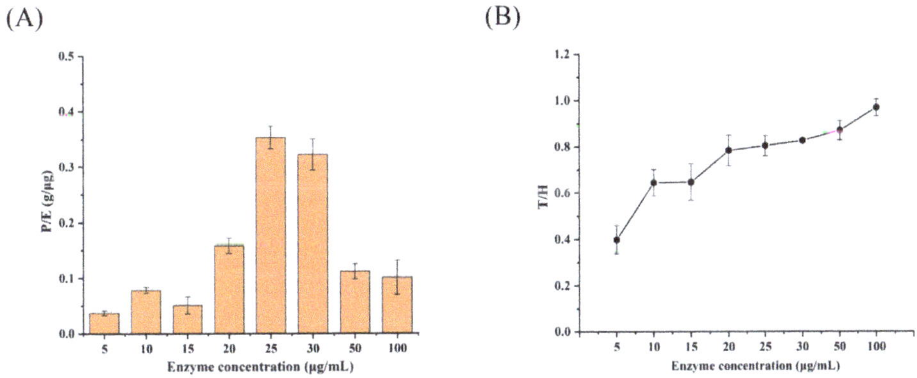

Figure 7. Effect of enzyme concentration on the P/E and T/H of the recombinant LS; P/E is the ratio of levan production and enzyme concentration. (**A**) Effect of enzyme concentration on P/E of Psor-LS. (**B**) Effect of enzyme concentration on T/H of Psor-LS. All of the values are the mean of triplicate experiments.

3.9. Biological Production of Psor-LS

The biotransformation process of Psor-LS is shown in Figure 8A. Rapid sucrose consumption was shown in the first 1.5 h. The consumption rate slowed down in the next 1.5 h. After 3 h, the sucrose concentration was almost unchanged, consuming at a very low rate. When the reaction reaches equilibrium, the maximum conversion ratio of the transfructosylated product to sucrose was 29.2% at 3 h (Figure 8B). Like Cedi-LS, the product produced by Psor-LS decreased slowly after 3 h. By comparison, the LSs from *B. methylotrophicus* SK 21.002 and *B. licheniformis* NS032 could also produce levan effectively, and their conversion ratios were 33 and 25%, respectively [7,39].

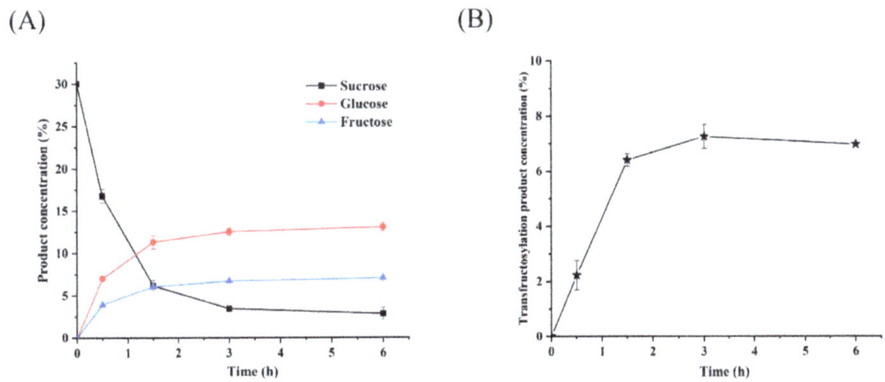

Figure 8. Biotransformation process and conversion ratio of recombinant LS. (**A**) The variation in sucrose, glucose, and fructose concentration in the biotransformation process of Psor-LS. (**B**) The variation in conversion ratio in the biotransformation process of Psor-LS. All of the values are the mean of triplicate experiments.

As reported, the T/H of Cedi-LS was 1.3 at 65 °C, higher than that of Psor-LS (0.8). T/H is considered to reflect the transfructosylation ability of LS and continue to the product distribution of LS. For instance, the *B. subtilis* LS showed higher T/H (2.7) at 0.1 U/mL enzyme concentration and produced HMW levan. When the enzyme concentration was 1 U/mL, the enzyme produced LMW levan with lower T/H (1.0) [15]. As shown in Figure 8A, the residual fructose in the Psor-LS system is significantly higher than that in Cedi-LS, consistent with its relatively lower T/H value. Meanwhile, the residual sucrose and glucose in Psor-LS system are lower than that in Cedi-LS, which indicates that Psor-LS has higher glucose utilization than Cedi-LS.

3.10. *Effect of Temperature on the Product Distribution of Cedi-LS and Psor-LS*

Many factors were considered to be potential reasons for the product specificity of LSs, such as sucrose concentration [5] and enzyme concentration [15]. The temperature could also affect the product specificity [17]. However, how temperature could affect the product specificity of LS remains unclear. To investigate the effect of temperature on the product distribution of LS, we reduced the reaction temperature from 65 to 35 °C. At 35 °C, the product conversion ratios of Cedi-LS and Psor-LS were 41.4 and 37.1%, respectively. The T/H values of the two enzymes were 2.3 and 1.0. Moreover, the viscosity of the reaction solution of Cedi-LS was increased when the temperature was decreased to 45 and 35 °C. At 2 h reaction time, the solution showed a "gel-similar" phenomenon (Figure 9C), which is much different from that of the Psor-LS solution (Figure 9D).

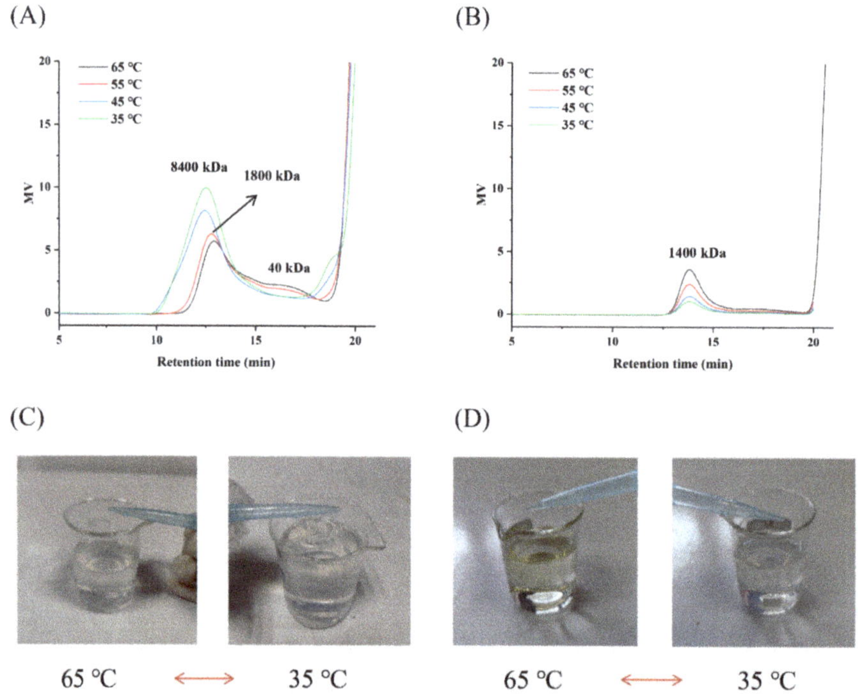

Figure 9. Effect of temperature on the product distribution and solution status of the recombinant LSs. (**A**) Effect of temperature on the product distribution of Cedi-LS. (**B**) Effect of temperature on the product distribution of Psor-LS. (**C**) Effect of temperature on the reaction mixture of Cedi-LS. (**D**) Effect of temperature on the reaction mixture of Psor-LS.

To further determine the possible change in product distribution at different temperatures, the reaction mixture components were analyzed by HPGFC in detail (Figure 9A,B). At its optimal temperature of 65 °C, the Cedi-LS could simultaneously produce FOS, LMW (4.1×10^4 Da), and HMW (1.8×10^6 Da) levan in the reaction mixture. On the contrary, the Psor-LS specifically produced FOS and HMW (1.4×10^6 Da) levan without LMW levan. When the temperature was decreased, the levan produced by Cedi-LS showed a higher M_W that reached 8.4 × 10^6 Da at 35 °C. Simultaneously, a low temperature results in higher production of HMW levan. Therefore, the increase in M_W and production of HMW levan were supposed to result in a higher viscosity of the reaction solution, as shown in Figure 9C. The low temperature increased the HMW levan in many LSs [40]. For instance, the LS from *R. aquatilis* ATCC 33071 mainly produced FOS at 55–60 °C, while it synthesized HMW levan (1 × 10^6 Da) at low temperature (37 °C) [41]. The production of HMW levan increased obviously at low temperatures (4 °C) in *Z. mobilis* LS [17]. On the contrary, the production of HMW levan in Psor-LS decreased as the temperature decreased. This means that lower temperatures promoted the synthesis of FOS in Psor-LS. As far as the authors are concerned, this is the first LS that prefers to produce FOSs rather than HMW levan when the temperature is decreased. Moreover, the M_W of the levan from Psor-LS was not changed, indicating that the temperature has different effects on the product distribution of Cedi-LS and Psor-LS.

3.11. Product Purification and Analysis

The residual enzyme in the mixture was removed by Sevage reagent, and the polysaccharide was separated by multiple ethanol precipitation. When the final ethanol concentration was 60%, the polysaccharide produced from Psor-LS was obtained. The obtained precipitate was vacuum freeze-dried for 48 h to remove moisture altogether (Figure 10A). The products of Psor-LS were identified as β-(2, 6) levan and levan-type FOSs by the NMR analysis (Figure S1). The ^{1}H spectrograms were compared with those of *L. reuteri* LTH5448 LS and *L. jensenii* IS [9,38]. Meanwhile, the ^{13}C chemical shifts reported for biosynthesized levan are compared in Table S2. The result revealed that the polysaccharide synthesized by Psor-LS was β-(2, 6) levan.

(A)

(B)

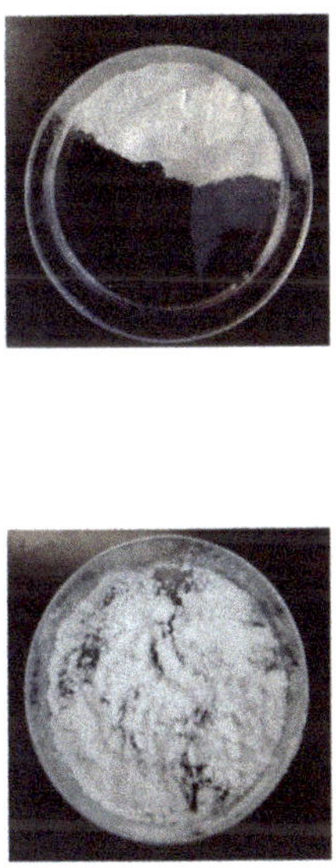

(C)

Enzyme	FOS	LMW-levan	HMW-levan
Cedi-LS	✓	✓	✓
Psor-LS	✓	✗	✓

(D)

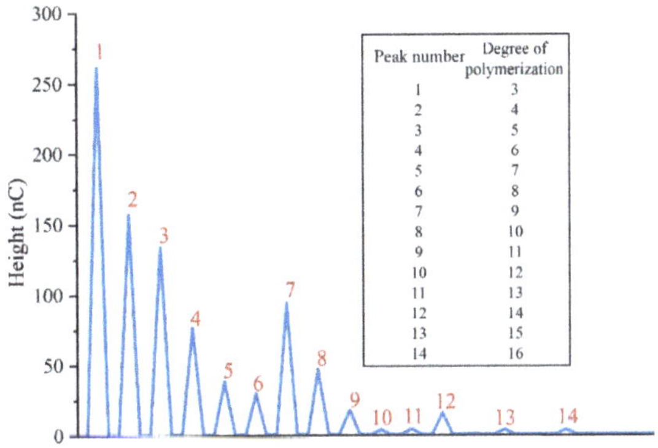

Figure 10. Products from the recombinant LSs. (**A**) The purified HMW levan from Psor-LS. (**B**) The purified FOS from Psor-LS. (**C**) The component analysis of products from Cedi-LS and Psor-LS. (**D**) The HPIC spectrogram of purified FOS from Psor-LS.

The M_W of levan synthesized by Psor-LS was 1.4 × 10^6 Da (65 °C). Generally, the M_W of HMW levan produced by LS from different microorganisms were different, such as the

LSs from *T. sakaeratensis* ($1.0-6.8 \times 10^5$ Da) [42], *A. diazotrophicus* SRT4 (2.0×10^6 Da) [43] and *Bacillus aryabhattai* (5.3×10^7 Da) [44]. FOS was purified by activated carbon chromatography to remove the sucrose and monosaccharides and dried in a freeze dryer for 48 h. The purified FOS is shown in Figure 10B. The purity of FOS produced by Psor-LS with DP $\leq$ 16 could reach more than 90%, but quantitative comparison cannot be carried out due to the significant loss in the purification process (Figure 10D).

4. Conclusions

In this study, a novel thermostable LS from *P. orientalis* was identified. The Psor-LS retained 46% of its initial activity at 55 °C for 9 h and 50% at 45 °C for 60 h. Meanwhile, there are noticeable differences in the product distribution between the Cedi-LS and Psor-LS. The M_W of levan synthesized by Cedi-LS was increased from 1.8×10^6 Da (65 °C) to 8.4×10^6 Da (35 °C). On the contrary, the decrease in temperature did not significantly affect the product distribution of Psor-LS. At 65 °C, the Psor-LS would specifically produce FOSs and HMW levan without LMW levan. Notably, at 35 °C, the reaction equilibrium of Psor-LS from sucrose (30%) was 37%, and a certain amount of FOS (DP $\leq$ 16) was obtained among them.

Supplementary Materials: The following supporting information can be downloaded at: https://www.mdpi.com/article/10.3390/polym15061435/s1, Figure S1: Nuclear magnetic resonance (NMR) analysis spectrogram of the products from Psor-LS; Table S1: The genbank number and sequnce length of different microbial FSs; Table S2: ^{13}C chemical shifts reported for biosynthesized levan [7,45].

Author Contributions: C.G.: Supervision. X.Z.: Data analysis, writing—original draft. D.N.: Software. W.Z.: Methodology and software. W.X.: Writing—review and editing. W.M.: Co-Supervision, project administration. All authors have read and agreed to the published version of the manuscript.

Funding: This work was supported by the National Natural Science Foundation of China (22278183), the National Key R&D Program of China (2022YFD2101400), the Natural Science Foundation of Jiangsu Province (BK20210463), the Special Fund from Post-doctor Innovation Research Program of Shandong Province (SDCX-ZG-202203049), and the Independent Projects for Young Scholars at Jiangnan University (JUSRP122011).

Institutional Review Board Statement: Not applicable.

Informed Consent Statement: Not applicable.

Data Availability Statement: Data presented in this study are available on request from the corresponding author.

Conflicts of Interest: The authors declare no conflict of interest.

References

1. Lammens, W.; Le Roy, K.; Schroeven, L.; Van Laere, A.; Rabijns, A.; Van den Ende, W. Structural insights into glycoside hydrolase family 32 and 68 enzymes: Functional implications. *J. Exp. Bot.* **2009**, *60*, 727–740. [CrossRef] [PubMed]
2. Belghith, K.; Dahech, I.; Belghith, H.; Mejdoub, H. Microbial production of levansucrase for synthesis of fructooligosaccharides and levan. *Int. J. Biol. Macromol.* **2012**, *50*, 451–458. [CrossRef] [PubMed]
3. Ozimek, L.; Kralj, S.; van der Maarel, M.; Dijkhuizen, L. The levansucrase and inulosucrase enzymes of *Lactobacillus reuteri* 121 catalyse processive and non-processive transglycosylation reactions. *Microbiology* **2006**, *152*, 1187–1196. [CrossRef] [PubMed]
4. Anwar, M.; Kralj, S.; Pique, A.; Leemhuis, H.; van der Maarel, M.; Dijkhuizen, L. Inulin and levan synthesis by probiotic *Lactobacillus gasseri* strains: Characterization of three novel fructansucrase enzymes and their fructan products. *Microbiology* **2010**, *156*, 1264–1274. [CrossRef]
5. Caputi, L.; Nepogodiev, S.; Malnoy, M.; Rejzek, M.; Field, R.; Benini, S. Biomolecular Characterization of the Levansucrase of *Erwinia amylovora*, a Promising Biocatalyst for the Synthesis of Fructooligosaccharides. *J. Agric. Food Chem.* **2013**, *61*, 12265–12273. [CrossRef]
6. Vigants, A.; Kruce, R.; Bekers, M.; Zikmanis, P. Response of *Zymomonas mobilis* levansucrase activity to sodium chloride. *Biotechnol. Lett.* **1998**, *20*, 1017–1019. [CrossRef]
7. Zhang, T.; Li, R.; Qian, H.; Mu, W.; Miao, M.; Jiang, B. Biosynthesis of levan by levansucrase from *Bacillus methylotrophicus* SK 21.002. *Carbohydr. Polym.* **2014**, *101*, 975–981. [CrossRef]

8. Paul, A.; Samaddar, N.; Dutta, D.; Bagchi, A.; Chakravorty, S.; Chakraborty, W.; Gachhui, R. Mercuric Ion Stabilizes Levansucrase Secreted by *Acetobacter nitrogenifigens* Strain RG1(T). *Protein J.* **2011**, *30*, 262–272. [CrossRef]
9. Ni, D.; Xu, W.; Bai, Y.; Zhang, W.; Zhang, T.; Mu, W. Biosynthesis of levan from sucrose using a thermostable levansucrase from *Lactobacillus reuteri* LTH5448. *Int. J. Biol. Macromol.* **2018**, *113*, 29–37. [CrossRef]
10. Gao, S.; Qi, X.; Hart, D.; Gao, H.; An, Y. Expression and Characterization of Levansucrase from *Clostridium acetobutylicum*. *J. Agric. Food Chem.* **2017**, *65*, 867–871. [CrossRef]
11. Ragab, T.; Shalaby, A.; Awdan, S.; El-Bassyouni, G.; Salama, B.; Helmy, W.; Esawy, M. Role of levan extracted from bacterial honey isolates in curing peptic ulcer: In vivo. *Int. J. Biol. Macromol.* **2020**, *142*, 564–573. [CrossRef]
12. Esawy, M.; Amer, H.; Gamal-Eldeen, A.; El Enshasy, H.; Helmy, W.; Abo-Zeid, M.; Malek, R.; Ahmed, E.; Awad, G. Scaling up, characterization of levan and its inhibitory role in carcinogenesis initiation stage'. *Carbohydr. Polym.* **2013**, *95*, 578–587. [CrossRef]
13. Esawy, M.; Ahmed, E.; Helmy, W.; Mansour, N.; El-Senousy, W.; El-Safty, M. Production of levansucrase from novel honey *Bacillus subtilis* isolates capable of producing antiviral levans. *Carbohydr. Polym.* **2011**, *86*, 823–830. [CrossRef]
14. Gamal, A.; Abbas, H.; Abdelwahed, N.; Kashef, M.; Mahmoud, K.; Esawy, M.; Ramadan, M. Optimization strategy of *Bacillus subtilis* MT453867 levansucrase and evaluation of levan role in pancreatic cancer treatment. *Int. J. Biol. Macromol.* **2021**, *182*, 1590–1601. [CrossRef] [PubMed]
15. Raga-Carbajal, E.; Carrillo-Nava, E.; Costas, M.; Porras-Dominguez, J.; Lopez-Munguia, A.; Olvera, C. Size product modulation by enzyme concentration reveals two distinct levan elongation mechanisms in *Bacillus subtilis* levansucrase. *Glycobiology* **2016**, *26*, 377–385. [CrossRef]
16. Kirtel, O.; Menendez, C.; Versluys, M.; Van den Ende, W.; Hernandez, L.; Oner, E. Levansucrase from Halomonas smyrnensis AAD6(T): First halophilic GH-J clan enzyme recombinantly expressed, purified, and characterized. *Appl. Microbiol. Biotechnol.* **2018**, *102*, 9207–9220. [CrossRef]
17. Santos-Moriano, P.; Fernandez-Arrojo, L.; Poveda, A.; Jimenez-Barbero, J.; Ballesteros, A.; Plou, F. Levan versus fructooligosaccharide synthesis using the levansucrase from *Zymomonas mobilis*: Effect of reaction conditions. *J. Mol. Catal. B Enzym.* **2015**, *119*, 18–25. [CrossRef]
18. Bradford, M. A rapid and sensitive method for the quantitation of microgram quantities of protein utilizing the principle of protein-dye binding. *Anal. Biochem.* **1976**, *72*, 248–254. [CrossRef]
19. Kuhn, R.C.; Filho, F.M. Purification of fructooligosaccharides in an activated charcoal fixed bed column. *New Biotechnol.* **2010**, *27*, 862–869. [CrossRef]
20. Abraham, M.; Murtola, T.; Schulz, R.; P'alla, S.; Smith, J.; Hessa, B.; Lindahl, E. GROMACS: High performance molecular simulations through multi-level parallelism from laptops to supercomputers. *SoftwareX* **2015**, *1*, 19–25. [CrossRef]
21. Maier, J.; Martinez, C.; Kasavajhala, K.; Wickstrom, L.; Hauser, K.; Simmerling, C. ff14SB: Improving the Accuracy of Protein Side Chain and Backbone Parameters from ff99SB. *J. Chem. Theory Comput.* **2015**, *11*, 3696–3713. [CrossRef] [PubMed]
22. Hon, J.; Borko, S.; Stourac, J.; Prokop, Z.; Zendulka, J.; Bednar, D.; Martinek, T.; Damborsky, J. EnzymeMiner: Automated mining of soluble enzymes with diverse structures, catalytic properties and stabilities. *Nucleic Acids Res.* **2020**, *48*, W104–W109. [CrossRef] [PubMed]
23. Waterhouse, A.; Bertoni, M.; Bienert, S.; Studer, G.; Tauriello, G.; Gumienny, R.; Heer, F.; de Beer, T.; Rempfer, C.; Bordoli, L.; et al. SWISS-MODEL: Homology modelling of protein structures and complexes. *Nucleic Acids Res.* **2018**, *46*, W296–W303. [CrossRef] [PubMed]
24. Liu, Q.; Yu, S.; Zhang, T.; Jiang, B.; Mu, W. Efficient biosynthesis of levan from sucrose by a novel levansucrase from *Brenneria goodwinii*. *Carbohydr. Polym.* **2017**, *157*, 1732–1740. [CrossRef]
25. Jadaun, J.; Narnoliya, L.; Agarwal, N.; Singh, S. Catalytic biosynthesis of levan and short-chain fructooligosaccharides from sucrose-containing feedstocks by employing the levansucrase from *Leuconostoc mesenteroides* MTCC10508. *Int. J. Biol. Macromol.* **2019**, *127*, 486–495. [CrossRef]
26. Sangiliyandi, G.; Chandra Raj, K.; Gunasekaran, P. Elevated temperature and chemical modification selectively abolishes levan forming activity of levansucrase of *Zymomonas mobilis*. *Biotechnol. Lett.* **2004**, *21*, 179–182. [CrossRef]
27. Ni, D.; Kirtel, O.; Yin, D.; Xu, W.; Chen, Q.; Oner, E.; Mu, W. Improving the catalytic behaviors of *Lactobacillus*-derived fructansucrases by truncation strategies. *Enzyme Microb. Technol.* **2021**, *149*, 109857. [CrossRef]
28. Ni, D.; Zhang, S.; Kirtel, O.; Xu, W.; Chen, Q.; Oner, E.; Mu, W. Improving the thermostability and catalytic activity of an inulosucrase by rational engineering for the biosynthesis of microbial inulin. *J. Agric. Food Chem.* **2021**, *69*, 13125–13134. [CrossRef]
29. Klaewkla, M.; Pichyangkura, R.; Charoenwongpaiboon, T.; Wangpaiboon, K.; Chunsrivirot, S. Computational design of oligosaccharide producing levansucrase from *Bacillus licheniformis* RN-01 to improve its thermostability for production of levan-type fructooligosaccharides from sucrose. *Int. J. Biol. Macromol.* **2020**, *160*, 252–263. [CrossRef]
30. Ben Ammar, Y.; Matsubara, T.; Ito, K.; Izuka, M.; Limpaseni, T.; Pongsawasdi, P.; Minamiura, N. Characterization of a thermostable levansucrase from *Bacillus* sp TH4-2 capable of producing high molecular weight levan at high temperature. *J. Biotechnol.* **2002**, *99*, 111–119. [CrossRef]
31. Salama, B.; Helmy, W.; Ragab, T.; Ali, M.; Taie, H.; Esawy, M. Characterization of a new efficient low molecular weight *Bacillus subtilis* NRC 16 levansucrase and its levan. *J. Basic Microbiol.* **2019**, *59*, 1004–1015. [CrossRef]

32. Inthanavong, L.; Tian, F.; Khodadadi, M.; Karboune, S. Properties of *Geobacillus Stearothermophilus* levansucrase as potential biocatalyst for the synthesis of levan and fructooligosaccharides. *Biotechnol. Prog.* **2013**, *29*, 1405–1415. [CrossRef]
33. Li, S.; Yan, Y.; Zhou, Z.; Yu, H.; Zhan, Y.; Zhang, W.; Chen, M.; Lu, W.; Ping, S.; Lin, M. Single amino acid residue changes in subsite −1 of levansucrase from *Zymomonas mobilis* 10232 strongly influence the enzyme activities and products. *Mol. Biol. Rep.* **2011**, *38*, 2437–2443. [CrossRef]
34. Xu, W.; Ni, D.; Yu, S.; Zhang, T.; Mu, W. Insights into hydrolysis versus transfructosylation: Mutagenesis studies of a novel levansucrase from *Brenneria* sp. EniD312. *Int. J. Biol. Macromol.* **2018**, *116*, 335–345. [CrossRef]
35. El-Refai, H.; Abdel-Fattah, A.; Mostafa, F. Enzymic synthesis of levan and fructo-oligosaccharides by *Bacillus circulans* and improvement of levansucrase stability by carbohydrate coupling. *World J. Microbiol. Biotechnol.* **2009**, *25*, 821–827. [CrossRef]
36. Ortiz-Soto, M.; Porras-Dominguez, J.; Rodriguez-Alegria, M.; Morales-Moreno, L.; Diaz-Vilchis, A.; Rudino-Pinera, E.; Beltran-Hernandez, N.; Rivera, H.; Seibel, J.; Lopez Munguia, A. Implications of the mutation S164A on *Bacillus subtilis* levansucrase product specificity and insights into protein interactions acting upon levan synthesis. *Int. J. Biol. Macromol.* **2020**, *15*, 898–908. [CrossRef]
37. Okuyama, M.; Serizawa, R.; Tanuma, M.; Kikuchi, A.; Sadahiro, J.; Tagami, T.; Lang, W.; Kimura, A. Molecular insight into regioselectivity of transfructosylation catalyzed by GH68 levansucrase and beta-fructofuranosidase. *J. Biol. Chem.* **2021**, *296*, 100398. [CrossRef]
38. Ni, D.; Chen, Z.; Xu, W.; Zhang, W.; Mu, W. Efficient production of inulin and oligosaccharides using thermostable inulosucrase from *Lactobacillus jensenii*. *Int. J. Biol. Macromol.* **2020**, *165*, 1250–1257. [CrossRef]
39. Kekez, B.; Gojgic-Cvijovic, G.; Jakovljevic, D.; Stefanovic Kojic, J.; Markovic, M.; Beskoski, V.; Vrvic, M. High levan production by *Bacillus licheniformis* NS032 using ammonium chloride as the sole nitrogen source. *Appl. Biochem. Biotechnol.* **2015**, *175*, 3068–3083. [CrossRef]
40. Oner, E.; Hernandez, L.; Combie, J. Review of Levan polysaccharide: From a century of past experiences to future prospects. *Biotechnol. Adv.* **2016**, *34*, 827–844. [CrossRef]
41. Kang, S.; Lee, J.; Park, Y.; Lee, C.; Kim, S.; Chang, B.; Kim, C.; Seo, J.; Rhee, S.; Jung, S.; et al. Secretory production of *Rahnella aquatilis* ATCC 33071 levansucrase expressed in Escherichia coli. *Int. J. Biol. Macromol.* **2004**, *14*, 1232–1238.
42. Aramsangtienchai, P.; Kongmon, T.; Pechroj, S.; Srisook, K. Enhanced production and immunomodulatory activity of levan from the acetic acid bacterium, *Tanticharoenia sakaeratensis*. *Int. J. Biol. Macromol.* **2020**, *163*, 574–581. [CrossRef] [PubMed]
43. Hernandez, L.; Arrieta, J.; Betancourt, L.; Falcon, V.; Madrazo, J.; Coego, A.; Menendez, C. Levansucrase from *Acetobacter diazotrophicus* SRT4 is secreted via periplasm by a signal-peptide-dependent pathway. *Curr. Microbiol.* **1999**, *39*, 146–152. [CrossRef] [PubMed]
44. Nasir, A.; Sattar, F.; Ashfaq, I.; Lindemann, S.; Chen, M.-H.; Van den Ende, W.; Öner, E.; Kirtel, O.; Khaliq, S.; Ghauri, M.; et al. Production and characterization of a high molecular weight levan and fructooligosaccharides from a rhizospheric isolate of *Bacillus aryabhattai*. *LWT* **2020**, *123*, 109093. [CrossRef]
45. Xavier, J.R.; Ramana, K.V. Optimization of Levan Production by Cold-Active Bacillus licheniformis ANT 179 and Fructooligosaccharide Synthesis by Its Levansucrase. *Appl. Biochem. Biotechnol.* **2017**, *181*, 986–1006. [CrossRef]

polymers MDPI

Article

Phormidium ambiguum and *Leptolyngbya ohadii* Exopolysaccharides under Low Water Availability

Isabela C. Moia [1], Sara B. Pereira [2,3], Paola Domizio [1], Roberto De Philippis [1] and Alessandra Adessi [1,*]

1 DAGRI—Department of Agriculture, Food, Environment and Forestry, University of Florence, Via Maragliano 77, 50144 Firenze, Italy
2 i3S—Instituto de Investigação e Inovação em Saúde, Universidade do Porto, Rua Alfredo Allen, 208, 4200-135 Porto, Portugal
3 IBMC—Instituto de Biologia Celular e Molecular, Universidade do Porto, Rua Alfredo Allen, 208, 4200-135 Porto, Portugal
* Correspondence: alessandra.adessi@unifi.it

Abstract: Cyanobacteria can cope with various environmental stressors, due to the excretion of exopolysaccharides (EPS). However, little is known about how the composition of these polymers may change according to water availability. This work aimed at characterizing the EPS of *Phormidium ambiguum* (Oscillatoriales; Oscillatoriaceae) and *Leptolyngbya ohadii* (Pseudanabaenales; Leptolyngbyaceae), when grown as biocrusts and biofilms, subject to water deprivation. The following EPS fractions were quantified and characterized: soluble (loosely bound, LB) and condensed (tightly bound, TB) for biocrusts, released (RPS), and sheathed in *P. ambiguum* and glycocalyx (G-EPS) in *L. ohadii* for biofilms. For both cyanobacteria upon water deprivation, glucose was the main monosaccharide present and the amount of TB-EPS resulted was significantly higher, confirming its importance in these soil-based formations. Different profiles of monosaccharides composing the EPSs were observed, as for example the higher concentration of deoxysugars observed in biocrusts compared to biofilms, demonstrating the plasticity of the cells to modify EPS composition as a response to different stresses. For both cyanobacteria, both in biofilms and biocrusts, water deprivation induced the production of simpler carbohydrates, with an increased dominance index of the composing monosaccharides. The results obtained are useful in understanding how these very relevant cyanobacterial species are sensitively modifying the EPS secreted when subject to water deprivation and could lead to consider them as suitable inoculants in degraded soils.

Keywords: EPS monosaccharidic composition; water deprivation; soil restoration; dehydration; biocrusts; biofilms

Citation: Moia, I.C.; Pereira, S.B.; Domizio, P.; De Philippis, R.; Adessi, A. *Phormidium ambiguum* and *Leptolyngbya ohadii* Exopolysaccharides under Low Water Availability. *Polymers* **2023**, *15*, 1889. https://doi.org/10.3390/polym15081889

Academic Editor: Paschalis Alexandridis

Received: 27 January 2023
Revised: 31 March 2023
Accepted: 12 April 2023
Published: 14 April 2023

1. Introduction

Cyanobacteria are a group of prokaryotic microorganisms found in fresh and marine waters and soils [1] and represent the first colonizers in drylands [2]. They have been studied for contributing to agricultural productivity in a process called "cyanobacterization", which consists of cyanobacteria inoculation in the soil to provide soil structural stability and biofertilizer properties [3–5]. The selection of feasible species for a given environmental condition, allied to a cheaper biomass production and dispersion strategy, represents a new biotechnological approach to increase fertility in agricultural soils.

This biotechnological potential is due to the formation of biocrusts, which are composed of microorganisms aggregated to soil particles [6,7]. Due to the excretion of exopolysaccharides (EPS), the biocrust-forming cyanobacteria allow: (i) the increase in soil organic carbon pool, and stimulation of exoenzyme activity in the soil [8]; (ii) water retention and infiltration, which have a positive correlation with the total carbohydrate content of the biocrust [9,10]; (iii) the retention of metabolites and nutrients, thus preventing molecules from leaching down to the subcrustal layer [11]; and (iv) the release of vitamins, amino acids and phytohormones in the soil, which act as biofertilizers [12,13].

The rather complex monosaccharidic composition usually provides to the cyanobacterial EPS an amphiphilic character, determining their functions in soil [14–16]. For example, the hydrophobic character of some fractions of the polymer, mainly due to the presence of deoxysugars, such as rhamnose and fucose, may favor the adhesion of the cyanobacterial filaments to solid substrates and the formation of soil aggregates [5,14,17]. On the other hand, the negatively charged, hydrophilic fractions of the EPS, characterized by a large presence of uronic acids and sulfate groups, are involved in binding minerals, nutrients and water molecules, thus favoring the survival of the microbial community residing in the biocrust [18,19]. These EPS features provide sustainment to the biocrusts and can enable cyanobacteria to cope with periods of desiccation [8,20,21].

The EPS versatility and their protection role confer resistance also for biofilms to reach ecological success [22]. For instance, the EPS hydrophobic/hydrophilic character was reported for biofilm-forming cyanobacteria isolated from wastewater treatment plant and brackish lagoon [23]. These authors reported that the high uronic acids proportions could be successfully exploited for different biotechnological applications. Moreover, Keshari et al. (2021) [24] showed that the cyanobacterium *Scytonema mille*, which has a thick sheath (polysaccharide anchored to the cell wall) was the major microorganism in the biofilms on the Buddha statue in India, indicating that the sheath may support the organism to cope with desiccation [24].

The abiotic conditions of the environment and the cyanobacterium species also influence the molecular size distribution of the EPS [9]. For example, the presence of exopolysaccharides with low molecular weight (MW) was more frequently present when the filamentous cyanobacterium *Leptolyngbya ohadii*, grown as biocrust, was subjected to high water availability [17]. This pattern might be related to the increase in the activity of enzymes that decompose the higher MW to smaller MW polymers [17]. Additionally, it was reported that *Phormidium ambiguum* liquid culture (where it experiences optimal nutritional conditions) and biocrusts produced polymers with MW distributed heterogeneously through the EPS fractions [25,26]. Therefore, the chemical and macromolecular characteristics of exopolysaccharides can be modulated according to soil type [25], nutrient supply [27], and water availability [17,21].

Recently, it has been shown that *L. ohadii*, isolated from biocrusts of the Negev desert [6,28], can revive after hydration/dehydration cycles [29,30]. The genus *Phormidium* has been found in arid areas, where hot deserts are present [31–33]. These areas are susceptible to wind erosion, due to the changes in physicochemical properties and decrease in the stability of the soil, that accelerates land degradation [34,35]. Therefore, it is important to evaluate cyanobacteria tolerating long periods of drought in order to stabilize and increase the strength of the soil. For instance, *P. ambiguum* was shown to perform an essential role in the aggregation of soil particles, improving soil structure and stability, and indicates it to be a good candidate to be used in soil restoration in arid areas [25,36]. However, little is known about these cyanobacteria growth and EPS composition and role upon low water availability when cultivated in different modes of life (sandy soil biocrusts and biofilms liquid cultures) and about the possible differences of their EPS in biofilms or biocrusts. This work aimed to characterize *P. ambiguum* and *L. ohadii* upon water deprivation in terms of the amount, monosaccharidic composition, and MW distribution of the soluble soil EPS fraction (loosely bound EPS, LB-EPS) and of the condensed soil EPS fraction (tightly bound EPS, TB-EPS) in biocrust-forming conditions, and of the RPS and anchored polysaccharides (either sheath or glycocalyx EPS) in biofilm-forming conditions. Understanding these differences could reveal the diversity of existing mechanisms to tolerate harsh environmental conditions, such as water-deprivation, determining relevant information for potential biotechnological applications.

2. Materials and Methods

2.1. Cyanobacteria Origin and Growth Conditions

Two filamentous non-heterocystous cyanobacteria, previously reported to be present in the arid soil cyanobacterial communities [30,37], were selected to evaluate the effect of water deprivation stress on their EPS secretion: the sheathed *Phormidium ambiguum* Gomont NIES-2121, purchased at NIES Collection, Japan, and originally isolated from an African soil [26], and *Leptolyngbya ohadii* provided by the Department of Plant and Environmental Sciences of the Hebrew University of Jerusalem, Israel, originally isolated in the Negev desert [30]. The latter cyanobacterium produces an EPS fraction, not structured similar to a sheath, that remains attached to the filament when grown in liquid culture hereafter referred to as glycocalyx EPS (G-EPS) [10]. Both cyanobacteria, maintained in the laboratory of the Department of Agriculture, Food, Environment, and Forestry (DAGRI) of the University of Florence (Italy), were grown in flasks containing liquid BG-11 medium [38], at 25 °C, under continuous illumination of 15 μmol photons m^{-2} s^{-1} and continuous stirring at 100 rpm until the stationary phase. These cultures were used for biomass inoculation described below.

2.2. Inoculation and Biocrust Sampling

In these experiments, the cyanobacterial strains were inoculated (nine replicates for each cyanobacterium) in microcosms of Petri dishes, 150 mm (diameter) × 20 mm (depth), containing 300 g of an autoclaved commercial dried silica sand (VAGA s.r.l., Pavia, Italy), with a granulometry of 0.3–0.6 mm. Before inoculation the cultures were centrifuged at different speeds to optimize their sedimentation. *P. ambiguum* was centrifuged at 7000× *g* and *L. ohadii* was centrifuged at 4000× *g*, for 20 min, at room temperature and the supernatant was discarded. To remove any remaining culture medium, the pellet filaments were resuspended in sterile distilled water, centrifuged again at the same speeds, and the supernatant discarded. The pellet was resuspended in sterile distilled water in a volume to provide enough inoculum to disperse the resuspended filaments spirally on the microcosms [10]. Each microcosm was inoculated with 160 mg (dry weight) of cyanobacterial biomass. The chlorophyll *a* amount inoculated in each microcosm corresponded to 2.54 μg/g sand and 2.9 μg/g sand for *P. ambiguum* and *L. ohadii*, respectively. Inoculated microcosms were maintained inside a Plexiglass incubator with controlled temperature at 25 °C and continuous light intensity of 30 μmol photons m^{-2} s^{-1} to stimulate growth. The microcosms were daily watered with 35 mL of sterilized distilled water (enough volume to wet the entire surface of the biocrust).

After 4 weeks, all microcosms formed a visible crust on the sand substrate and three microcosms (N = 3) of each cyanobacterium were randomly collected. The remaining six microcosms were kept in incubation for 5 more days under two different conditions: three (N = 3) were subjected to water deprivation stress by stopping the watering and leading to the drying of the biocrusts, while three (N = 3) the watering continued and worked as controls (Figure S1). After this additional period (totaling 33 days) the biocrusts in all six microcosms were collected. All the formed crust was collected separated from the sand substrate and gently homogenized with a sterilized spatula. After homogenization, the crusts were weighted and used to determine chlorophyll *a* content, EPS amount, monosaccharidic composition, and MW distribution as described below.

2.3. Biofilm Formation in Liquid Culture Medium

Cultures of each cyanobacterium were inoculated in 1 L Pyrex Erlenmeyer flasks containing 500 mL of BG-11 culture medium [38] (nine replicates for each cyanobacterium). After inoculation, each replicate had an initial dry weight of 0.1 mg mL^{-1}. The chlorophyll *a* amount inoculated in each flask corresponded to 9.65 μg mL^{-1} and 6.82 μg mL^{-1} for *P. ambiguum* and *L. ohadii*, respectively. The flasks were maintained, without agitation, inside a Plexiglass incubator with controlled temperature at 25 °C and continuous light intensity of 30 μmol photons m^{-2} s^{-1} to stimulate growth. After 4 weeks, 3 biofilms trials

(N = 3) of each cyanobacterium were randomly collected. The remaining 6 biofilms trials were kept in incubation for 5 more days, under two different conditions: 3 (N = 3) were subjected to water deprivation stress by removing the liquid medium and maintaining the biofilm attached to the glass bottom of the Erlenmeyer flasks, while 3 (N = 3) were kept with the liquid medium (Figure S1). After this additional period (totaling 33 days) the 6 biofilms were collected. Samples were used to determine chlorophyll *a* content, EPS amount, monosaccharidic composition, and MW distribution as described below.

2.4. Biocrusts and Biofilm Characterization

2.4.1. Growth Measurements through Chlorophyll *a*

For biocrusts, chlorophyll *a* content was determined according to the method as reported in Castle et al. (2011) [39]. All biofilms, after resuspension of the water deprived samples in saline solution (150 mM NaCl in sterile distilled water), were homogenized with a sterilized glass rod and collected with serological pipette. The extraction was performed according to the method as reported in Ritchie (2006) and Yéprémian et al. 2016 [40,41]. Briefly, 1 g of homogenized crust and 2 mL of biofilm homogenate were collected into screw-cap vials. The biofilms were centrifuged at 3800× *g* for 10 min, and the supernatant discarded. Next, the weighted crust and the pelletized biofilm were treated with 5 mL of ethanol, at 80 °C, for 5 min. Samples were incubated in the dark, at 4 °C for 30 min before being centrifuged at 3800× *g* for 15 min. The supernatant was recovered and determined by measuring the absorbance (A) at 665 nm. Chlorophyll *a* content was calculated according to previous study [40]:

$$\text{Chlorophyll } a\ [\mu g/g\ \text{crust}] = (11.9035 \times A_{665} \times \text{Ve}) \times (\text{g } crust^{-1}) \times \text{L} \quad (1)$$

$$\text{Chlorophyll } a\ [\mu g/mL] = (11.9035 \times A_{665} \times \text{Ve}) \times (\text{Vs}^{-1}) \times \text{L} \quad (2)$$

where Ve is the volume of ethanol (mL), Vs is the volume of sample (mL), and L is the path length (cm).

2.4.2. EPS Isolation, Quantification and Characterization

The EPSs were extracted from the biocrusts in two different fractions according to previous works [17,19,42]. The one easily released into the sand substrate and more water soluble, referred to as loosely bound EPS (LB-EPS), and the one more condensed, firmly attached to cells and sand particles, referred to as tightly bound EPS (TB-EPS). LB-EPS were recovered by resuspending the biocrusts in distilled water and incubating them at room temperature for 15 min. Next, samples were centrifuged at 3800× *g* at 8 °C for 30 min and the LB-EPS-containing supernatants collected. This extraction was repeated three times for each sample. TB-EPS were recovered by treating the biocrust pellet resulting from the LB-EPS extraction with 0.1 M Na_2EDTA for 16 h at room temperature. Next, samples were centrifuged at 3800× *g* at 8 °C for 30 min and the TB-EPS-containing supernatants collected. This extraction was repeated three times for each sample, the last two extractions performed for 120 min each.

Regarding the biofilms, EPS were extracted from the glass-rod homogenized biofilms cultures and centrifuged at 3800× *g* for 30 min at room temperature. The RPS were isolated from the supernatant while the pelleted cells were saved for sheath or G-EPS extraction. For that, the supernatants were concentrated by evaporation using an orbital evaporator at 35 °C. Next, the RPS were precipitated by mixing with two volumes of cold (4 °C) isopropyl alcohol and incubated at 4 °C for 8 h. After centrifuging at 4000× *g* for 15 min, the EPS pellets were resuspended in distilled water. For sheath or G-EPS extraction, the pelleted cyanobacterial cells obtained after the centrifugation of the culture biofilms were washed with 5 mL of 1.5% NaCl solution. After removing the 1.5% NaCl solution, the pellets were resuspended in 5 mL of sterile distilled water at 80 °C for 1 h [19]. After centrifuging at

4000× g for 30 min, the sheath-containing supernatants for *P. ambiguum* and the G-EPS for *L. ohadii* were collected.

The RPS and TB-EPS extracts were confined in dialysis membranes (12–14 kDa MW cut off, Medicell International London) and dialyzed against distilled water for 24 h, with two changes of water. All the EPS extracts were quantified by the phenol-sulphuric acid assay method [43]. For the determination of the monosaccharidic composition, the samples were hydrolyzed in 2 N trifluoroacetic acid (TFA), for 120 min at 120 °C. The tubes containing the samples were then dipped in cool water and the samples were evaporated in an orbital evaporator at 35 °C. Subsequently, samples were suspended in HPLC-grade water and evaporated again, repeating this step for one more time. Finally, monosaccharide composition was analyzed with a Dionex ICS-2500 ion exchange chromatograph (Dionex, United States) equipped with an ED50 pulsed amperometric detector operating with a gold working electrode (Dionex) and a CarboPac PA1 column of 250 mm length and 4.6 mm internal diameter (Dionex, Sunnyvale, CA, USA). Eluents were HPLC-grade water (A), 0.185 M NaOH (B), and 0.488 M sodium acetate (C), at a flow rate of 1 mL min^{-1}. Single sugars were identified and quantified based on the retention time of reference standards. Results were expressed in molar ratio.

Molecular size distribution of the EPSs was analyzed using a Varian ProStar HPLC chromatograph (Varian, CA, USA) equipped with a refractive index detector and two columns for Size Exclusion Chromatography (Polysep-GFC-P6000 and 4000, Phenomenex, CA, USA) connected in series. Samples were analyzed with runs of 70 min and with HPLC-grade water as eluent, at a flow rate of 0.4 mL min^{-1}. Dextran (Sigma-Aldrich, Burlington, MA, USA) at different MWs (2, 1.1, 0.41, 0.15, and 0.05 M Da) were used as reference standards. To obtain the % of the different MW classes the ratio between each peak area and the total area under the curve was calculated and the resulting % area was assigned to the corresponding size class according to the retention time of the peak output.

2.5. Data Analysis

Possible differences in chlorophyll *a* and EPSs amount among the two cyanobacterial strains grown in biocrusts and biofilm-cultures were analyzed using one-way analysis of the variance (ANOVA) at 95% of the significance, followed by Tukey post hoc test. Variables were previously checked for normality and homogeneity of variance using the Shapiro–Wilk and Levene's test, respectively. To correlate parameters, linear regression analyses were performed, and r^2 and p values are reported. For statistical analysis of the monosaccharidic profiles, Student's t test was used to compare the relative amount of each monosaccharide in the control and water deprivation replicates. Furthermore, the number of monosaccharides, diversity, dominance, and equitability indices of sugar residues of the different strains and conditions (sheath, G-EPS, and RPS in liquid cultures, and LB-EPS and TB-EPS in biocrusts) were compared. Shannon diversity index was calculated using the percentiles of a bootstrap distribution with 9999 repetitions. All statistical analysis was performed using Past 4.09 software.

3. Results

3.1. Effect of Water Deprivation on Chlorophyll a and EPS Production in Sandy Microcosms

The growth of the cyanobacterial strains was evaluated by measuring the chlorophyll *a* content. Chlorophyll *a* content decreased significantly ($p < 0.05$) during the water deprivation period for both cyanobacteria (Figure 1). The content of chlorophyll *a* decreased in a negative correlation with the EPSs contents (Figure 2), presenting an r^2 of 0.87, 0.84, 0.89, and 0.77, respectively; $p < 0.05$, for *P. ambiguum* LB-EPS, *P. ambiguum* TB-EPS, *L. ohadii* LB-EPS, and *L. ohadii* TB-EPS, respectively. The significantly higher amount of chlorophyll a obtained for *P. ambiguum* control (i.e., 4 weeks watered plus 5 days watered), represented twice the amount of chlorophyll *a* produced by *L. ohadii* control (Figure 1).

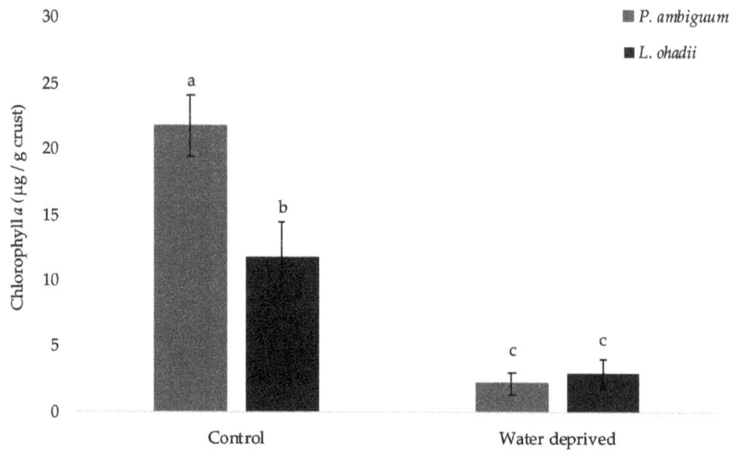

Figure 1. Chlorophyll *a* content in cyanobacterial biocrusts (values represent the mean of $N = 3$, error bars represent SD). Different letters represent significant differences ($p < 0.05$).

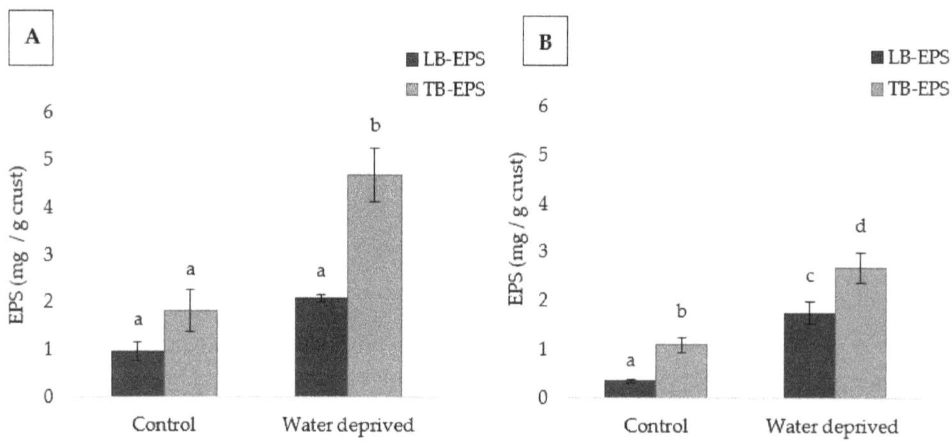

Figure 2. EPSs contents in *P. ambiguum* (**A**) and *L. ohadii* (**B**) biocrusts (values represent the mean of $N = 3$, error bars represent SD). Different letters represent significant differences ($p < 0.05$) in each graph.

The amount of LB-EPS, the loosely bound polymers, that are weakly attached to cells and sediments, and the TB-EPS, the tightly bound polymers, which have stronger bounds to cells and sediments, changed according to the cyanobacterial strain and growth condition (Figures 2 and S2). For *P. ambiguum*, the only significant difference observed was the increase in the TB-EPS fraction in water deprivation compared to the TB-EPS from the control and all the LB-EPS fractions. On the other hand, for *L. ohadii*, in LB-EPS from water deprived condition was significantly higher compared to those of the control. For this strain, the TB-EPS amount was significantly higher than the LB-EPS amount in each condition. The TB-EPS after water deprivation resulted significantly the highest value among all the EPSs extracted for both cyanobacteria (Figure 2).

When comparing the amount of LB-EPS in each condition for both cyanobacteria, those from *P. ambiguum* in control showed significantly higher value than those of *L. ohadii* in the same condition (Figure S3). After the water deprivation period, the LB-EPSs of these strains were not statistically different. On the contrary, while *P. ambiguum* TB-EPS control was not statistically different from *L. ohadii* TB-EPS control, after the water deprivation period, *P. ambiguum* TB-EPS showed significantly higher value than *L. ohadii* TB-EPS.

3.2. Changes in Monosaccharidic Composition and Molecular Weight Distribution of Microcosms EPSs

The EPSs were analyzed in terms of monosaccharide composition, showing differences in its relative abundances between the control and water deprived microcosms. The EPS fractions of *P. ambiguum* were mainly composed of glucose (Figures 3 and S2). The LB-EPS after water deprivation had the highest relative amount of this sugar among the EPSs extracted from this cyanobacterium biocrust. On the other hand, the LB-EPS in the control microcosms had a higher diversity and was also composed by significantly higher molar ratios of uronic acids (Figure 3A,C and Table S1). Moreover, fucose and rhamnose were detected in LB-EPS control and water deprived. The TB-EPS of water deprived *P. ambiguum* microcosms showed lower diversity than the control TB-EPS of this cyanobacterium (Table S1). When comparing the LB-EPS and TB-EPS only in water deprivation conditions, the first fraction showed lower diversity than the latter (Table S1). Galactose was found only after water deprivation in LB-EPS and in both conditions in TB-EPS (Figure 3A,C). Additionally, fucose was detected in the control and water deprived TB-EPS.

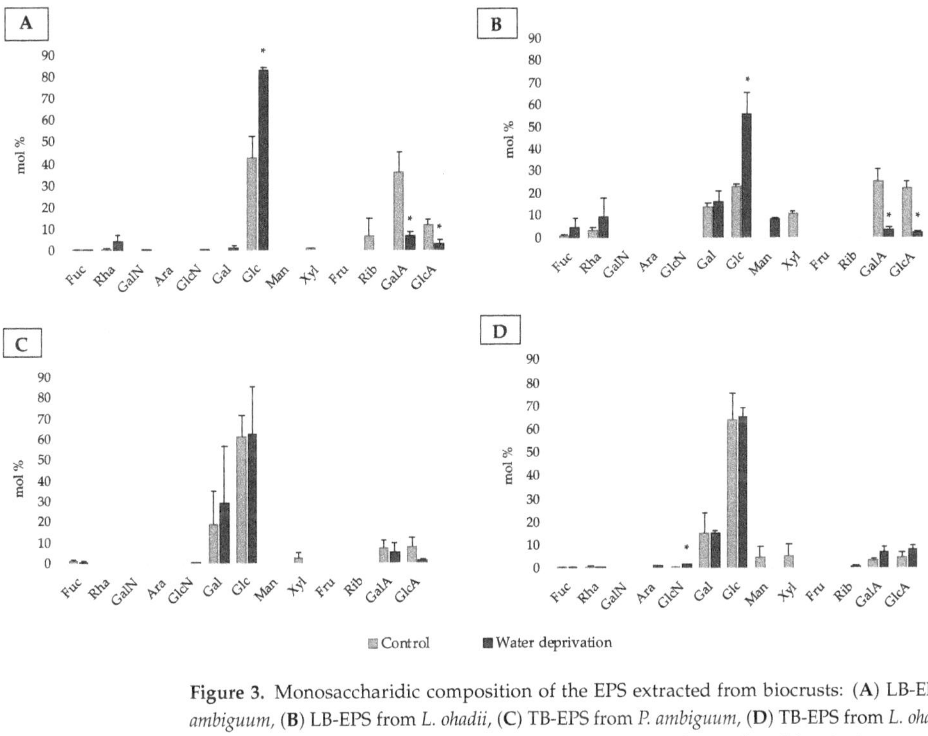

Figure 3. Monosaccharidic composition of the EPS extracted from biocrusts: (**A**) LB-EPS from *P. ambiguum*, (**B**) LB-EPS from *L. ohadii*, (**C**) TB-EPS from *P. ambiguum*, (**D**) TB-EPS from *L. ohadii*. Molar percentages (%) of single sugars are represented (expressed as moles of the single monosaccharide divided by the total amount of moles of monosaccharides in the EPS × 100). Symbol *, when present, indicates significant differences between the control and watered-deprived period in each monosaccharide. Fuc, fucose; Rha, rhamnose; GalN, galactosamine; Ara, arabinose; GlcN, glucosamine; Gal, galactose; Glc, glucose; Man, mannose; Xyl, xylose; Fru, fructose; Rib, ribose; GalA, galacturonic acid; GlcA, glucuronic acid.

After 4 weeks watered, *L. ohadii* LB-EPS and TB-EPS were mainly composed of glucose (Figure S2). The control LB-EPS and TB-EPS were mainly composed of glucose, galactose, and uronic acids (Figure 3B,D). The control TB-EPS showed higher diversity than the TB-EPS extracted after water deprivation (Table S1). For this cyanobacterium, the TB-EPS of water-deprived microcosms showed the highest molar percentage of glucose, followed by galactose. This EPS fraction after water deprivation condition had higher dominance than the control TB-EPS, the same profile observed for the LB-EPS. When comparing the LB-EPS and TB-EPS only in water deprivation conditions, the first fraction showed higher diversity than the second fraction. The TB-EPS demonstrated higher dominance compared with LB-EPS, the same profile shown by *P. ambiguum* water-deprived biocrusts. Moreover, fucose and rhamnose were present in both control and water-deprived EPS fractions.

All the EPS extracted were analyzed in terms of molecular weight (MW) distribution. The results revealed the presence of molecules ranging from 50 kDa to 2 MDa. The *P. ambiguum* LB-EPS and TB-EPS were composed predominantly of MW molecules of 410 kDa, except LB-EPS in water deprivation stress that also showed molecules higher than 2 MDa. The *L. ohadii* LB-EPS after the watered periods was predominantly composed by molecular weight molecules of 1.1 MDa–410 kDa and after a water deprivation of 1.1 MDa. The TB-EPSs extracted from this cyanobacterium were mainly constituted of molecules with MW ranging between 2 MDa–1.1 MDa (80% of chromatogram area) and a small fraction of molecules with MW ranging between 410 kDa–150 kDa (15% of chromatogram area) for all the conditions tested (data not shown).

3.3. *Effects of Water Deprivation in Chlorophyll a and EPS Production in Biofilms*

For the controls and the water-deprived biofilms, no cyanobacterium produced significantly more chlorophyll *a* than the other (Figure 4). Therefore, no significant correlation could be attributed between the EPSs contents and chlorophyll a ($p > 0.05$).

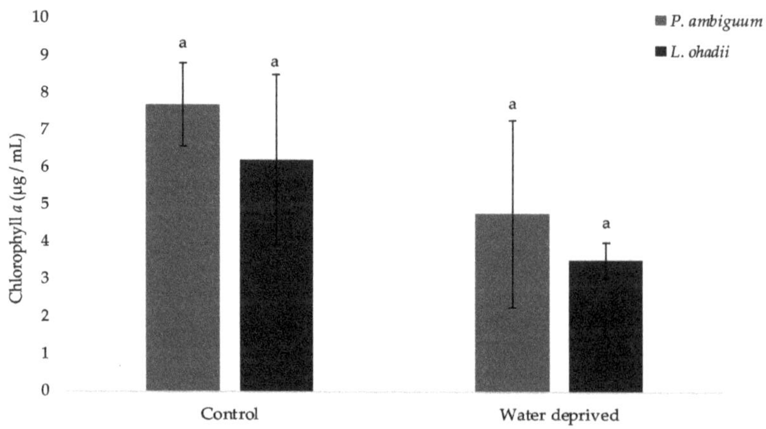

Figure 4. Chlorophyll *a* content in cyanobacterial biofilms (values represent the mean of $N = 3$, error bars represent SD). Different letters represent significant differences ($p < 0.05$).

For both cyanobacteria, the amount of RPS and sheath or G-EPS did not show significant differences between the control and the water-deprived biofilms (Figure 5A,B). For *P. ambiguum*, the sheath EPS amount was significantly higher, under all conditions, than the RPS, indicating that the sheath gives the major contribution to the total carbohydrates of this cyanobacterium (Figures 5A and S4B) in biofilm forming conditions. For *L. ohadii* under water deprivation stress, the G-EPS was significantly higher than the RPS, but it was not significantly higher than the control RPS (Figure 5B). Moreover, when comparing the RPS between the cyanobacteria in each condition, *L. ohadii* RPS after water deprivation stress resulted significantly higher (Figure S5A). On the other hand, the sheath EPS resulted significantly higher than G-EPS in water deprivation stress, indicating how sheath EPS could contribute to *P. ambiguum* as protection in biofilms (Figure S5B).

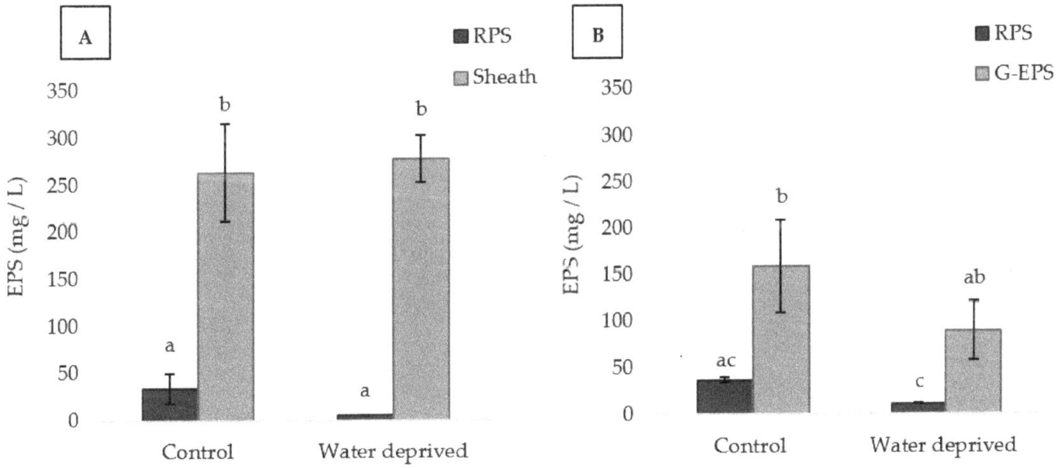

Figure 5. EPSs contents in *P. ambiguum* (**A**) and *L. ohadii* (**B**) biofilms (values represent the mean of $N = 3$, error bars represent SD). Different letters represent significant differences in each graph ($p < 0.05$).

3.4. Changes in Monosaccharidic Composition and Molecular Weight Distribution of Biofilms EPSs

The monosaccharidic composition revealed differences in biofilm-cultures, showing different profiles between the strains, the conditions, and comparing them to the biocrusts. The RPS after water deprivation were mainly composed of glucose and uronic acids (Figure 6A,B). After water deprivation, only glucuronic acid demonstrated significantly higher molar ratio than the control for *P. ambiguum* RPS. Unlike *P. ambiguum* LB-EPS, the molar ratio of glucose in *P. ambiguum* RPS did not show significant differences between the control and water deprivation. The RPS after water deprivation showed higher equitability than the RPSs extracted from the control biofilms, conferring the highest diversity, despite the smaller number of monosaccharides (Table S2). The sheath EPS after water deprivation showed lower diversity than the RPS after water deprivation and the sheath EPS control, due to the higher dominance of glucose. This moiety also demonstrated significantly higher molar ratio than the control in the sheath EPS (Figure 6C, Table S2).

For all the conditions, the RPSs demonstrated higher diversity than the G-EPS in *L. ohadii* biofilm (Table S2). For this cyanobacterium RPS, the galacturonic and glucuronic acids were significantly higher in water deprivation compared to the control. A similar profile was found in *P. ambiguum* biofilm, in which these sugars dominated the RPS after water deprivation. Unlike *L. ohadii* LB-EPS, the molar ratio of glucose in *L. ohadii* RPS after water deprivation was significantly lower compared to the control. In addition, unlike *P. ambiguum* biofilm, the *L. ohadii* RPS in water deprivation showed lower diversity compared with the RPS control. The latter resulted in the highest number of monosaccharides and the highest equitability, leading to the highest diversity among all the *L. ohadii* EPS fractions. Finally, glucose and galactose were the monosaccharides mostly present in G-EPSs (Figure 6D). The G-EPS after water deprivation showed lower diversity than the RPS after water deprivation and the G-EPS control, due to the higher dominance of glucose as observed in the sheath EPS after water deprivation (Figure 6B–D, Table S2).

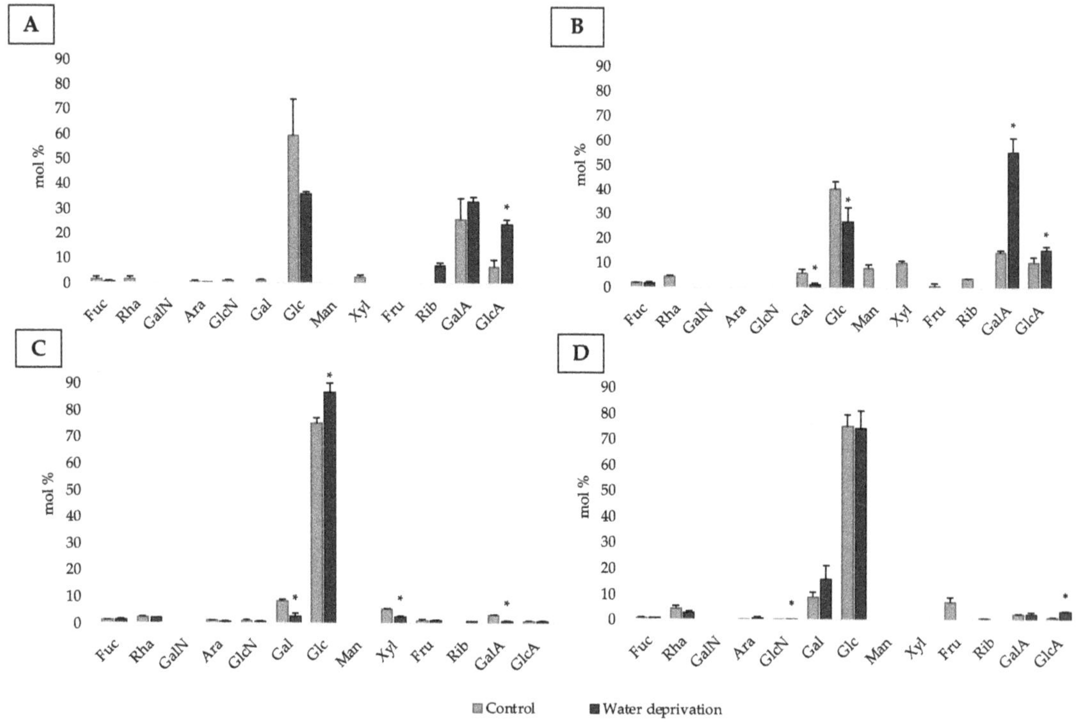

Figure 6. Monosaccharidic composition of the EPS extracted from biofilms: (**A**) RPS from *P. ambiguum*, (**B**) RPS from *L. ohadii*, (**C**) Sheath EPS, (**D**) G-EPS. Molar percentages (%) of single sugars are represented (expressed as moles of the single monosaccharide divided by the total amount of moles of monosaccharides in the EPS × 100). Symbol *, when present, indicates significant differences between the control and the water-deprived period in each monosaccharide. Fuc, fucose; Rha, rhamnose; GalN, galactosamine; Ara, arabinose; GlcN, glucosamine; Gal, galactose; Glc, glucose; Man, mannose; Xyl, xylose; Fru, fructose; Rib, ribose; GalA, galacturonic acid; GlcA, glucuronic acid.

The EPS extracted from biofilms were analyzed in terms of molecular weight (MW) distribution. The MW distribution of the RPSs extracted were not detectable and data are not shown. The sheath EPSs of *P. ambiguum* were predominantly composed of molecules lower than 50 kDa, while the *L. ohadii* G-EPSs were constituted of molecules having an apparent MW between 410 kDa and 50 kDa (data not shown).

4. Discussion

Filamentous cyanobacteria have been studied for their potential to survive water deprivation [29,44,45]. Previous studies demonstrated that the profile of exopolysaccharides is heterogeneously dependent on the strain [17,26,42]. In this study, two different strains were cultivated as biocrusts and biofilms, showing different responses to water deprivation stress.

4.1. EPS and Growth Profile in Biocrust-Forming Cyanobacteria

When the two cyanobacteria were incubated in microcosms, the amount of chlorophyll *a* decreased under water deprivation, indicating that five days deprivation of water had an impact in photosynthesis. The photosynthetic apparatus has been reported to respond to water deprivation conditions [30,31,46–48]. These authors reported that the phycobilisomes in the desiccated state lose their organized structure, and after re-wetting, photosynthetic activity was reactivated. Previous study identified genes encoding proteins involved in photosynthesis and chlorophyll *a* synthesis whose transcription was induced by re-wetting the biocrust of *Microcoleus vaginatus* [49]. Additionally, previous work reported that the dry phase induces an osmotic stress, which regulates genes and decreases the photosynthesis activity. Therefore, this agrees with our observation of a decrease in chlorophyll *a* induced by water deprivation [50,51].

Cultures of bacteria and cyanobacteria have been reported to increase EPS production under water deprived conditions, suggesting that resources were allocated to EPS production in response to the stress [52–54]. This may explain the apparent contradiction between the increase in EPS and the decrease in photosynthesis. In this work, the amount of the EPS changed with the water availability: *P. ambiguum* TB-EPS amount was significantly higher only under water deprivation. It was previously reported that this cyanobacterium promotes higher increase in TB-EPS compared to other cyanobacteria (i.e., *Scytonema javanicum*) [26]. In the present work, it also promoted significantly higher TB-EPS amounts compared to *L. ohadii* biocrusts under water deprivation (Figure S3B), corroborating with previously reports that *P. ambiguum* has a more prominent role of aggregation of soil particles and improvement of soil strength, due to the synthesis of large amounts of TB-EPS [25,26,36]. This could be an advantage for this cyanobacterium to deal with periods of limited water availability. The significant increase in the amount of LB-EPS of *L. ohadii* under the water deprivation stress condition corroborates previous findings [17] that the amount of LB-EPS in *L. ohadii* biocrusts increases when incubated with lower amounts of water. This suggests that the amount of water influences the amount of LB-EPS for this cyanobacterium. On the other hand, in contrast to previous reports that the amount of *L. ohadii* TB-EPS was not influenced by water availability, in the present study the amount of TB-EPS was higher under water deprivation condition, being significantly higher than the other EPSs fractions quantified for *L. ohadii* biocrust justifying their large presence in biocrusts that are subject to frequent dry and wet cycles [55].

Not only the amount of EPSs can be modulated by the harsh environmental conditions to which cyanobacteria were submitted, but also their monosaccharidic composition [8]. The higher relative abundance of glucose in the LB-EPS and TB-EPS of both cyanobacterial biocrusts, and the increase in the molar ratio of this sugar in water deprivation conditions is in agreement with previous studies that reported the higher abundance of glucose in sand biocrusts of these cyanobacteria under low water additions, mainly in LB-EPS [17,25,26]. Moreover, galactose was another sugar found in biocrusts that present higher molar ratios compared to other sugars. It has been reported to be highly produced by *P. ambiguum* TB-EPS in sand soils [25], confirming the observations on *P. ambiguum* biocrusts reported here. These data corroborate previous findings that, in nutrient limited conditions often found in biocrusts from drylands, more essential monosaccharides are required by the cyanobacteria [56], which could explain the increase in the dominance index in water deprivation stress condition.

Despite the dominance of glucose, cyanobacterial EPS are also characterized by the presence of uronic acids and deoxysugars, such as fucose and rhamnose [57–59]. In this work, uronic acids were present in significantly higher percentage after the watered periods and decreased after the water deprivation period. The uronic acids have hydrophilic character involved in the chelation of minerals, nutrients, and water molecules when present [18] and have been detected mostly under the highest water availability in *L. ohadii* LB-EPS [17]. On the contrary, fucose and rhamnose have hydrophobic character increasing cell capacity to adhere to solid surfaces [5,19]. Though the mol% of these monosaccharides is probably too low, in our results, to observe a significant difference in polymer hydrophobicity, these data still show the plasticity of the cells to modify EPSs to respond to the surrounding environment performing with the amphiphilic character of the macromolecules. In addition, the sugar amount and composition were different between the strains and EPS fractions, showing how these polymers are heterogeneously dependent on the cyanobacteria and abiotic factors.

The molecular weight distribution of EPS is another feature that deserves further investigations about its response to water deprivation. In this work, the TB-EPSs extracted were mainly composed of molecules ranging between 2 MDa–410 kDa of apparent MW. The biocrust stability was previously reported to be provided by the TB-EPS, which was composed of MW in the range 410 kDa–50 kDa, but by higher abundance of MW in the range 2 MDa–1.1 MDa [25,26,36]. Moreover, the LB-EPSs were also represented by >2 MDa–1.1 MDa molecules after water deprivation, while after the watered periods molecules in the range 1.1 MDa–410 kDa were present. The presence of high MW molecules could also be due to molecule aggregation after the extraction. However, the analysis confirms the presence of large polymers, which can contribute to the viscosity of the EPS. The presence of such high MW molecules under low moisture could indicate a higher water retention during periods of drought [17,25,60].

4.2. Biocrusts and Biofilms EPSs, Similarities and Differences

The sheath EPS is important for mechanical and physicochemical stability [61–63]. On stones and monuments, the presence of biofilm-forming sheathed cyanobacteria has already been reported [18,64]. The thick sheath protects the cyanobacteria, helping it to survive in low nutrient environment [22], high temperatures, and water deprivation conditions [24,63]. This EPS structure has been documented to be thicker in stressed cells [65] and strongly connects sand grains [60,66]. Taking that into consideration, it is possible that the *P. ambiguum* TB-EPS might be similar to the sheath EPS, due to the similar role performed by the two EPS fractions. These findings were also reported by previous work which documented the relation between TB-EPS and sheath EPS produced by Schizothrix cf. delicatissima [42]. Regarding the *L. ohadii* TB-EPS and G-EPS some similarities in the monosaccharidic composition were observed. Both fractions have predominantly abundances of glucose and galactose that were not significantly different between the control and water-deprived samples. This could indicate that also the TB-EPS might derive from the G-EPS, as suggested by previous authors who reported that the dominant presence of glucose in *L. ohadii* G-EPS combined with other sugars are related with stronger sand stabilizing capability [10], a feature related to TB-EPS [8,19].

In sand biocrusts, the LB-EPS is released and remains in the biocrusts, with no loss of polysaccharides as in water-deprived biofilms. Furthermore, after water deprivation, rhamnose was detected in both cyanobacteria LB-EPSs, but not in the RPSs, showing the relevance of this hydrophobic monosaccharide in water-deprived biocrusts. This is particularly important for demonstrating the functionality of LB-EPS in contributing to surface hydrophobicity [5]. The presence of uronic acids in *P. ambiguum* and *L. ohadii* RPS was already observed by previous studies [10,26], but here it was demonstrated that even in water deprivation these sugars are predominantly present. In this stressful condition, the RPS fraction from the cyanobacteria tested in this work showed a different relative amount of these sugars from the LB-EPSs. While in the latter significantly increased the amount of glucose and reduced the amount of uronic acids, in the RPSs the amount of glucose significantly decreased while the amount of uronic acids increased compared to their controls. This suggests that the water-deprived condition for biofilms may stimulate intracellular enzymes that oxidize glucose to form glucuronic acid, which in turn is converted to galacturonic acid [67–69], since the cyanobacteria may be undergoing an oxidation process imposed by the absence of culture medium [70–73]. This behavior shows how the environment can modulate the monosaccharidic profile. Therefore, without the sand substrate, the relative abundance of the sugars changed after water deprivation. Moreover, the diverse composition of monosaccharides and MW distribution between the cyanobacteria supports the existence of different response to the surrounding environment in normal or under stress conditions [74].

5. Conclusions

The present work provided important indications regarding the defense mechanisms of two cyanobacteria, namely, *P. ambiguum* and *L. ohadii,* to water deprivation in biocrusts and in biofilm formation. The strains showed EPS synthesis capability even in water deprived conditions, demonstrating a distinct production pattern of EPS. Particularly the production of TB-EPS was significantly increased in biocrusts after the water deprivation, suggesting the importance of this fraction in slowing down desiccation. In biofilms under water deprivation, the sheath EPS was significantly more abundant than the RPSs and G-EPS, suggesting that it may perform a crucial role in stress tolerance. The different profile of monosaccharidic composition demonstrated the plasticity of the cells to modify EPS in order to respond different stresses, as reported here by the higher concentration of deoxysugars in biocrusts compared to biofilms. For the two cyanobacteria evaluated in this work, water-deprived biofilms and biocrusts tend to produce simpler carbohydrates and increase their dominance index. The knowledge of different cyanobacteria biological mechanisms to cope with harsh environmental conditions is important to select the best strain for the use in the restoration of dryland soils and the improvement of ecosystems services.

Supplementary Materials: The following supporting information can be downloaded at: https://www.mdpi.com/article/10.3390/polym15081889/s1, Figure S1: biocrusts and biofilms control and water deprived; Figure S2: biocrust features after 4 weeks watered; Figure S3: LB-EPS and TB-EPS contents ; Figure S4: biofilm features after 4 weeks with culture medium; Figure S5: RPS, sheath and G-EPS contents; Table S1: diversity indices of the biocrusts EPS; Table S2: diversity indices of the biofilms EPS; Table S3: sugar total mols and mols of each monosaccharide in each EPS fraction for *P. ambiguum* biocrusts; Table S4: sugar total mols and mols of each monosaccharide in each EPS fraction for *L. ohadii* biocrusts; Table S5: sugar total mols and mols of each monosaccharide in each EPS fraction for *P. ambiguum* biofilms; Table S6: sugar total mols and mols of each monosaccharide in each EPS fraction for *L. ohadii* biofilms.

Author Contributions: Conceptualization, I.C.M. and A.A.; methodology, I.C.M., A.A., S.B.P. and P.D.; formal analysis, I.C.M.; writing—original draft preparation, I.C.M.; writing—review and editing, A.A., S.B.P., P.D. and R.D.P.; supervision, R.D.P. and A.A. All authors have read and agreed to the published version of the manuscript.

Funding: This research received no external funding.

Institutional Review Board Statement: Not applicable.

Data Availability Statement: Not applicable.

Conflicts of Interest: The authors declare no conflict of interest.

References

1. Gaysina, L.A.; Saraf, A.; Singh, P. Cyanobacteria in diverse habitats. In *Cyanobacteria*; Mishra, A.K., Tiwari, D.N., Rai, A.N., Eds.; Academic Press: Cambridge, MA, USA, 2019; pp. 1–28.
2. Abinandan, S.; Subashchandrabose, S.R.; Venkateswarlu, K.; Megharaj, M. Soil microalgae and cyanobacteria: The biotechnological potential in the maintenance of soil fertility and health. *Crit. Rev. Biotechnol.* **2019**, *39*, 981–998. [CrossRef]
3. Lan, S.; Zhang, Q.; Wu, L.; Liu, Y.; Zhang, D.; Hu, C. Artificially accelerating the reversal of desertification: Cyanobacterial inoculation facilitates the succession of vegetation communities. *Environ. Sci. Technol.* **2014**, *48*, 307–315. [CrossRef] [PubMed]
4. Jafarpoor, A.; Sadeghi, S.H.; Darki, B.Z.; Homaee, M. Changes in hydrologic components from a mid-sized plots induced by rill erosion due to cyanobacterization. *ISWCR* **2022**, *10*, 143–148. [CrossRef]
5. Rossi, F.; Mugnai, G.; De Philippis, R. Cyanobacterial biocrust induction: A comprehensive review on soil rehabilitation-effective biotechnology. *Geoderma* **2022**, *415*, 115766. [CrossRef]
6. Hagemann, M.; Henneberg, M.; Felde, V.J.M.N.L.; Berkowicz, S.M.; Raanan, H.; Pade, N.; Felix-Henningsen, P.; Kaplan, A. Cyanobacterial populations in biological soil crusts of the northwest Negev desert, Israel—effects of local conditions and disturbance. *FEMS Microbiol.* **2017**, *93*, fiw228. [CrossRef]
7. Weber, B.; Belnap, J.; Büdel, B.; Antoninka, A.J.; Barger, N.N.; Chaudhary, V.B.; Darrouzet-Nardi, A.; Eldridge, D.J.; Faist, A.M.; Ferrenberg, S.; et al. What is a biocrust? A refined, contemporary definition for a broadening research community. *Biol. Rev.* **2022**, *97*, 1768–1785. [CrossRef] [PubMed]

8. Chen, L.; Rossi, F.; Deng, S.; Liu, Y.; Wang, G.; Adessi, A.; De Philippis, R. Macromolecular and chemical features of the excreted extracellular polysaccharides in induced biological soil crusts of different ages. *Soil Biol. Biochem.* **2014**, *78*, 1–9. [CrossRef]
9. Colica, G.; Li, H.; Rossi, F.; Li, D.; Liu, Y.; De Philippis, R. Microbial secreted exopolysaccharides affect the hydrological behavior of induced biological soil crusts in desert sandy soils. *Soil Biol. Biochem.* **2014**, *68*, 62–70. [CrossRef]
10. Mugnai, G.; Rossi, F.; Chamizo, S.; Adessi, A.; De Philippis, R. The role of grain size and inoculum amount on biocrust formation by *Leptolyngbya ohadii*. *Catena* **2020**, *184*, 104248. [CrossRef]
11. Swenson, T.L.; Couradeau, E.; Bowen, B.P.; De Philippis, R.; Rossi, F.; Mugnai, G.; Northen, T.R. A novel method to evaluate nutrient retention by biological soil crust exopolymeric matrix. *Plant Soil* **2018**, *429*, 53–64. [CrossRef]
12. Hussain, A.; Hasnain, S. Phytostimulation and biofertilization in wheat by cyanobacteria. *J. Ind. Microbiol. Biotechnol.* **2011**, *38*, 85–92. [CrossRef] [PubMed]
13. Gr, S.; Yadav, R.K.; Chatrath, A.; Gerard, M.; Tripathi, K.; Govindsamy, V.; Abraham, G. Perspectives on the potential application of cyanobacteria in the alleviation of drought and salinity stress in crop plants. *J. Appl. Phycol.* **2021**, *33*, 3761–3778. [CrossRef]
14. Zampieri, R.M.; Adessi, A.; Caldara, F.; Codato, A.; Furlan, M.; Rampazzo, C.; De Philippis, R.; La Rocca, N.; Dalla Valle, L. Anti-inflammatory activity of exopolysaccharides from *Phormidium* sp. ETS05, the most abundant cyanobacterium of the therapeutic Euganean thermal muds, Using the Zebrafish Model. *Biomolecules* **2020**, *10*, 582. [CrossRef]
15. Guo, Y.; Furrer, J.M.; Kadilak, A.L.; Hinestroza, H.F.; Gage, D.J.; Cho, Y.K.; Shor, L.M. Bacterial extracellular polymeric substances amplify water content variability at the pore scale. *Front. Environ. Sci.* **2018**, *6*, 93. [CrossRef]
16. Nishanth, S.; Bharti, A.; Gupta, H.; Gupta, K.; Gulia, U.; Prasanna, R. Cyanobacterial extracellular polymeric substances (EPS): Biosynthesis and their potential applications. In *Microbial and Natural Macromolecules Synthesis and Applications*, 1st ed.; Das, S., Dash, H.R., Eds.; Academic Press: London, UK, 2021; pp. 349–369.
17. Mugnai, G.; Rossi, F.; Felde, V.J.M.N.L.; Colesie, C.; Büdel, B.; Peth, S.; Kaplan, A.; De Philippis, R. The potential of the cyanobacterium *Leptolyngbya ohadii* as inoculum for stabilizing bare sandy substrates. *Soil Biol. Biochem.* **2018**, *127*, 318–328. [CrossRef]
18. Rossi, F.; Micheletti, E.; Bruno, L.; Adhikary, S.P.; Albertano, P.; De Philippis, R. Characteristics and role of the exocellular polysaccharides produced by five cyanobacteria isolated from phototrophic biofilms growing on stone monuments. *Biofouling* **2012**, *28*, 215–224. [CrossRef]
19. Rossi, F.; Mugnai, G.; De Philippis, R. Complex role of the polymeric matrix in biological soil crusts. *Plant Soil* **2018**, *429*, 19–34. [CrossRef]
20. Tamaru, Y.; Takani, Y.; Yoshida, T.; Sakamoto, T. Crucial role of extracellular polysaccharides in desiccation and freezing tolerance in the terrestrial cyanobacterium *Nostoc commune*. *Appl. Environ. Microbiol.* **2005**, *71*, 7327–7333. [CrossRef]
21. Xu, H.; Raanan, H.; Dai, G.; Oren, N.; Berkowicz, S.; Murik, O.; Kaplan, A.; Qiu, B. Reading and surviving the harsh conditions in desert biological soil crust: The cyanobacterial viewpoint. *FEMS Microbiol. Rev.* **2021**, *45*, fuab036. [CrossRef]
22. Zammit, G. Phototrophic biofilm communities and adaptation to growth on ancient archaeological surfaces. *Ann. Microbiol.* **2019**, *69*, 1047–1058. [CrossRef]
23. Di Pippo, F.; Ellwood, N.T.W.; Gismondi, A.; Bruno, L.; Rossi, F.; Magni, P.; De Philippis, R. Characterization of exopolysaccharides produced by seven biofilm-forming cyanobacterial strains for biotechnological applications. *J. Appl. Phycol.* **2013**, *25*, 1697–1708. [CrossRef]
24. Keshari, N.; Das, S.K.; Adhikary, S.P. Colonization and survival of a stress tolerant cyanobacterium on a heritage monument of Santiniketan, India. *Int. Biodeterior. Biodegrad.* **2021**, *164*, 105294. [CrossRef]
25. Chamizo, S.; Adessi, A.; Mugnai, G.; Simiani, A.; De Philippis, R. Soil type and cyanobacteria species influence the macromolecular and chemical characteristics of the polysaccharidic matrix in induced biocrusts. *Microb. Ecol.* **2019**, *78*, 482–493. [CrossRef] [PubMed]
26. Chamizo, S.; Adessi, A.; Torzillo, G.; De Philippis, R. Exopolysaccharide features influence growth success in biocrust-forming cyanobacteria, moving from liquid culture to sand microcosms. *Front. Microbiol.* **2020**, *11*, 568224. [CrossRef] [PubMed]
27. Brüll, L.P.; Huang, Z.; Thomas-Oates, J.E.; Paulsen, B.S.; Cohen, E.H.; Michaelsen, T.E. Studies of polysaccharides from three edible species of *Nostoc* (cyanobacteria) with different colony morphologies: Structural characterization and effect on the complement system of polysaccharides from *Nostoc commune*. *J. Phycol.* **2000**, *36*, 871–881. [CrossRef]
28. Hagemann, M.; Henneberg, M.; Felde, V.J.; Drahorad, S.L.; Berkowicz, S.M.; Felix-Henningsen, P.; Kaplan, A. Cyanobacterial diversity in biological soil crusts along a precipitation gradient, Northwest Negev Desert, Israel. *Microb. Ecol.* **2015**, *70*, 219–230. [CrossRef]
29. Murik, O.; Oren, N.; Shotland, Y.; Raanan, H.; Treves, H.; Kedem, I.; Keren, N.; Hagemann, M.; Pade, N.; Kaplan, A. What distinguishes cyanobacteria able to revive after desiccation from those that cannot: The genome aspect. *Environ. Microbiol.* **2017**, *19*, 535–550. [CrossRef]
30. Raanan, H.; Oren, N.; Treves, H.; Berkowicz, S.M.; Hagemann, M.; Pade, N.; Keren, N.; Kaplan, A. Simulated soil crust conditions in a chamber system provide new insights on cyanobacterial acclimation to desiccation: Simulation of BSC conditions and acclimation. *Environ. Microbiol.* **2016**, *18*, 414–426. [CrossRef] [PubMed]
31. Chen, L.; Yang, Y.; Deng, S.; Xu, Y.; Wang, G.; Liu, Y. The response of carbohydrate metabolism to the fluctuation of relative humidity (RH) in the desert soil cyanobacterium *Phormidium tenue*. *Eur. J. Soil Biol.* **2012**, *48*, 11–16. [CrossRef]

32. Hu, C.; Liu, Y.; Paulsen, B.S.; Petersen, D.; Klaveness, D. Extracellular carbohydrate polymers from five desert soil algae with different cohesion in the stabilization of fine sand grain. *Carbohydr. Polym.* **2003**, *54*, 33–42. [CrossRef]
33. Johansen, J.R. Cryptogamic crusts of semiarid and arid lands of North America. *J. Phycol.* **1993**, *29*, 140–147. [CrossRef]
34. Burezq, H. Combating wind erosion through soil stabilization under simulated wind flow condition—Case of Kuwait. *Int. Soil Water Conserv. Res.* **2020**, *8*, 154–163. [CrossRef]
35. Dregne, H.E. Land degradation in the drylands. *Arid Land Res. Manag.* **2002**, *16*, 99–132. [CrossRef]
36. Chamizo, S.; Mugnai, G.; Rossi, F.; Certini, G.; De Philippis, R. Cyanobacteria inoculation improves soil stability and fertility on different textured soils: Gaining insights for applicability in soil restoration. *Front. Environ. Sci.* **2018**, *6*, 49. [CrossRef]
37. Roncero-Ramos, B.; Muñoz-Martín, M.A.; Cantón, Y.; Chamizo, S.; Rodríguez-Caballero, E.; Mateo, P. Land degradation effects on composition of pioneering soil communities: An alternative successional sequence for dryland cyanobacterial biocrusts. *Soil Biol. Biochem.* **2020**, *146*, 107824. [CrossRef]
38. Rippka, R.; Deruelles, J.; Waterbury, J.B.; Herdman, M.; Stainier, R.Y. Generic assignments, strain histories and properties of pure cultures of cyanobacteria. *J. Gen. Microbiol.* **1979**, *111*, 1–61. [CrossRef]
39. Castle, S.C.; Morrison, C.D.; Barger, N.N. Extraction of chlorophyll *a* from biological soil crusts: A comparison of solvents for spectrophotometric determination. *Soil Biol. Biochem.* **2011**, *43*, 853–856. [CrossRef]
40. Ritchie, R.J. Consistent sets of spectrophotometric chlorophyll equations for acetone, methanol and ethanol solvents. *Photosynth. Res.* **2006**, *89*, 27–41. [CrossRef] [PubMed]
41. Yéprémian, C.; Catherine, A.; Bernard, C.; Congestri, R.; Elersek, T.; Pilkaityte, R. Chlorophyll *a* extraction and determination. In *Handbook of Cyanobacterial Monitoring and Cyanotoxin Analysis*, 1st ed.; Meriluoto, J., Spoof, L., Codd, G.A., Eds.; John Wiley & Sons, Ltd.: Chichester, UK, 2016; pp. 331–334.
42. Mugnai, G.; Rossi, F.; Felde, V.J.M.N.L.; Colesie, C.; Büdel, B.; Peth, S.; Kaplan, A.; De Philippis, R. Development of the polysaccharidic matrix in biocrusts induced by a cyanobacterium inoculated in sand microcosms. *Biol. Fertil. Soils* **2018**, *54*, 27–40. [CrossRef]
43. Dubois, M.; Gilles, K.A.; Hamilton, J.K.; Rebers, P.A.; Smith, F. Colorimetric method for determination of sugars and related substances. *Anal. Chem.* **1956**, *28*, 350–356. [CrossRef]
44. Oren, N.; Raanan, H.; Murik, O.; Keren, N.; Kaplan, A. Dawn illumination prepares desert cyanobacteria for dehydration. *Curr. Biol.* **2017**, *27*, R1056–R1057. [CrossRef]
45. Lin, C.; Wu, J. Tolerance of soil algae and cyanobacteria to drought stress. *J. Phycol.* **2014**, *50*, 131–139. [CrossRef]
46. Bar Eyal, L.; Eisenberg, I.; Faust, A.; Raanan, H.; Nevo, R.; Rappaport, F.; Krieger-Liszkay, A.; Sétif, P.; Thurotte, A.; Reich, Z.; et al. An easily reversible structural change underlies mechanisms enabling desert crust cyanobacteria to survive desiccation. *BBA Bioenerg.* **2015**, *1847*, 1267–1273. [CrossRef]
47. Bar Eyal, L.; Ranjbar Choubeh, R.; Cohen, E.; Eisenberg, I.; Tamburu, C.; Dorogi, M.; Ünnep, R.; Appavou, M.-S.; Nevo, R.; Raviv, U.; et al. Changes in aggregation states of light-harvesting complexes as a mechanism for modulating energy transfer in desert crust cyanobacteria. *Proc. Natl. Acad. Sci. USA* **2017**, *114*, 9481–9486. [CrossRef]
48. Sakamoto, T.; Wei, Y.; Yuasa, K.; Nishiyama, Y. Recovery of photosynthesis after long-term storage in the terrestrial cyanobacterium *Nostoc commune*. *J. Gen. Appl. Microbiol.* **2022**, *68*, 169–174. [CrossRef] [PubMed]
49. Rajeev, L.; da Rocha, U.; Klitgord, N.; Luning, E.G.; Fortney, J.; Axen, S.D.; Shih, P.M.; Bouskill, N.J.; Bowen, B.P.; Kerfeld, C.A.; et al. Dynamic cyanobacterial response to hydration and dehydration in a desert biological soil crust. *ISME J.* **2013**, *7*, 2178–2191. [CrossRef] [PubMed]
50. Liu, Y.; Liu, K.; Ai, Y.; Jiang, H.; Gao, X.; Qiu, B. Differential display analysis of cDNA fragments potentially involved in *Nostoc flagelliforme* response to osmotic stress. *J. Appl. Phycol.* **2012**, *24*, 1487–1494. [CrossRef]
51. Lan, S.; Wu, L.; Zhang, D.; Hu, C. Desiccation provides photosynthetic protection for crust cyanobacteria *Microcoleus vaginatus* from high temperature. *Physiol. Plant.* **2014**, *152*, 345–354. [CrossRef] [PubMed]
52. Roberson, E.B.; Firestone, M.K. Relationship between desiccation and exopolysaccharide production in a soil *Pseudomonas* sp. *Appl. Environ. Microbiol.* **1992**, *58*, 1284–1291. [CrossRef]
53. Wu, S.; Yu, K.; Li, L.; Wang, L.; Liang, W. Enhancement of exopolysaccharides production and reactive oxygen species level of *Nostoc flagelliforme* in response to dehydration. *Environ. Sci. Pollut. Res.* **2021**, *28*, 34300–34308. [CrossRef]
54. Han, P.P.; Shen, S.G.; Guo, R.J.; Zhao, D.X.; Lin, Y.H.; Jia, S.R.; Yan, R.R.; Wu, Y.K. ROS is a factor regulating the increased polysaccharide production by light quality in the edible cyanobacterium *Nostoc flagelliforme*. *J. Agric. Food Chem.* **2019**, *67*, 2235–2244. [CrossRef] [PubMed]
55. Román, J.R.; Roncero-Ramos, B.; Rodríguez-Caballero, E.; Chamizo, S.; Cantón, Y. Effect of water availability on induced cyanobacterial biocrust development. *Catena* **2021**, *197*, 104988. [CrossRef]
56. Mager, D.M.; Thomas, A.D. Extracellular polysaccharides from cyanobacterial soil crusts: A review of their role in dryland soil processes. *J. Arid Environ.* **2011**, *75*, 91–97. [CrossRef]
57. Kehr, J.C.; Dittmann, E. Biosynthesis and function of extracellular glycans in cyanobacteria. *Life* **2015**, *5*, 164–180. [CrossRef] [PubMed]
58. Laroche, C. Exopolysaccharides from microalgae and cyanobacteria: Diversity of strains, production strategies, and applications. *Mar. Drugs* **2022**, *20*, 336. [CrossRef]

59. Uhliariková, I.; Matulová, M.; Capek, P. Structural features of the bioactive cyanobacterium *Nostoc* sp. exopolysaccharide. *Int. J. Biol. Macromol.* **2020**, *164*, 2284–2292. [CrossRef]
60. Rossi, F.; De Philippis, R. Role of cyanobacterial exopolysaccharides in phototrophic biofilms and in complex microbial mats. *Life* **2015**, *5*, 1218–1238. [CrossRef]
61. Pereira, S.; Zille, A.; Micheletti, E.; Moradas-Ferreira, P.; De Philippis, R.; Tamagnini, P. Complexity of cyanobacterial exopolysaccharides: Composition, structures, inducing factors and putative genes involved in their biosynthesis and assembly. *FEMS Microbiol. Rev.* **2009**, *33*, 917–941. [CrossRef]
62. Svircev, Z.; Markovic, S.B.; Stevens, T.; Codd, G.A.; Smalley, I.; Simeunovic, J.; Obreht, I.; Dulic, T.; Pantelic, D.; Hambach, U. Importance of biological loess crusts for loess formation in semi-arid environments. *Quat. Int.* **2013**, *296*, 206–215. [CrossRef]
63. Hoiczyk, E. Structural and biochemical analysis of the sheath of *Phormidium uncinatum*. *J. Bacteriol.* **1998**, *180*, 3923–3932. [CrossRef]
64. Keshari, N.; Adhikary, S.P. Characterization of cyanobacteria isolated from biofilms on stone monuments at Santiniketan, India. *Biofouling* **2013**, *29*, 525–536. [CrossRef] [PubMed]
65. Olsson-Francis, K.; Watson, J.; Cockell, C. Cyanobacteria isolated from the high-intertidal zone: A model for studying the physiological prerequisites for survival in low Earth orbit. *Int. J. Astrobiol.* **2013**, *12*, 292–303. [CrossRef]
66. Tirkey, J.; Adhikary, S.P. Cyanobacteria in biological soil crusts of India. *Curr. Sci.* **2005**, *89*, 515–521.
67. Borg, A.J.E.; Dennig, A.; Weber, H.; Nidetzky, B. Mechanistic characterization of UDP-glucuronic acid 4-epimerase. *FEBS J.* **2021**, *288*, 1163–1178. [CrossRef] [PubMed]
68. Campbell, R.E.; Mosimann, S.C.; van De Rijn, I.; Tanner, M.E.; Strynadka, N.C. The first structure of UDP-glucose dehydrogenase reveals the catalytic residues necessary for the two-fold oxidation. *Biochemistry* **2000**, *13*, 7012–7023. [CrossRef]
69. Ramachandran, S.; Fontanille, P.; Pandey, A.; Larroche, C. Gluconic acid: A review. *Food Technol. Biotechnol.* **2006**, *44*, 185–195.
70. Kranner, I.; Birtic, S. A modulating role for antioxidants in desiccation tolerance. *Integr. Comp. Biol.* **2005**, *45*, 734–740. [CrossRef]
71. Oliveira, P.; Martins, N.M.; Santos, M.; Couto, N.A.; Wright, P.C.; Tamagnini, P. The *Anabaena* sp. PCC 7120 exoproteome: Taking a peek outside the box. *Life* **2015**, *5*, 130–163. [CrossRef]
72. Shirkey, B.; Kovarcik, D.P.; Wright, D.J.; Wilmoth, G.; Prick, T.F.; Helm, R.F.; Gregory, E.M.; Potts, M. Active Fe-containing superoxide dismutase and abundant *sodF* mRNA in *Nostoc commune*(cyanobacteria) after years of desiccation. *J. Bacteriol.* **2000**, *182*, 189–197. [CrossRef]
73. Wada, N.; Sakamoto, T.; Matsugo, S. Multiple roles of photosynthetic and sunscreen pigments in cyanobacteria focusing on the oxidative stress. *Metabolites* **2013**, *3*, 463–483. [CrossRef]
74. Delattre, C.; Pierre, G.; Laroche, C.; Michaud, P. Production, extraction and characterization of microalgal and cyanobacterial exopolysaccharides. *Biotechnol. Adv.* **2016**, *34*, 1159–1179. [CrossRef] [PubMed]

polymers

MDPI

Review

Microbial Exopolysaccharide Composites in Biomedicine and Healthcare: Trends and Advances

Vishal Ahuja [1,2], Arvind Kumar Bhatt [3], J. Rajesh Banu [4], Vinod Kumar [5], Gopalakrishnan Kumar [6], Yung-Hun Yang [7,8] and Shashi Kant Bhatia [7,8,*]

1 University Institute of Biotechnology, Chandigarh University, Mohali 140413, Punjab, India
2 University Centre for Research & Development, Chandigarh University, Mohali 140413, Punjab, India
3 Department of Biotechnology, Himachal Pradesh University, Shimla 171005, Himachal Pradesh, India
4 Department of Life Sciences, Central University of Tamil Nadu, Thiruvarur 610005, Tamil Nadu, India
5 Centre for Climate and Environmental Protection, School of Water, Energy and Environment, Cranfield University, Cranfield MK43 0AL, UK
6 Institute of Chemistry, Bioscience and Environmental Engineering, Faculty of Science and Technology, University of Stavanger, P.O. Box 8600 Forus, 4036 Stavanger, Norway
7 Department of Biological Engineering, College of Engineering, Konkuk University, Seoul 05029, Republic of Korea
8 Institute for Ubiquitous Information Technology and Applications, Seoul 05029, Republic of Korea
* Correspondence: shashibiotechhpu@gmail.com

Abstract: Microbial exopolysaccharides (EPSs), e.g., xanthan, dextran, gellan, curdlan, etc., have significant applications in several industries (pharma, food, textiles, petroleum, etc.) due to their biocompatibility, nontoxicity, and functional characteristics. However, biodegradability, poor cell adhesion, mineralization, and lower enzyme activity are some other factors that might hinder commercial applications in healthcare practices. Some EPSs lack biological activities that make them prone to degradation in ex vivo, as well as in vivo environments. The blending of EPSs with other natural and synthetic polymers can improve the structural, functional, and physiological characteristics, and make the composites suitable for a diverse range of applications. In comparison to EPS, composites have more mechanical strength, porosity, and stress-bearing capacity, along with a higher cell adhesion rate, and mineralization that is required for tissue engineering. Composites have a better possibility for biomedical and healthcare applications and are used for 2D and 3D scaffold fabrication, drug carrying and delivery, wound healing, tissue regeneration, and engineering. However, the commercialization of these products still needs in-depth research, considering commercial aspects such as stability within ex vivo and in vivo environments, the presence of biological fluids and enzymes, degradation profile, and interaction within living systems. The opportunities and potential applications are diverse, but more elaborative research is needed to address the challenges. In the current article, efforts have been made to summarize the recent advancements in applications of exopolysaccharide composites with natural and synthetic components, with special consideration of pharma and healthcare applications.

Keywords: biopolymers; exopolysaccharides; EPS composites; healthcare; food

Citation: Ahuja, V.; Bhatt, A.K.; Banu, J.R.; Kumar, V.; Kumar, G.; Yang, Y.-H.; Bhatia, S.K. Microbial Exopolysaccharide Composites in Biomedicine and Healthcare: Trends and Advances. *Polymers* **2023**, *15*, 1801. https://doi.org/10.3390/polym15071801

Academic Editor: Alexander Malkin

Received: 24 February 2023
Revised: 27 March 2023
Accepted: 3 April 2023
Published: 6 April 2023

1. Introduction

Exopolysaccharides (EPSs) are natural biopolymers synthesized by microorganisms and are secreted for various respective functions, such as defense, biofilm formation, pathogenicity, structure, adhesion, etc. [1]. EPSs are long-chain biomolecules with molecular weights ranging from 10 to 30 kD, and are produced during the late exponential and or stationary phase of microbial growth. EPSs are produced in response to environmental stress conditions, such as pH and temperature, and exposure to heavy metals or inhibitors, etc. [2,3]. Biochemically, EPSs are carbohydrate polymers composed of glucose, galactose, and rhamnose, accompanied by non-carbohydrate moieties such as proteins, enzymes, nucleic acid, etc. The exact composition may vary with microorganisms and growth conditions.

Based on the biochemical structure and composition, EPSs can be categorized into homo-exopolysaccharides (HOEPSs) and hetero-exopolysaccharides (HEEPSs) [4]. HOEPSs are composed of one type of monosaccharide, such as α-D-glucans, β-D-glucans, fructans, and polygalactans, interlinked with α-1-6, α-1-3, β-1-2, β-1-3, β-2-6, and β-2-1 linkage among the subunits, depending upon the monomeric units. Dextran is one of the well-known examples of HOEPSs, which is made of glucose interlinked with α-1-6 glucoside linkage. In contrast, HEEPSs are comprised of different sugar monomers, along with their respective derivatives and non-carbohydrate moieties. Succinoglycan is one of HEEPSs present in bacterial biofilm. It is comprised of saccharide-based oligomers in which sugar molecules are derivatized with acyl, pyruvate, and succinic acid [4–6]. Both classes are further recognized as subgroups based on the dominant sugar residues [4]. In recent years, EPSs have gained wide attention for their cost-effective production and diverse applications as pharma and healthcare products, cosmeceutical, nutraceutical, functional food, and biocontrol agents in agriculture [7], oil recovery in petroleum industries [8], heavy metal removal [9], drug delivery, and tissue regeneration and repair [10]. Microbial exopolysaccharides have applications in various sectors, as depicted in Figure 1.

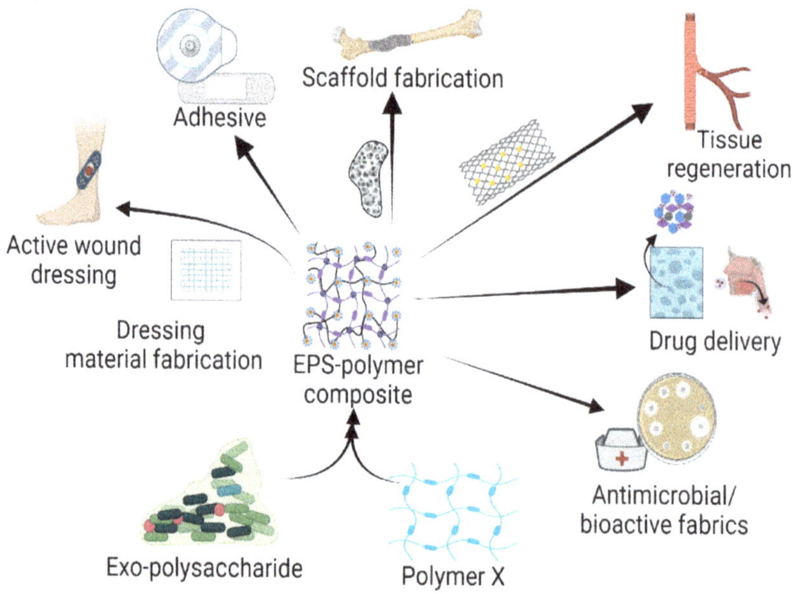

Figure 1. Exopolysaccharide composites with natural and synthetic polymers and their application in the health sector.

Most exopolysaccharides have gained attention in healthcare due to their nontoxicity, and biocompatibility, but dextran is the only known exopolysaccharide that has been commercialized in healthcare as a plasma volume compensator [6,11]. Alongside that, the majority of applications in healthcare are still under trial. Prasher et al. [12] used a dextran derivative, i.e., acetylated dextran, as a drug delivery vehicle for the treatment of respiratory disease due to biodegradability, pH sensitivity, high encapsulation efficacy, and the ability to cross the mucosal layer. Yahoum et al. [13] encapsulated metformin hydrochloride in xanthan gum microspheres and found a sustainable release of metformin hydrochloride from the microsphere. The eyes are one of the most sensitive parts of the body that need special care. EPSs have proven safe and biocompatible for biological systems, which allows their application in ophthalmic formulations. Khare et al. [14] evaluated an ophthalmic solution comprising gellan-gum-based nanosuspension with posaconazole for fungal keratitis. The main bottleneck of using native EPS molecules in commercial products is solubility in different media, bioavailability, degradation, etc. Xanthan, a HEEPS comprising a glucose backbone along with trisaccharide side chains, has poor thermal stability and electrical conductivity and is prone to microbial contamination [15,16]. Curdlan has high immunomodulatory potential, good gelling ability, and thermal stability, but is disadvantaged by the issue of solubility in water [17]. Hyaluronic acid has high water retention, but poor mechanical stability [18,19]. EPS has been blended with other natural biomolecules or synthetic polymers to improve and control the functional features of base EPS molecules, such as solubility, antimicrobial potential, mechanical strength, water retention, etc. The approach has proven beneficial, as a hyaluronic acid (HA) composite has higher mechanical strength and stability. HA composites with poly(Ne-acryloyl L-lysine) have shown a double network structure and is used for the fabrication of a more physiologically relevant 3D in vitro model for breast cancer [20,21]. Similarly, a HA composite with a self-assembling peptide carrying an IKVAV adhesive motif was used for the fabrication of a scaffold for breast cancer [22], and composites with silk fibroin–gelatin and heparan sulfate was used for the scaffold fabrication for cholangiocarcinoma [23]. A chitosan–curdlan composite is characterized by important features of both EPS components, as chitosan forms a fibrillary scaffold, and curdlan provides support to mesenchymal cell adhesion and promotes bone growth [24].

EPS composites have shown a way to overcome the bottlenecks of native EPS molecules by offering 3D-stable and porous architecture, offering more support for cell adhesion and proliferation, and an overexpression of enzymes governing wound healing and mineralization. However, there are a lot of factors impeding the commercialization of composite materials. Among the major challenges, stability and activity after prolonged storage must be addressed for commercialization apart from the cost of the product. The current review summarizes the major applications of EPS composites in healthcare, as well as research challenges.

2. Microbes Producing Exopolysaccharides

Exopolysaccharides are produced by diverse microorganisms, including yeasts, fungi, and bacteria, utilizing various raw materials (Figure 2). Besides the native physiological role, microbial EPSs also have diverse applications in many industries. Several efforts have been made by various researchers for the production of EPSs (Table 1). In comparison to yeasts and other fungi, probiotic bacteria are commonly used due to non-pathogenicity and categorization of generally regarded as safe (GRAS) [25]. It has been found that carbon source has a direct relation with EPS production and composition. A study on 20 strains of *Lactobacillus paracasei* revealed that with a change in carbon source, not only does EPS yield, but it also influences the monosaccharide composition. The yield of EPS was increased by 115% with optimized carbon sources, including fructose, glucose, galactose, lactose, mannose, and trehalose [26]. Among different carbon sources, including sucrose, maltose, lactose, glycerol, and sorbitol, maximum EPS production has been achieved with maltose by *Candida guilliermondii* and *Candida famata* (0.505 and 0.321, respectively) [27]. Among 156 lactic acid bacteria isolated from healthy young children's feces, the maximum EPS production, i.e., 59.99 g/L, was reported from *Weissella confusa* VP30 after 48 h in growth media containing 10% sucrose [28].

The use of commercial-grade sugars/substrates has a direct impact on product cost, and is one of the factors responsible for high production costs. Hence low-cost waste materials, such as lignocellulosic residues and wastewater, are preferred as raw materials for EPS production. The carbohydrate and organic fractions present in the waste can be used by microorganisms. It opens up the opportunity to reduce product costs, along with waste management. Da Silva et al. [29] have compared coconut shells, cocoa husks, and sucrose for xanthan gum production by *Xanthomonas campestris* pv. *campestris* IBSBF 1866 and 1867. The study revealed that xanthan gum yields were higher, i.e., 4.48 g/L and 3.89 g/L, in the case of cocoa husk by *Xanthomonas* strains 1866 and 1867, respectively, but the apparent viscosity was higher than sucrose, i.e., 181.88 mPas over cocoa husk, with a viscosity of 112.06 mPas.

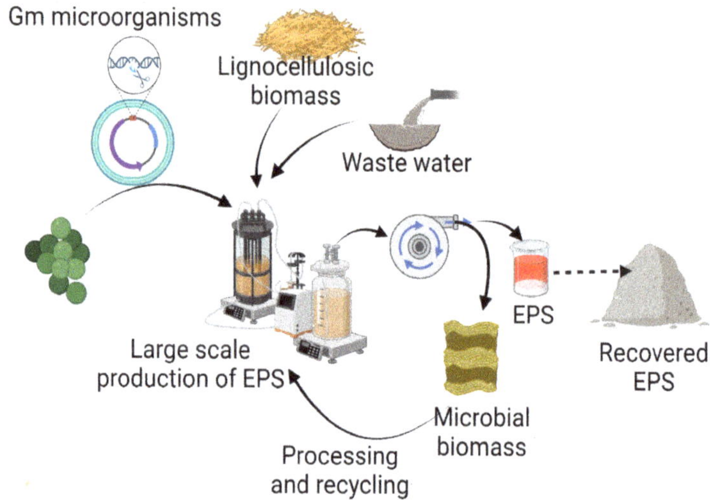

Figure 2. Microbial production of EPS using waste, and its recovery.

Choi et al. [30] used spent media wastewater (originated from kimchi fermentation) for EPS production using *Leuconostoc mesenteroides* WiKim32. Under optimal conditions, the maximum EPS productivity was 7.7–9.0 g/L, with a conversion of 38.6–45.1%. The EPSs were nontoxic and exhibited thermal tolerance and antioxidant activity. Pan et al. [31] optimized dextran production from *Leuconostoc pseudomesenteroides* XG5 using an L9 (33) orthogonal test. Under optimal conditions (sucrose 100 g/L, pH 7.0, 25 °C, at 100 rpm for 36 h), the maximum dextran yield of 26.02 g/L and 40.07 g/L were recorded at a laboratory scale and fed-batch fermentation, respectively. The EPS was also exhibiting water-holding capacity and antioxidant activity. It reduced the chewiness and hardness of yogurt, but the resilience increased during the 14 days of storage. Product cost is one of the main obstacles in commercialization, hence process economics is one of the major factors that must be assessed. Integration of multiple processes might improve process economics, as well as environmental adaptability. *Sphingobium yanoikuyae* was evaluated for the coproduction of EPS and polyhydroxyalkanoates (PHAs) using lignocellulosic hydrolysate. Hydroxymethyl furfural (HMF), one of the hydrolysis byproducts, improved the consumption of glucose and xylose during fermentation. The optimum C/N ratio of 5 resulted in the maximum EPS production of 3.24 ± 0.05 g/L, however, a further increase in the C/N ratio (30) favored PHB accumulation (38.7 ± 0.08% *w*/*w*) [32]. Biomass hydrolysate is usually accompanied by phenolics, furfurals, and HMF, which also act as fermentation inhibitors and hinder microbial action. Some of the methods, including activated carbon-based adsorption and membrane filtration, have been proposed for hydrolysate detoxification [33,34]. Removal of inhibitors might improve microbial action and product yield. Bhatia et al. [32] have compared the potential of various raw and detoxified hydrolysates for EPS production. In comparison to raw hydrolysate, detoxified biomass hydrolysate showed increased EPS production and maximum EPS production was reported with detoxified pine biomass hydrolysate, i.e., 2.83 ± 0.03 g/L. Table 1 summarizes the recent efforts for EPS production using various microbes (yeasts, bacteria, and fungi), utilizing various pure carbon and organic waste materials.

Table 1. Microbial production of EPS from various substrates.

EPS	Organism	Substrate	Growth Conditions	Working Volume	EPS Yield	Key Achievements	Reference
Dextran (α-D-gluco pyranosyl moieties interlinked with α-(1,6) linkage and have α-(1,2)/α-(1,3)/α-(1,4) branching) 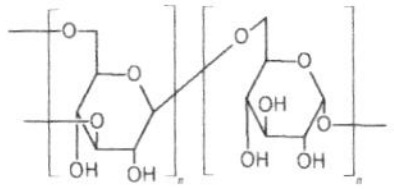	*Leuconostoc mesenteroides* SF3	10% sucrose	Temp 25 °C; pH 6; incubation period 16 h; inoculum 24 h old, 10% inoculum with cell concentration of 10^8 cells/mL.	100 mL	23.8 ± 4 g/L	Water absorption capacity 361.8% ± 3.1; oil absorption capacity 212.0% ± 6.7; emulsion activity 58.3% ± 0.7.	[35]
	Lactobacillus spp.	15% sucrose	Temp 30 °C; pH 7; incubation period 24 h; inoculum 24 h old 4%, growth conditions aerobic.	100 mL	5.8 mg/mL	*Lactobacillus* strains were isolated from the human vagina and infant stool.	[36]
	Leuconostoc pseudomesenteroides DSM20193	Brewers' spent grain	Initial cell concentration 6.0 Log cfu/g; temp 25 °C; period 24 h.	1000 g	1.2 g/100 g	EPS production is accompanied by mannitol; no dextran production without a starter (commercial granulated sugar).	[37]
	Leuconostoc pseudomesenteroides XG5	100 g/L sucrose	Temp 25 °C; pH 7.0; mixing rate 20 rpm; time period 60 h; inoculum 2%.	35 L	26.02 g/L dextran	Protein content in EPS reduced when extracted with EDTA or NaOH+formal-dehyde.	[31]
	Weissella confusa A16	Brewers' spent grain	Initial cell concentration 6.0 log cfu/g; temp 25 °C; period 24 h.	1000 g	1.1 g/100 g	No mannitol production was observed, but a starter was required for EPS production.	[37]
Curdlan (Type HOEPS, unbranched; molecular weight 5.3×10^4–2×10^6 Da; components glucosyl residues inter-connected with β-D-(1→3) bonds) 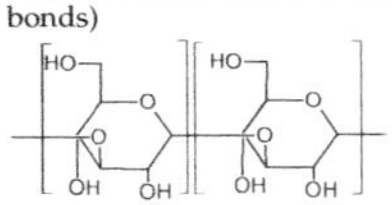	*Agrobacterium* sp. IFO 13140	50 g/L	Synthetic medium; temp 30 °C; mixing rate 150 rpm; period 5 d; pH 7.	100 mL	-	Water holding capacity and oil holding capacity 64% and 98% higher in comparison to commercial curdlan.	[38]
	Bacillus cereus PR3	10% starch	Synthetic medium; period 96 h.	100 mL	20.88 g/L	Anti-oxidant activity increased with curdlan.	[39]
	Agrobacterium sp. ATCC 31749	Asparagus spear bottom part juice	Synthetic medium; temp 30 °C; mixing rate 200 rpm; period 168 h.	100 mL	40.2 g/L	Curdlan production is higher with sucrose in comparison to mineral salt.	[40]
Xanthan (Type HEEPS; components backbone made of D-glucose unit linked with β-1,4-glycosidic bonds and side chain trisaccharide; side chain comprised of mannose, glucuronic acid, and mannose, terminal mannose with pyruvic acid residues; molecular weight 2.0×10^6–2.0×10^7 Da) 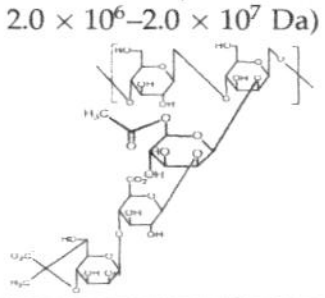	*Xanthomonas campestris*	20 g/L glucose	Stainless steel supported biofilm reactor; period 27 h; synthetic medium; mixing rate 180 rpm.	150 mL	3.47 ± 0.71 g/L	Use of biofilm reactor increased the xanthan recovery.	[41]
	Xanthomonas campestris	20 g/L glucose	Polyethylene supported biofilm reactor; period 78.5 h; synthetic medium; mixing rate 180 rpm.	150 mL	3.21 ± 0.68 g/L	Biofilm reactor increased the glucose consumption.	[41]

Table 1. *Cont.*

EPS	Organism	Substrate	Growth Conditions	Working Volume	EPS Yield	Key Achievements	Reference
Gellan gum (Type HEEPS; components backbone made up of β-d-glucose, l-rhamnose, and d-glucuronic acid along with acetate and glycerate attached to glucose) 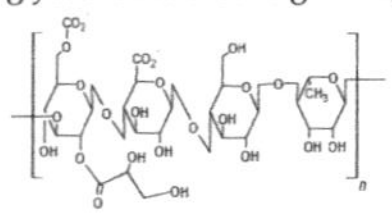	*Sphingomonas pseudosanguinis* (Accession No. GI:724472387)	80 g/L biodiesel-derived waste glycerol	pH 7; synthetic medium; temp 30 °C; mixing rate 200 rpm; inoculum 10%; period 7 days; 0.5 vvm.	3 L	51.6 g/L	At lower concentrations, glycerol is consumed completely at all pHs, but at a higher concentration, it is not exhausted completely.	[42]
	Sphingomonas yabuuchiae (GI:724472388)				52.6 g/L		[42]
EPS Br42 was found to be a heteropolysaccharide consisting of glucose and galacturonic acid with a molecular weight of about 286 kDa.	*Brevibacillus borstelensis* M42	2% glucose	Period 60 h.	-	1.88 g/L	Water-holding capacity 510 ± 0.35%, oil-holding 374 ± 0.12% and swelling capacities 146.6 ± 5.75%.	[43]
EPS K1T-9 (EPS type HEEPS; components glucose and galacturonic acid; molecular weight 207 kDa.	*Neorhizobium urealyticum* sp. nov.	Glucose 5 g/L	Zobell's marine broth; pH 7; temp 28 °C; mixing rate 150 rpm; inoculum size 5 mL/100 mL; period 72 h.	Working volume 400 mL	3.38 g/L	Water holding 356 ± 0.8%, oil holding 697 ± 1% (coconut oil); 317 ± 1.3% (olive oil), swelling capacity 200 ± 1.1%.	[44]

Preparation of composites for healthcare applications needs high purity, therefore the downstream processing becomes an inseparable part of processing after fermentation. After production recovery, the identification of structural and chemical characteristics is necessary for further applications. The characterization of EPSs is quantitative, as well as qualitative. EPSs are mainly comprised of carbohydrates conjugated with other biomolecules, hence the basic characterization techniques employed are a colorimetric estimation and use spectrophotometry [45]. For carbohydrate estimation 'Dinitrosalicylic Acid Reagent' is one of the common methods which quantify the reducing sugars [46]. Similarly, for protein, Bradford's dye-binding method [47] and Lowry's method [48] are used. Besides basic characterization with colorimetric methods, Fourier transform infrared spectroscopy (FTIR) is employed to detect the available functional groups and structural functionalities of EPSs. For the detailed structures of EPSs, nuclear magnetic resonance (NMR) and mass spectra are used. In NMR, the sample is dissolved in deuterated solvents for quantification with respect to internal standards [49,50]. The mass spectrum of EPS provides a monosaccharide composition. For analysis, EPSs are hydrolyzed with acid hydrolysis, followed by silylation derivatization. The derivatives are detected and identified by gas chromatography-mass spectrometry [49]. The biological potential of EPS has followed the general procedure for antimicrobial, antioxidant, anti-inflammatory and other activities [45].

3. Microbial Exopolysaccharide Composites and Their Applications

Exopolysaccharides are natural biopolymers exhibiting diverse applications, variation in structure, adaptability, and the presence of different functional groups. These polymers have shown their suitability as drug carriers and medical sealants. Even with polymeric nature and biocompatibility, these polymers have also shown some weak points. Rapid degradation, hydrophilic nature, low mechanical and tensile strength, and stress tolerance, particularly for scaffold preparation, restrict their application. Some EPSs lacks bioactivity itself, which suggests the addition of multiple drug compounds to add bioactivity for in vivo application. Exopolysaccharide composites have shown a higher potential than native EPSs due to the presence of secondary polymer molecules that offer hybrid characteristics of both (Table 2). The higher mechanical strength with different functional characteristics contributes to diverse functional groups, and is responsible for physiological and chemical properties, contributed to by the members of the composite. To date, EPS composites have been prepared with natural, as well as synthetic, polymers. The selection of secondary polymers relies upon the type of application such as drug loading and release, tissue support and engineering, or ex vivo applications such as a sealant (Figure 3).

Piola et al. [51] prepared a composite hydrogel with gelatin and xanthan gum to support the growth of human skin cells. The composite was printed with CellInk Inkredible 3D printer, using glutaraldehyde solution as a crosslinker. The printed hydrogel was compatible and suitable for the growth of human keratinocytes, as well as fibroblast. Alvel et al. [52] also prepared a composite hydrogel of xanthan with Konjac glucomannan, which also focused on wound healing. On the other hand, a 3D scaffold for tissue engineering was prepared with alginate–gellan gum [53] and methacrylated gellan gum [54].

The applications of EPS composites with various polymers have been summarized below:

3.1. Exopolysaccharide Composites with Natural Materials

EPS itself is a natural biopolymer that is combined with other natural biomolecules, including proteins, enzymes, lipids, carbohydrates, etc., and thus in most cases, the composite has healthcare-associated applications, including medical sealants and scaffolds in tissue recovery and repair. The major advantages of EPS composites with natural polymers are bioactivity, biodegradability, and biocompatibility. As both the components are natural in origin, there will be a negligible possibility for inflammation and rejection. Alongside that, these composites will not have any toxic effect on the host, but may also act as an antimicrobial formulation for targeting microorganisms. Most biological materials can also be produced in bulk amounts using microorganisms, hence the cost of the products is also reduced along with the higher availability. The addition of other polymers may also add a specific type of bioactivity, such as antimicrobial, anti-inflammatory, and even catalytic activity to the composite, which widens the possible application in different fields [55–57].

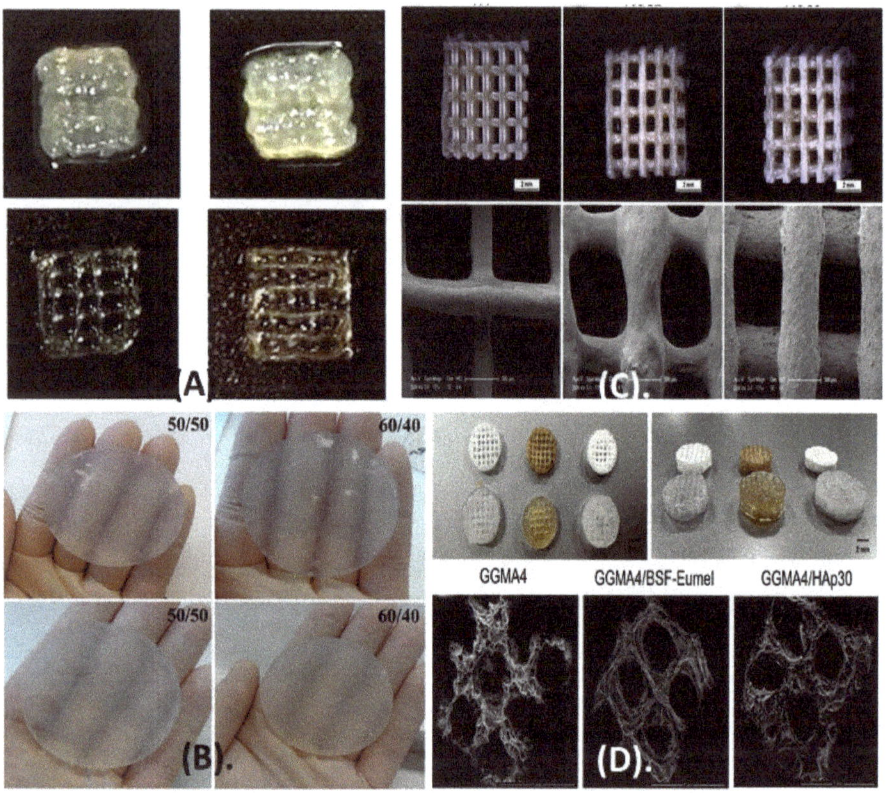

Figure 3. Composite scaffold prepared from different exopolysaccharides (**A**). Xanthan gum–gelatin [51]; (**B**). Xanthan gum–Konjac glucomannan [52]; (**C**). Alginate–gellan gum composites [53] and (**D**). Composite methacrylated gellan gum [54].

3.1.1. Cellulose Composites with Natural Polymers

Cellulose is a linear structural polysaccharide comprised of glucan chains connected with cellobiose residues via β-1,4-glycosidic linkage. These structures are packed in forms of microfibrils, kept together with hydrogen bonds and Van der Waals interactions. Depending upon the packaging, it exhibits different degrees of polymerization. It is usually present in the secondary cell wall of plants, but is also present in some bacteria, including members of the Acetobacter, Agrobacterium, Azotobacter, Alcaligenes, Pseudomonas, Rhizobium, and Sarcina genera [58]. The main function of cellulose is to provide strength to structure [59], however, cellulose does not exhibit antimicrobial or similar bioactivity in itself [60], but its high mechanical strength and stress tolerance nature make it suitable for the fabrication of scaffold and dressing material when used together with some bioactive material. Various polymers have been exploited to form composites with EPSs, which improve their strength and water retention capacity and also aid in antimicrobial properties.

Ojagh et al. [61] produced cellulose with *Gluconacetobacter xylinus* by static fermentation. The cellulose and diethylaminoethyl cellulose were derivatized to carboxymethyl cellulose and carboxymethylated diethylaminoethyl cellulose, respectively, and processed for composite preparation. Briefly, cellulose is a carbohydrate that is rich in hydroxyl groups, which makes it negatively charged. However, similar charges on both components make interactions repulsive. Derivatization of cellulose to carboxymethylated diethylaminoethyl cellulose (CMDEAEC) has both hydroxyls, as well as amine groups in its structure. In contrast to cellulose, CMDEAEC has a net positive charge, and thus the higher amount of CMDEAEC can be loaded on BC in composite formation. The composite of bacterial cellulose–carboxymethyl cellulose has a higher drug loading capacity and swelling ratio than the composite of bacterial cellulose–carboxymethylated diethylaminoethyl cellulose, and even native bacterial cellulose itself. The drug release follows the Higuchi and Korsmeyer–Peppas models. The model is most suitable to describe the release of drugs from the polymer matrix hydrogel. The model suggests that drug release increases from the carrier with time. Methylene blue is a positively charged molecule and the presence of cellulose supports its binding to the composite. The work has also shown that chemical modification adds unique properties to EPSs, improves functionality, and enhances the interaction with another polymer to form a stable composite. Bacterial cellulose was functionalized by derivatization with two active agents, i.e., glycidyl trimethylammonium chloride and glycidyl hexadecyl ether. These agents act upon hydroxyl functional groups of glucose by a heterogeneous reaction, which simultaneously deprotonates the hydroxyl group, as well as the addition of epoxides. It was observed that mere derivatization reduced the bacterial population of *Staphylococcus aureus* 6538PTM and *Escherichia coli* (Migula) ATCC® 8739TM almost by half (53% and 43%, respectively) within 24 h upon direct contact. However, the derivative did not have any cytotoxic effect in terms of morphology and viability upon keratinocytes (HaCaT cell line) and almost 90–100% viability was recorded after 6 days of direct contact. Modified hydrogel has shown an equitant wound closure rate in an in vitro scratch assay, with complete coverage of the wound area after 5 days [60]. It was suggested that the addition of epoxide to cellulose adds antimicrobial properties to cellulose that are also exhibited by the composite.

Composite hydrogels are made by attaching TEMPO-modified nanocrystalline cellulose to the methacrylated gelatin backbone. The human adipose-derived mesenchymal stem cells were encapsulated within the composite and cultured in normal and osteogenic media for 14 days, and the expression of valve interstitial cell phenotypes was observed. The encapsulated cells have lowered alpha-smooth muscle actin expression, while the expression of vimentin and aggrecan increased. Cells cultured in osteogenic media have a reduced expression of osteogenic genes (Runx2 and osteocalcin) that support resistance to calcification. With the composite, a tall and self-standing tubular structure was constructed with a composite hydrogel that also sustained cell viability for possible application in cardiovascular systems [62]. The addition of cellulose to chitosan has improved the mechanical strength, porosity, cytocompatibility, and drug release rate for use in bone tissue engineering. The ternary complex of alginate, chitosan, and bacterial cellulose was used for the preparation of scaffolds using a hydroxyapatite/D-glucono-δ-lactone complex-based gelling system. The composite-based scaffold can adapt to 3D morphology and is stabilized with extensive cross-linking. The smaller size of pores supports the attachment and growth of tissue, along with the required mechanical integrity. Alginate controls the swelling behaviors of the scaffold by intermolecular hydrogen bonds and prevents the degradation of the composite. It has shown high protein adsorption and release potential, along with commendable cytocompatibility, that supported its application in tissue engineering [63]. The polymer supported the 3D structure and allowed for the proliferation of cells, increasing the possible application opportunities in the reconstruction of complex and large organ structures.

3.1.2. Dextran Composites with Natural Polymers

Dextran is a non-toxic homoexopolysaccharide, mostly produced by lactic acid bacteria. It is made up of glucose subunits interconnected with α-(1$\rightarrow$6) bonds, and forms a linear backbone. However, liner chains also have D-glucose side chains as branches attached with α-(1→4), α-(1→3), or α-(1→2) linkage. Dextran exhibited diversity in molecular weights that range from 40–70 kDa. Dextran is water soluble, increases viscosity, and has a tendency to expand in the presence of water, contributing to the volume expansion in the case of plasma volume compensation, gelling agent in foods and pharmaceuticals [6,11]. It is also used in the preparation of molecular sieves while purifying macromolecules [20]. The solubility and nontoxicity of dextran have a big role in its application for drug delivery and preparation of wound dressing materials for biological systems.

Dextran was oxidized with periodate-assisted oxidation, followed by a reaction with chitosan hydrochloride. The composite did not have any cytotoxicity and offered minimal swelling in phosphate-buffered saline, along with good adhesiveness. The optimum adhesive strength of the composite, i.e., 200–400 gf/cm^2, was 4–5-fold higher than commercial fibrin glue. The burst strength of the composite was 400–410 mm of Hg, which makes it suitable to be used as a medical sealant to control bleeding during surgical procedures. It also works under high blood pressure conditions. Both, adhesiveness and hemostat were assessed and found to be suitable in a rabbit liver injury model [64]. Besides medical adhesiveness, dextran has shown commendable suitability for in vivo applications, such as drug delivery and wound healing. Dextran composites alone can contribute to both adhesiveness and drug-carrying capacity. Even though a number of drugs are available for wound dressing and tissue repair, in some cases, injury worsens due to failure of wound dressing, mainly attributed to poor bioavailability, hydrophobic nature of drugs, and high level of reactive oxygen species. Dextran-based composites have offered the solution for all these challenges of rapid tissue repair. The drug delivery and bioavailability can be improved further when used in the nano form due to increased surface area. Andrabi et al. [65] have fabricated a composite nano-hydrogel with gelatin, oxidized dextran, and curcumin and cerium oxide-based nano-formulation. Two types of dextran derivatives have been used, i.e., alkylated dextran with 1-bromohexadecane (O-hexadecyl-dextran) and oxidized dextran. Curcumin nanoparticles were synthesized with alkylated dextran, while, on the other hand, cerium oxide nanoparticles were loaded onto the gelatin hydrogel with oxidized dextran. Altogether, these formulations form a hybrid nano-hydrogel that is used for wound healing. Here, alkylated dextran was amphiphilic in nature, which maintains the bioavailability of the drug (curcumin). Cerium oxide acts as an antioxidant and controls the reactive oxygen species and other free radicals. Hydrogel has shown a prolonged and consistent release of the drug, with ~63% release in 108 h. It also promotes cell migration to induce wound healing, along with antioxidant and in vivo anti-inflammatory activity (~39%).

Poor wound healing raises concerns in cases of noncompressible injuries due to a higher rate of blood loss. These injuries mostly occur from gunshots or injuries with sharp objects, such as knives, etc. In the absence of timely repair of the injury, higher blood loss and trauma-assisted mortality have been observed. Cryogels were prepared from oxidized dextran, chitosan, collagen, and polydopamine nanoparticles. In the formulation, dextran was used as a crosslinker, while chitosan was a hemostatic agent. The cryogels have an intensively branched and interconnected macroporous structure, with a high capacity to absorb blood or water. In vitro assessment suggested a very high coagulation potential due to strong procoagulant ability, and high adhesion potential for fibrinogens and blood cells. Moreover, it can activate platelets and associate intrinsic pathways. It reduced bleeding time and blood loss [66]. The dextran–thyme magnesium-doped hydroxyapatite composite is prepared by mixing salt precipitate with dextran and essential oil solution (Figure 4). The composites have shown antimicrobial activity and have considerable results against *Staphylococcus aureus, Enterococcus faecalis, E. coli, Pseudomonas aeruginosa,* and *Candida albicans.* The composites have the potential for application as antimicrobial coating materials [67].

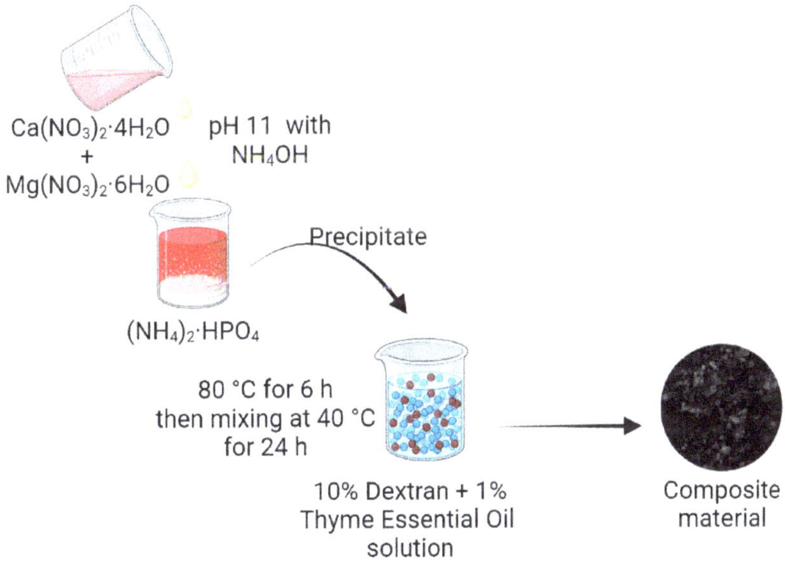

Figure 4. Preparation of antimicrobial coatings with dextran, thyme essential oil, magnesium salt, and hydroxyapatite, as suggested by Iconaru et al. [67].

3.1.3. Xanthan Composites with Natural Polymers

Xanthan is a hetero-exopolysaccharide, comprising glucose, mannose, and glucuronic acid. It is produced and secreted by *Xanthomonas campestris*. Biochemically, it is made of repeated units of pentasaccharides interconnected with β 1, 4-linked d-glucose backbone. The chain also has a substitution with trisaccharide side-chain linkage. The monosaccharide composition represents β-D-glucose, α-D-mannose, and α-D-glucuronic acid in a 2:2:1 ratio. Xanthan is known for its high viscosity, emulsion stabilization, and shear-thinning activity. Xanthan is non-toxic and biocompatible, but it is disadvantaged by poor electrical conductivity and stability of heat [20]. In comparison to macrostructures, the application of nanomaterials increases the interactions at a molecular level, has a higher surface area, and reduces the drug dosage required due to the higher efficiency of drug delivery in the nano form [68,69]. The application of EPS composites as nanomaterials might improve the performance, hence efforts have been made to incorporate the nanomaterials into the composite.

Nanocomposites comprising sodium alginate and xanthan gum, reinforced with cellulose nanocrystals and/or halloysites, are highly porous, with an extensive pore network. Nanotubes and nanocrystals have uniform dispersion and partial orientation within a composite. Composites with nanocrystals and nanotubes have porosity ranging from $91.7 \pm 0.81\%$ to $88.5 \pm 0.64\%$, and water uptake capacity ranging from $14.73.7 \pm 0.46$ g/g to 11.34 ± 0.32 g/g. The composite is thermally stable and has high compressive strengths of 91.1 ± 1.2 kPa to 114.4 ± 0.6 kPa in dry form, and 9.0 ± 0.8 kPa to 10.6 ± 0.8 kPa in wet form. The composite has high cytocompatibility for MC3T3-E1 osteoblastic cells, and the viability of the cells increased with nanotube components in the composite. Both nitrocellulose and nanotubes increased the mechanical stability of the composite, along with conducive bioactivity, including higher cell adhesion and proliferation. The compressive strength of the composites was higher than alginate alone, as well as the alginate-xanthan gum blend, i.e., 91 kPa and 80.7 kPa. The blending of xanthan gum lowered the stiffness, which was further improved by the addition of nanocrystals and nanotubes, with the maximum strength reaching 114.4 kPa [70]. Mechanical strength and adhesion are important parameters for wound dressing material. Hydrogels of xanthan gum and Konjac glucomannan blend have shown significant firmness and cohesiveness when both polymers are used in an equal ratio, i.e., 1:1. All the blends prepared with components of 1:1 and 2:3 ratios (glucomannan: xanthan gum) have a water content of more than 99% and water contact angle of less than 90°. This feature is also important, as it prevents dehydration in wounds and supports healing by improving cell adhesion. In vitro analysis suggested that even after 72 h of contact, the hydrogel did not affect the morphology and viability, and supported the proliferation of fibroblasts. The hydrogel also provides cell migration during the healing phase that aided in rapid wound healing, which led to the full coverage of the wound site after 12 h, while in the control, it took 24 h [52].

Medical adhesives have been used in device and tape manufacturing. Biocompatible adhesive might offer a durable and user-friendly alternative for application. Feng et al. [71] have developed medical-grade adhesives with soybean protein sources at different concentrations of xanthan gum. The composite comprising 0.5% xanthan and 0.5% soybean proteins have shown a maximal adhesion strength of 321 kPa, which was 2.6-fold higher than the control, i.e., SP alone. The addition of xanthan increased the viscosity of the composite and improved the hydrogen bond. On a molecular basis, the addition of xanthan reduced the α-helix content and increased the β-sheet content in the protein secondary structure. The composite has shown a compact and viscous surface that supports adhesion. Sometimes, bone injuries need support and artificial implants for regeneration and recovery. Zia et al. [72] have prepared a polymer composite with xanthan gum, chitosan, and nanohydroxyapatite and polyelectrolyte complex. Osteo-conductive tri-composite scaffolds were prepared for osteo-regeneration. The composite scaffold, containing polymer electrolyte and hydroxyapatite in the ratio of 1:1, exhibited a commendable porous structure with a compressive strength swelling ability with slower degradation rates. Analysis in simulated body fluid, xanthan gum, chitosan, and nano-hydroxy apatite have apatite-like surface structures. In vitro interaction studies revealed that the nanocomposite scaffold with chitosan and nanohydroxyapatite supported cellular viability, attachment, and proliferation of MG-63 cells.

Polymeric porous scaffolds were made from chitosan and xanthan, along with 5% hydroxyapatite and brushite, for advanced mesenchymal stem cells. Composite scaffolds have an amorphous structure phase, having major bands of amide for chitosan and xanthan. Scaffolds have porous structures with calcium phosphate fillers. The elasticity modulus was higher for composite scaffolds with brushite than with hydroxyapatite and composite alone. Composites alone have a higher cell viability than scaffold–calcium phosphate, having acceptable cell viability. The composite with any calcium phosphate forms had a higher inflammatory response after 48 h, while scaffold + hydroxyapatite with mesenchymal stem cells had the lowest inflammatory cell number. It was clear from the work that calcium phosphate improved mechanical strength, but lowered cell viability. The toxic effect was countered by the addition of mesenchymal stem cells [73]. Now, the era of 3D printing has started, but the fabrication of biological structures needs supportive and efficient materials with a high mechanical strength and elasticity modulus. Piola et al. [51] developed crosslinked 3D-printable hydrogel with gelatin and xanthan gum, especially for wound dressing. The results have suggested that both gelatin and xanthan gum are important for composite formation, but xanthan concentration is crucial for the printability of the composite, as 1–1.2% of xanthan is required to attain printability, irrespective of the concentration of gelatin; however, 1.2% concentration was optimum.

3.1.4. Pullulan Composites with Natural Polymers

Pullulan, an unbranched homopolysaccharide, is composed of triose units made from three glucose units interconnected by α-1,4 glycosidic bonds. These maltotriose units are linked with each other by α-1,6 glycosidic bonds. It is produced by *Aureobasidium pullulans* from starch, and secreted out [74,75]. The α-bond contributes to its aqueous solubility and flexibility. Overall, it is a biodegradable, nontoxic, and biocompatible polymer [76]. The characteristics are suitable for hydrogels and hybrid polymer fabrication, but the main drawbacks are its high-water solubility and hygroscopic nature. In addition, it has poor support for cell adhesion and proliferation, and is poorly osteogenic, mainly attributed to its hydrophilicity [77,78].

For making bone grafts, mechanical strength and degradation profile are important parameters. Usually, a hydrogel has a porous structure with intense networking. Pullulan is one of the non-immunogenic biopolymers that can be used for scaffold fabrication. Pullulan–dextran composite scaffolds, together with interfacial polyelectrolyte complexation fibers, have an improved adhesion for cells in comparison to pullulan. Further addition of extracellular proteins proved beneficial for cell adhesion and growth. The composite scaffold induced endothelial cell growth and followed the zero-order release kinetics for bovine serum albumin, as well as vascular endothelial growth factor [79]. Hydrogels are made from pullulan and covered with 5% hydroxyapatite nano-crystals and 3% poly(3-hydroxybutyrate). Making composites with fillers, such as butyrate, increased the compressive modulus of the composite scaffold by 10 times. The adhesion support for cells was improved by the presence of hydroxylapatite nanocrystals [78].

Posterior segment eye diseases need invasive intravitreal-injection-assisted treatment. However, prolonged injections become painful and can be countered with the use of an efficient drug delivery system. Kicková et al. [80] prepared pullulan–dexamethasone conjugates for sustainable drug delivery. Dexamethasone was loaded onto pullulan in a 1:20 ratio, and attached with pullulan via hydrogen bonds. The composite particles were stable at a wider range of temperatures, from 4 to 37 °C, at a physiological pH. At pH 5.0, these composites were acting like lysosomes and released the drug slowly, as 50% of the drug was released in the vitreous, while at pH 5.0, this occurred across a 2-day period. In vitro evaluation of biocompatibility showed no signs of toxicity on the retinal pigment epithelial cell line (ARPE-19).

Sustainable drug release from the composites has also opened the path for designing matrices and scaffolds for wound healing. The research work of Chen et al. [81] has proved that dressing materials have a critical role in wound healing time, pattern, and cellular response on wound healing, from in-depth histologic and histopathologic analysis in mice models. The dressing materials made from collagen hydrogel followed a human-like wound-healing model in mice. In comparison to the control, the hydrogel of pullulan–collagen induced rapid wound closure and healing, healing with less dense and shorter collagen fibers with random alignment.

Baron et al. [82] also used a pullulan composite for wound healing and drug delivery. Hydrogels made from oxidized pullulan and dopamine have possible applications in hemostatic wound dressing. Cryogels were prepared by (hemi)acetal and Schiff base bonds between dialdehyde pullulan and dopamine. Two types of formulation were prepared, i.e., PD1 by pullulan derivatives attached with dopamine derivatives and PD2 by adsorbed dopamine on the scaffold. PD1 is loaded with 20% dopamine, while PD2 has only 1.14% carried within the polysaccharide network. The porosity was maximum in the scaffold alone (80.41%) and reduced after loading; however, the network density was maximal in PD1 (47.1%). The water adsorption capacity was minimum for the crude scaffold (31.41%) due to the smooth texture and the maximum PD1 (59.01%). The PD1 hydrogel exhibited fast swelling initially, followed by stabilization. In comparison to the crude scaffold and PD2, PD1 has the highest mechanical stability of 558 N. The in vitro hemolytic analysis suggested a high rate of hemolysis in PD2, i.e., 99.04%, followed by 7.12% in PD1 and a minimum for the crude scaffold (0.15%).

3.1.5. Levan Composites with Natural Polymers

Levan and gellan are exopolysaccharides extensively used in the food industry. Levan is a bioactive, user-friendly, non-toxic homo-exopolysaccharide produced by microorganisms, as well as plants. Among microorganisms, *Aerobacter levanicum*, *Bacillus subtilis*, *B. polymyxa*, *Corynebacterium laevaniformans*, *Pseudomonas* sp., and *Streptococcus* sp. are among some common producers of levan. It is one of the natural fructans, made of fructose subunits interlinked by β-(2→6) glycosidic linkages, along with side chains attachment via β-(2→1) linkage to the main backbone. It is commonly used as a thickening/gelling agent, encapsulating material, and alternative to petrochemicals in medical applications [83]. Levan is soluble in water, as well as in oil. In industries, it is used as gum, sweetener, flavor carrier, surface finishing additive, gelling agent, emulsifier, cryoprotector, and osmoregulator. Its bioactivity includes antitumor and antihyperlipidemic and radioprotector activities [84].

Levan has an immunomodulatory effect on the host, hence it is commonly used as a drug carrier, as well as drug coating material. Bovine serum albumin was used as a model compound, and was encapsulated in levan nanoparticles to assess carrying and release behavior. The nanoparticle has surface charges ranging from +4.3 mV to +7.6 mV, varying in particle size, from 200 nm to 537 nm. With the increase in size, the encapsulation capacity of nanoparticles also increased from 49.3% to 71.3%. Both types of particles have shown controlled release of BSA during in vitro analysis [85]. Besides drug delivery, levan has also been used as tissue filler. Due to its flexibility, an injectable biofiller was designed with levan, together with carboxymethyl cellulose and Pluronic F127, for the regeneration of soft tissue. The hydrogel offered an elastic modulus higher than hyaluronic acid hydrogel. The presence of interconnected pores also makes it suitable to be used as a filler. In vitro analysis has confirmed that levan improved cell proliferation and collagen synthesis in human dermal fibroblast cells without any cytotoxicity. The levan hydrogel was stable over 2 weeks in vivo, which was higher than the Pluronic F127 hydrogel or hyaluronic acid hydrogel alone. Apart from this, the levan hydrogel has also shown anti-wrinkle activity in wrinkle model mice, which was also higher in comparison to the hyaluronic acid hydrogel [86]. The hydrolysed derivative of the levan polysaccharide is prepared for nanoparticle synthesis by the electro-hydrodynamic atomization (EHDA) technique, and loaded with resveratrol (encapsulation efficiency 13.8 $\pm$ 1.3%). The drug-loaded nanoparticles have followed first-order kinetics in resveratrol release at different pHs, as higher loading is accompanied by a more gradual release. It showed a burst release mechanism, as 65–70% of the drug was released initially, followed by slow release. There was no sign of toxicity during in vitro assessment with human dermal fibroblast cell lines (PCS-201-012) [87]. Activation of metalloproteinase supports the healing of injured tissue. The application of levan in biomedical dressing sealants is limited, as it gets removed in a wet environment. The addition of catechol improved the adhesion strength of the levan composite to 42.17 $\pm$ 0.24 kPa (>3 times of fibrin glue), that persists even in a wet environment, and allowed for its application in hemostatic surgeries and wound healing. Besides adhesion, it also induced rapid blood clotting and healing of rat skin incisions. In comparison to levan, the composite has a lower endotoxin level [88]. The major challenge with levan is its cost and lower yield, thus the cost of composites and prepared products will also be high. For better market reach, bulk and cost-effective production is necessary.

3.1.6. Gellan Composites with Natural Polymers

On the other hand, gellan gum is a hetero exopolysaccharide composed of two β-D-glucose, β-D glucuronic acid, and α L-rhamnose subunits. It is an anionic, water-soluble polysaccharide, secreted by *Sphingomonas elodea*. Gellan gum is a strong gelling agent, able to form a gel at very even low concentrations. Based on the gel-forming ability, two types of gellan gums are available, i.e., low-acyl form that makes hard and brittle gel, and the second high-acyl form, that produces soft and elastic gels. It can be used to make low-viscosity suspensions, and as a gelling agent in industries [74]. It can offer diverse forms in gelling ability, but it lacks stability against stress and shear tolerance for tissue engineering. Zheng et al. [89] have blended gelatin with gellan gum to prepare injectable scaffolds applicable for skin regeneration. The aim was to cover irregularly shaped wounds followed by recovery. Gelatin and gellan gum had a synergetic effect on the composite hydrogel and exhibited both shear-thinning, as well as self-recovering abilities. Further addition of tannic acid induced rapid wound healing in mice model.

Some of the studies have also shown the application of biopolymer composites as a cell carrier in addition to drug compounds, as suggested in the case of retinal pigment epithelium cells. The deterioration and damage in retinal pigment epithelium cells lead to blindness that can be cured with the replacement of damaged cells with healthy ones. Kim et al. [90] have shown that hydrogels prepared from gellan gum and silk sericin can be used as a cell carrier also. Sericin improved the compressive strength of composite over gellan gum to support the growth and proliferation of cells. A composite of gellan gum with 0.5% sericin has shown compressive strength to 10 kPa and improved the gene expression associated with ARPE-19 cell proliferation.

3.2. *Exopolysaccharide Composites with Synthetic Polymers*

Similar to natural polymers and biomolecules, synthetic polymers have also been used to make a composite with exopolysaccharides, which improves the binding ability and woven strength of any fiber. The section summarized some of the composites of exopolysaccharides and synthetic polymers.

3.2.1. Dextran Composites with Synthetic Polymers

Dextran composites are prepared with synthetic polymers mainly to improve the water/solvent interaction behaviors, biodegradability, and mechanical strength. In the context of medical/healthcare-associated operations, polymers have higher strength or compounds with additional health benefits have a critical contribution. The fabrication of advanced fabric with biodegradability and inbuilt healing power might improve the treatment and revolutionize healthcare practices. A biocompatible dressing material fiber was prepared with electrospun poly(vinyl alcohol)–dextran, prepared from dextran and poly(vinyl alcohol) by using citric acid as a cross-linker, followed by loading with sodium ampicillin. The mechanical strength was mainly governed by the concentration of citric acid, as it determines the degree of cross-linking. However, swelling, adsorption of protein, and drug release were decreased, as the CA concentration increased, while high concentrations of dextran induced the proliferation of HFB-4 cells and offered higher antimicrobial activity. After 24–48 h of treatment, all the fabrics had the potential to accelerate the wound gap closure [91]. Kenawy et al. [91] used a composite of poly(vinyl alcohol)–dextran nanofibers for injury dressing materials. The composite was prepared by poly(vinyl alcohol) and dextran cross-linked with sodium ampicillin-loaded citric acid. The composite was electrospun for the fabrication of dressing material. The composite made with 10% PVA–10% dextran and 5% citric acid offered the best nanofiber suitable for dressing materials. The concentration of the cross-linker, i.e., citric acid, greatly influenced the characteristics, including mechanical stability, thermal stability, and water uptake. The composite nanofibers with a high concentration of dextran have encouraged the proliferation of HFB-4 cells.

3.2.2. Cellulose Composite with Synthetic Polymers

Bacterial cellulose is suitable for the fabrication of biopolymers, especially for healthcare-related applications, manufacturing scaffolds, implants, artificial blood vessels, and wound dressing materials, for wound and burn cases. The composite of bacterial cellulose with poly(vinyl alcohol) and hexagonal boron nitride was used for the preparation of a 3D scaffold for bone tissue engineering, using 3D printing technology. In composite bacterial cellulose, the major characteristics of the scaffold were determined. The addition of cellulose reduced the pore size in the composite and increased the viscosity to 81.3 mPa.s. Composites with bacterial cellulose have a lower tensile strength and the highest break in elongation, i.e., 93% was observed in the case of the composite prepared with 0.5% cellulose, 12% poly(vinyl alcohol), and 0.25% boron nitride. The biocompatibility assessment with human osteoblast cells suggested that at a lower concentration of cellulose, the viability of cells reduced by it further increased when the concentration of cellulose increased from

0.1/0.2% to 0.25 and 0.5%. The composite also offered a surface for cell adhesion and proliferation [92].

In a similar line, Zhang et al. [93] have fabricated macroporous hydrogels with dextran and polydopamine to be used as a carrier for antibiotics. Polydopamine affected the structure, as well as the functionality of the hydrogel, as with the increase in the concentration of dopamine, the pore size decreased and the surface area increased. The hydrogel has shown an increase in negative charge with the increase in the concentration of polydopamine in the composite, but the storage modulus and mechanical strength was increased. With the increase in dopamine content, the swelling ratio of hydrogel reduced and induced deswelling, which was due to the reduced pore size and higher interaction of dopamine with the internal structure of the hydrogel. The hydrogel lowered the viability of NIH3T3 cells slightly on the first day, but then no further cytotoxicity effect was reported. Hydrogels with a higher concentration of polydopamine were found to be suitable for drug loading and release. Hydrogels have shown up to 71% chlorhexidine acetate loading in 4.5 h, followed by its release of 12.58%, 16.06%, and 22.03% after 12, 24, and 48 h, respectively. The trend suggested that drug release was reduced with an increase in dopamine concentration. The work of Wang et al. [94] also emphasized the application of a cellulose composite for the preparation of artificial blood vessels with small diameters for thrombosis patients. A composite of poly(-caprolactone) and cellulose acetate was used to prepare nanofiber membranes for its further application in tubular scaffold fabrication, using different types of stainless-steel collectors. The composite scaffolds have water contact angles of 126.5° and 105.5°, which was increased by constructing a square-groove. In comparison to other collector mesh, scaffolds with a large mesh have 30% and 148% higher tensile strength over random-flat and tubular scaffolds, respectively. Similarly, the long mesh also offered 103% higher suture retention strength. Biocompatibility assessment revealed that the long mesh scaffold has 88% cell viability, and the blood coagulation index (BCI) was 5 min, which was around 89% of the standard value.

3.2.3. Xanthan Composites with Synthetic Polymers

Extracellular polysaccharides are one of the most suitable materials for the preparation of scaffolds, grafts, and dressing materials, as they have water retention capacity, significant mechanical and tensile strength, and adaptability similar to the natural organs and tissues. In addition, it also supports cell adhesion, growth, and proliferation.

Cell growth has a direct influence on the conductivity and impedance of the medium, which may be used as an indirect method to assess and monitor cell growth within polymeric scaffolds. A conducting polymer composite was designed for the preparation of porous scaffolds with poly(3,4-ethylenedioxythiophene) and xanthan gum and compared with PEDOT: polystyrene sulfonate scaffolds (Figure 5). The semisynthetic composite scaffold carries important characteristics, such as the conductivity of poly(3,4-ethylenedioxythiophene), along with the biocompatibility and mechanical strength of xanthan gum. Composite scaffolds have interconnected pores, with a size range of 10–150 μm, that can be tuned as required, and Young's module range was 10–45 kPa. The composite scaffold supports the cell growth of MDCK II eGFP and MDCK II LifeAct epithelial cells [95].

A series of composites were prepared with xanthan gum, citric acid, gelatin, glutaraldehyde, and HPLC-grade water, and evaluated for wound healing potential in rat skin wound models. All the composites have >90% of water-holding capacity. The presence of free and bound water was confirmed with FTIR studies, with inter- as well as intramolecular hydrogen bonds. Similar to water holding capacity, all the hydrogels have also shown significant wound healing capacity in deep second-degree skin burns in rats. After 20 days of application, a composite of xanthan gum–gelatin–glutaraldehyde and xanthan–citric acid and glutaraldehyde had a higher wound healing rate over others, as well as the control [96].

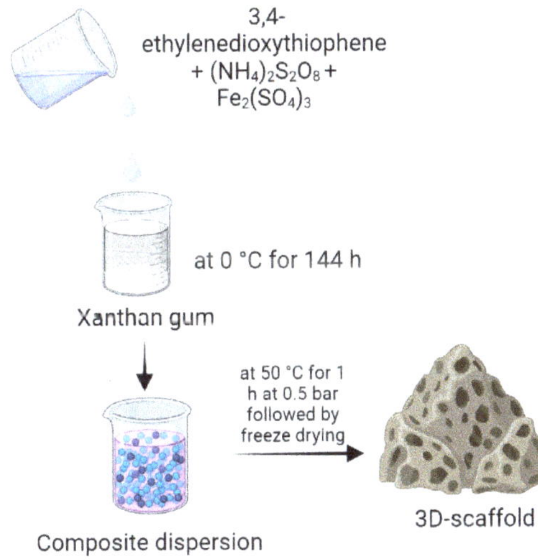

Figure 5. 3D scaffold prepared with xanthan gum and poly(3,4-ethylenedioxythiophene), as described by del Agua et al. [95].

Malik and et al. [97] prepared an oral, drug carrier composite with chitosan, xanthan gum, monomer 2-acrylamido-2-methylpropane sulfonic acid, and potassium persulfate by free radicle polymerization technique. The components were crosslinked with N′ N′-methylene bis-acrylamide. The composite was loaded with acyclovir as a model antiviral drug commonly prescribed in herpes simplex virus infections. The composite had a higher thermal stability and a porous structure that encapsulated the drug compound. The composite hydrogel has shown pH-dependent drug releasing behavior. Ilomuanya et al. [98] prepared a hybrid composite with antibacterial, antioxidative, and anti-inflammatory behavior for wound dressing. The composite was prepared with poly(vinyl alcohol) and xanthan gum/hypromellose/sodium carboxymethyl cellulose. Composites with silver nanoparticles have higher antimicrobial potential and have inhibition of >99% against wound pathogens such as *Acinetobacter baumannii*, *E. coli*, *K. pneumoniae*, *P. aeruginosa*, *S. aureus*, *S. epidermidis*, and *Candida albicans*. Maximum bactericidal effect was observed from hypromellose nanocomposite, with 99.9% growth reduction, within 1 h of application. Dressing material prepared with composites also has free radical scavenging behavior, along with a reduction in the inflammatory response in RAW 264.7 macrophages. Animal model studies have confirmed that the hypromellose-based nanocomposite has a higher wound-healing process.

3.2.4. Gellan Gum/Levan Composite with Synthetic Polymers

Composites with synthetic and natural components may have a higher applicability rate as natural polymers are biocompatible and biodegradable, but the addition of synthetic material may prolong the lifespan and modify the surface texture for the attachment of polymer. A levan-based composite was prepared with two different forms of levan, i.e., hydrolyzed and sulphated. Both hydrolyzed and sulfated levans were synthesized by microwave-assisted acid hydrolysis, with 5% acetic acid at 60% operating power for 60 s and mixing with chlorosulfonic acid for 24 h, respectively. The composite blend of 10% polycaprolactone (THF:DMF) and the aqueous solution of sulfated levan and hydrolysed levan in polyethylene oxide was used for coaxial electrospinning. The composite fiber has higher ultimate tensile strength and it increased with ShHL concentration. The composite increased the viability of L929 fibroblasts and HUVECs [99]. Adrover et al. [100] have prepared the composite beads for drug delivery with gellan gum, calcium ions ($CaCl_2$), and with/without synthetic clay laponite ionotropic gelation technique. The addition of laponite reduced the swelling degree of polymeric beads. Gellan gum and gellan-silicate composite beads were loaded with theophylline and cyanocobalamin for in vitro release behavior. Laponite controlled the drug entrapment efficiency, as well as the slower drug release into the gastric environment. Theophylline followed Fickian behavior for drug release, while cyanocobalamin release behavior was greatly influenced by the physical/chemical interaction of the composite and drug.

For bone regeneration, an ideal membrane is of utmost importance that not only induces cell proliferation followed by mineralization, but is also able to withstand stress. A composite of calcium–poly-γ-glutamic acid and glycerol with gellan gum was studied for bone regeneration. In composite membranes, calcium aggregates are distributed uniformly and act as commendable performers in protein adsorption, and bone cell proliferation with MG63 cells. The hydrogel has promising results for bone repair, and γ-PGA acts critically. The exopolysaccharide member acts as biocompatible material, and γ-PGA improved cell adhesion and proliferation. It also increased the secretion of alkaline phosphatase and induced mineralization. The third member, 'glycerol', enhanced mechanical strength, elongation at break, as well as diffusion rate. Both glycerol and γ-PGA delayed the degradation of the composite [101]. The main contribution of synthetic polymers in the composite is to improve the tensile and mechanical strength and delay the degradation, an important feature that is required to provide sufficient time for tissue repair and recovery.

3.2.5. Pullulan Composites with Synthetic Polymers

Drug-delivery-related applications required hydrophobic carriers and functional groups that can offer a site for attachments. Exopolysaccharides such as pullulan are soluble in an aqueous environment. However, these are sensitive to modification by hydrophobic functional groups, including the cholesterol functional group, which make them amphiphilic and ensure the drug release.

Cholesteryl-modified aminated pullulan polymers were prepared with cholesterol succinate and pullulan. Succinic anhydride cholesterol (0.2–0.6 g), 0.18 g dimethylaminoaniline, 0.35 g 1-ethyl-(3-dimethylaminopropyl) carbodiimide salt, and acid salt were dissolved in DMSO at room temperature, which activated succinic anhydride cholesterol. The activated solution was added to 5.6% amino pullulan solution in DMSO and mixed at 50 °C for 48 h, followed by cooling to room temperature. Anhydrous ethanol was added to the reaction liquid that precipitated the composite. The composition of different components affects the properties of the composite. With respect to different concentrations of cholesteryl substitution, particle size reduced from 178.0, 144.4, and 97.8 nm, with an increase in the extent of substitution. With an increase in substitution, the hydrophobicity of the pullulan derivative increased and the particle size reduced. Hydrophobicity also influenced the drug release, as derivatives with maximum hydrophobicity have the slowest drug release, i.e., 57.8%; and the lowest hydrophobicity have maximum drug release, i.e., 72.7% after 48 h. In contrast, the efficacy against lung cancer cells increased with a reduction in hydrophobicity [102]. Mommer et al. [103] have functionalized the pullulan by thiolactone-based activation that adds amines. Modified pullulan offered the possibility of forming a composite with amine-containing biological substrates. With respect to thiolactone substitution (2.5/5.0 mol%), hydrogels have different mesh sizes, i.e., 27.8 and 49.1 nm respectively. Cell proliferation studies were conducted with different cell lines (normal human dermal fibroblasts and hepatocytes) and it was reported that gelatin and H-Gly-Arg-Gly-Asp-Ser-OH (GRGDS) support cell proliferation, while H-Gly-Arg-Gly-Asp-Ser-OH as well as H-Gly-His-Lys-OH acetate salt improved the proliferation of hepatocytes (HepG2) by up to 10 folds over gelatin. A thermosensitive composite was prepared with pullulan, carrying pendant carboxymethyl groups and amphiphilic triblock copolymer, poloxamer 407, by grafting. The final composite has about 83.8% poloxamer grafted on the copolymer. The composite was highly flexible and elastic, which allowed for the copolymer to regain and recover the native structure after the removal of external force/stimuli. The gel sustained amoxicillin release over a period of 168 h. The composite can be used as a hydrogel for scaffold fabrication, as well as for drug delivery [104].

High water holding and moisture retention with mechanical properties are the main strengths of bacterial nanocellulose suitable for the preparation of wound dressing and fabrication of biomedical device production, but cellulose lacks antimicrobial and wound-healing capacity. On the other hand, pullulan can contribute to wound healing, and zinc oxide nanoparticles offered antibacterial properties. Aminoalkylsilane grafted bacterial cellulose membrane and established an interconnection between cellulose with pullulan–ZnO nanoparticle hybrid electrospun nanofibers. Dressing materials prepared from composites have better blood clotting performance over the control, i.e., BNC. The composite released ZnO and offered 5 logs of higher antibacterial activity than cellulose. Cytotoxicity analysis with L929 fibroblast cells suggested that the composite was safe for fibroblast cell proliferation. In the animal (rat) model, the composite material offered rapid healing and re-epithelialization rate with the formation of a small blood vessel and synthesis of collagen [105].

Table 2. Exopolysaccharides composites in biomedical applications.

EPS Composites and Derivatives	Product	Applications	Preparation	Composite Properties	References
Gelatin- penta methyl cyclo pentadienyl triphenylphosphine ruthenium chloride, and sodium persulfate	Hydrogel	Wound healing	• Gelatin solution was prepared in Milli-Q ultrapure water at 37 °C. The visible light photoinitiator, pentamethylcyclopentadienyl triphenylphosphine ruthenium chloride, and sodium persulfate (10 mM; Advanced BioMatrix, Inc., Sea Lion Pl, Carlsbad, CA, USA) were gently mixed with gelatin solution immediately before use	• The minimum gelling time for hydrogel was 3.79 s for hydrogel with 10% hydrogel and 1 mM photoinitiator with LED intensity of 30 mW/cm^2 • Burst pressure 126.20 mmHg • No toxicity against L929 fibroblast cells • Wound closure of 72.9% in 7 days without inflammation	[106]
Poly(vinyl alcohol)/Dextran-aldehyde	Hydrogel	Wound dressing	• PVA powder, mixed and swelled in water followed by homogenization at 90 °C for 2 h • PVA and dextran aldehyde solution frozen at −20 °C for 6 h • Lyophilize the sample to prepare PVA/DA hydrogel	• Fluid adsorption capacity increased with dextran content • Tensile strength decreased with the increase in dextran content • Wound coverage was higher/equal to commercial products i.e., at 45% on day 4 and 68% on day 8 with hydrogel and 40% and 65% with commercial products	[107]
Gelatin-pullulan Composite Nanofibers	Nanofibers	Tissue engineering	• Pullulan and gelatin solution were prepared in deionized water at 30 °C for 24 h • The mixture was electrospinned at room temperature at DC voltage of 20 kV and pumping rate of 0.5 mL/h	• Gelatin-pullulan composite prepared by mixing solutions on a magnetic stirrer at 30 °C for at least 24 h • The surface tension of all compositions was below 70 mm/m • Viscosity, conductivity, and surface tension increase with gelatin content but fiber diameter decreased • Gelatin content also lowered the elongation at the break	[108]
P3HB4HB/(GE + PVA)	Scaffold	Tissue engineering	• P3HB4HB solution in dichloromethane, used for electrospinning of core • Gelatin and PVA solution in deionized water, used in shell	• Pore size from 45–68 μm, porosity was 50–81% • Water contact angle 68–83° • Degradation increased with gelatine content • Lower gelatine content higher will be the cell adhesion	[109]

Table 2. *Cont.*

EPS Composites and Derivatives	Product	Applications	Preparation	Composite Properties	References
Polycaprolactone/gelatin	Scaffold	Diaphragmatic muscle reconstruction	• PCL mesh was designed with reinforcement of gelatin type A (porcine skin) and was prepared by 3D printing at 220 °C (400 μm diameter) at the rate of 8 mm/s	• Fibers were hydrophilic with a pore size of 1–100 μm • Mechanical strength 2.7–4 MPa • Highest degradation with fibers having the highest gelatine content • Fibre-supported adhesion and proliferation of cells	[110]
Gellan gum-egg shell membrane	Hydrogel	Regeneration of retinal pigment epithelium	• Eggshells (washed with 0.05 M sodium carbonate) • Cleaned samples were cryo-ground to powder (<100 μm) • Powder added to 2% gellan gum solution and 0.03 *w*/*v*% calcium chloride at 70 °C	• The compressive moduli of all combinations were 244 and 399 kPa • The addition of 4% egg shale membrane reduced the cross-linking and viscosity by 40% • Composite swelling also reduced by 30% • Composite has no cytotoxicity against retinal pigment epithelium	[111]
Hydroxyapatite-embedded levan	Hydrogel	Dermal filler improved collagen production and anti-wrinkle activity	• 5% levan, 17% Pluronic F127, and 0.5 wt% of carboxymethyl cellulose was dissolved in deionized water • Mixing at 4 °C for 24 h • Addition of hydroxyapatite to the solution	• 0.1–5% hydroxyapatite (HAp)-levan composite hydrogel to improve in vivo collagen production and anti-wrinkle efficacy • Increasing in HAp concentration reduced the in vitro stability and elasticity • Composite has improved cell proliferation for human dermal fibroblast cells in comparison to levan alone • The composite was stable over 8 week's period. • Composite with 1% HAp has excellent anti-wrinkle efficacy during 8 weeks by improving the collagen production	[112]
Alginate-gelatin	Hydrogel	Biomedical applications in wound dressing	• 4% aqueous solution of sodium alginate/gelatin at room temp and mixed for 4 h • Addition of 5% oxidized starch to the solution followed by degassing under vacuum	• Composite has maximum tenacity of 1.29 cn/dtex • Maximum elongation of 4.41% (composite with 16.7% gelatin) • Water absorption and retention increased by 19% (overall 335% and 311% respectively (composite with 16.7% gelatin)	[113]
Hydroxyapatite-chitosan-based hydrogels biomaterials loaded with metronidazole.	Hydrogel	Controlled drug delivery	• 1% Chitosan solution in acetic acid solution and 2.5% Metronidazole dispersion in deionized water was mixed for 1 h • Na_2HPO_4, $CaCl_2$ solution was mixed sequentially and pH adjusted to 10 with NH_4OH • After 1 h pH was adjusted to 7 and the addition of poly vinyl alcohol • After 1 h mixing at 70 °C, the mixture was frozen at −18 °C overnight	• Hydroxyapatite reduced the swelling ratio and prevent burst release with respect to pH variation • Drug release was 38–73% as compared to the control • High encapsulation efficiency (96%)	[114]

Table 2. *Cont.*

EPS Composites and Derivatives	Product	Applications	Preparation	Composite Properties	References
chitosan-gelatin scaffold loaded with aceclofenac	Scaffold	Controlled drug delivery	• 1% chitosan solution mixed with 3% (v/v) Tween 80 • 4% gelatin solution in distilled water was added to the chitosan solution at 37 °C. • 0.2% v/v glutaraldehyde added as cross-linker • 15 mg of Aceclofenac or [(2-{2,6-dichlorophenyl) amino} phenylacetooxyacetic acid] solution in methanol added dropwise to the CS-Tween 80	• Porosity-100% allows scaffold degradation and sustainable drug release • Best realised at pH: 6.8 (slightly acidic pH is good for drug release in osteopathic conditions) • Become liquid in external pressure and reform gel when pressure removed • No floccule in the scaffold and no aggregation behaviors	[115]
Curdlan-phosphorylated curdlan-ionic hydrogel-Metronidazole	Hydrogel	Controlled drug release	• 8% solution of curdlan and phosphorylated curdlan in NaOH • Addition of 1,4-butanediol diglycidyl ether drop wise • Sonication followed by lyophilisation	• Swelling ratio increased: 9 g/g to 16 g/g • Drug release increased from 50 to 90%	[116]
Hyaluronic acid-gelatin (0.5% HA-Ph + 5% gelatin-Ph)	Hydrogel	Adipose stem cells cultivation	• Gelatin-Ph and hyaluronic acid-Ph solution prepared in phosphate-buffered saline • The contents of [Ru(bpy)3]2+ and SPS in the solutions were fixed at 1.0 and 2.0 mM. hASCs were obtained from Lonza (Walkersville, MD, USA) and cultured in a growth medium for ASCs (PT-4505, Lonza) containing 10% (v/v) fetal bovine serum.	• Diffusion coefficient: 1.2–0.8 (10^{-10} m^2/s) • Young modulus: 1.0–1.1 kPa • Viscosity 80–90 mPa s	[117]
Dextran-Thyme Magnesium-Doped Hydroxyapatite	Coating	Antimicrobial coating	• $Ca(NO_3)_2{\cdot}4H_2O$ and $Mg(NO_3)_2{\cdot}6H_2O$ solution was added into $(NH_4)_2{\cdot}HPO_4$ by drop-by-drop • Precipitate was dispersed in 10% dextran and 1% thyme essential oil solution	• As coating material • Prevent biofilm formation by gram-positive and gram-negative bacteria and even *Candida*	[67]
Methacrylated gelatin-hyaluronic acid	Hydrogel scaffold	Tissue engineering	• Methylated gelatin and hyaluronic acid solutions were prepared in PBS at 37 °C • 0.5% (w/v) of Irgacure® 2959 photoinitiator was added to solution	• Hybrid hydrogel formed with modified gelatin and hyaluronic acid has a higher elastic modulus of 30 kPa, • Porosity was around 91% in comparison to the control	[118]
'Gelatin-hydroxy-phenyl propionic acid'-'hyaluronic acid tyramine'	Polymer network	Retinal ganglion cells replacement therapy	• Gtn-HPA and HA-Tyr composites were prepared via carbodiimide/active ester-mediated coupling reaction in phosphate buffer saline • H_2O_2 was added as a cross-linker	• Gel point 151–167 s • Elastic modulus 3101–4316 Pa • Shear 835–1072 Pa	[119]
Gellan Gum, Alginate and Nisin-Enriched Lipid Nanoparticles	Hydrogel	Wound recovery	• NSN-stearic acid nanoparticles fabricated by double emulsification/solvent evaporation method • Hydrogel was prepared by mixing gellan gum, alginate, and nisin-stearic acid nanoparticle	• Nisin-stearic acid-gellan gum-alginate nanoparticles were spherical • Average particle size 300 nm • Load size of 500 µg/mL was cytocompatible with L929 fibroblasts	[120]

Table 2. *Cont.*

EPS Composites and Derivatives	Product	Applications	Preparation	Composite Properties	References
Mg^{2+}-Gellan Gum/Poly-Acrylamide	Hydrogel	Wound healing	• 1.5 wt% GG + 1.5 mol/L acrylamide monomer + 0.015 mol/L N,N′-methylene dia crylamide + 5 mmol/L ammonium persulfate mixing at 80 °C • Cooled gel immersed in Mg^{+2} solution	• Mg^{2+}-GG/PAM hydrogel is 7.2 mol/m^3 • Tension strength 392 kPa • The controlled release rate of Mg^{2+} (in PBS and alkaline conditions) • Fibroblast cells have better proliferation over control • Complete wound closure without scare	[121]
Oxidized gellan gum + carboxy methyl chitosan	Hydrogel	Drug delivery and wound dressing	• Mixing of oxidized chitosan and gellan gum at room temp for 30 min • Addition of $CaCl_2$ at 37 °C	• Increase in concentration of gellan gum microsphere increased decreased the gelation time • Swelling rates were 35–34% (vary with components) • G′ of Gel made with cellulose: gellan in ratio of 1:1, 2:1, and 3:1 was 1750 Pa, 2260 Pa and 2110 Pa, respectively. • Drug release of 29.7% in 14 days	[122]
Gellan gum-alginate-calcium chloride	Hydrogel	Osteochondral repair	• 1% gellan gum solution in deionized water + 1.5% C alginate solution in deionized water + 1 mL $CaCl_2$ mixing for 20–30 min	• In osteochondral defect model with a diameter of 4.0 mm and depth 8.0 mm hydrogel-induced neovascularization in 4 weeks • Repair the defect at week 8	[123]

The composite preparation and applications in different fields are still not very well developed. Some of them are under screening and others have crossed trials in cell lines. The efforts have received wide attention, and for outstanding findings, patents have been awarded (Table 3).

Table 3. Patents awarded in microbial exopolysaccharide-related research.

Patent ID	EPS	Microorganisms	Application	Reference
WO2023036938A1	EPS	*Firmicutes*	Gene mutation in *Firmicutes* for EPS production	[124]
WO2023038519A1	Biopolymers	*aerobic granular sludge and anam- mox granular sludge*	Modification of biopolymers using polyols and polyacids	[125]
WO2023025235	EPS	*Lactic acid bacterium*	Treatment of sleeping disorder	[126]
WO2023020880	EPS	*Paenibacillus* strains	Production	[127]
RU2782953	-	*Paenibacillus polymyxa* strain essutm-2	Production	[128]
WO2023275343	Low-molecular-weight he800 exopolysaccharide derivatives	*Vibrio diabolicus* HE800	Anti-cancer	[129]
RU2784088	-	*Paenibacillus polymyxa* vsgutu-1	Production	[130]
WO2023014213	Cellulase	*Acetobacter aceti*	Production	[131]
CN115572690A	EPS	Lactic acid bacteria	Selection of EPS producing bacterial	[132]
CN115554197A	EPS	-	Soothing and repairing based cosmetic formulations	[133]
CN115505610A	EPS	*Chlorella pyrenoidosa*	Antitumor activity	[134]

Microbe-based polysaccharides are biocompatible, nonimmunogenic, biodegradable and most of these are USA FDA-approved and regarded as safe for human consumption [135]. According to GRAS notice 000099, the use of pullulan is allowed as an ingredient in tablets and capsules for dietary supplements [136]. The commercialization of such products still needs a long way to cover especially in the context of their behavior in various conditions to support their candidature.

4. Challenges and Future Prospective

Microbial EPS and associated composites have offered several advantages such as biocompatibility, biodegradability, and nontoxicity; however, there is still no product available at the commercial level. The commercialization of products prepared from EPS composites needs to address some of the common challenges, summarized below.

4.1. Strain Selection

The major fraction of EPS composites is microbial exopolysaccharides, which are produced by fermentation at various scales. Strain selection is one of the major problems associated with the large-scale production and commercialization of EPS and associated products. Several microorganisms have been screened for the production of EPS and some of them are either pathogenic or have low yield [137]. However, some of the microorganisms have offered compatible productivity without any pathogenicity, such as probiotic bacteria, including *Lactobacillus*, *Bifidobacterium*, and *Lactococcus*. In addition to yield, the tolerance to high salt concentrations (bile salt) and extremely acidic pH are other advantages that aided in the commercialization of products. Recent research has also proven that lactic acid bacteria also improved immunity and wound healing [49,138].

Genetic engineering of microorganisms is another possible strategy to improve productivity. However, the production of polysaccharides needs a combinatorial synthesis approach, as it is governed by multiple pathways. Cumulatively the pathways have the following three stages: nucleotide sugar precursors synthesis, extension, and synthesis of oligosaccharide units by glycosyltransferases and assembly of structural units to the final EPS, followed by export [139]. Some of the strategies have been summarized by Schmid et al. [140], but no report is available on the large-scale production of EPS with genetically engineered microorganisms. This may be due to the complex biosynthetic pathway governed by multiple gene systems. There is a need to work on this issue to increase EPS productivity with unique properties.

Production of exopolysaccharides, yield, and cost: Any microorganism can offer a higher possible production under optimal growth and production conditions. Before moving to higher-scale production, the identification of exopolysaccharides, their nature, and optimal conditions for maximum productivity are required [141]. As mentioned above, EPS production relies upon commercial-grade substrates with a higher cost, which also affects the product cost. A possible way out for the substrate is the utilization of low-cost lignocellulosic biomass from agricultural, as well as industrial sources, such as raw material. Previous research has used various agricultural biomass for pullulan production by *Aureobasidium pullulans*, i.e., hazelnut husk [142], sesame seed oil cake [143], corn cobs, and straw [144]. The process can be optimized via one factor or a statistical optimization approach. However, lignocellulosic materials, such as raw materials, need preprocessing such as hydrolysis, followed by detoxification. Acid hydrolysis is one of the most common methods used for sugar recovery in hydrolysate, but some byproducts, including furfurals and hydroxymethyl furfurals (from pentose and hexose), were also produced along with phenolics from lignin. These byproducts have shown an inhibitory effect on microbial growth and hence hydrolysate was detoxified by activated charcoal and membrane separation [33,145,146]. These additional steps for biomass processing and hydrolysate preparation increased the process, as well as product cost. The use of tolerant microorganisms might help overcome the challenges associated with inhibitors.

4.2. Downstream Processing

Downstream processing and product recovery is one of the major challenges, as it contributes to the majority of the product cost. Therefore, an efficient and low-cost downstream process is a prerequisite for the commercialization of exopolysaccharides and associated products [141]. The chemical nature and characteristics of exopolysaccharides, such as high viscosity and gel-forming ability, hinder their extraction. Some of the recovery approaches are trichloroacetic acid and proteases-assisted protein removal, dialysis, chromatography, and solvent-based precipitation used for EPS recovery. The ultra-grade of purity of a product is necessary for biomedical applications, hence the selection of the recovery process should be done as per the target product, as well as the physicochemical properties of EPS. Proteins are removed by protease/trichloroacetic acid to prevent contamination and cross-reaction. However, heat or use of concentrated acid damage the native EPS structure and change the integrity of bonds, branching, and sugar monomers. [147].

4.3. Composite-Forming Ability of EPS

Usually, EPSs are water-soluble (except a few, such as cellulose) and hydrophilic, which makes EPS prone to degradation in storage, as well as in the cellular environment. This also influences the selection of drug and secondary polymers, as components with similar charges usually make an unstable system due to repulsion. Due to the presence of hydroxyl and carboxylic functional groups, most of the EPS molecules are negatively charged, and thus loading of negatively charged groups and blending with negatively charged polymers is not possible due to electrostatic repulsion [137,148,149].

Derivatization of exopolysaccharides is one of the most common approaches to change the net charge, as well as improve interaction with other molecules, as observed in the case of cellulose, which is insoluble in water as well as organic solvents. The availability of the hydroxyl functional group makes it possible to modulate its chemical nature, including solubility, hydrophilicity/hydrophobicity, and mechanical strength either, by degradation or derivatization. Cellulose ether is the derivative prepared by replacing the hydroxyl group with the hydrocarbon group. It includes carboxymethyl, methyl, hydroxypropyl methyl, and hydroxyethyl derivatives of cellulose. Derivatization of cellulose has improved thermoplasticity, apart from hydrophilicity of derivative, in comparison to native cellulose [150]. Similar changes were also observed in gellan gum hydrogel used for muscular injury. The addition of laminin protein improved the muscular tissue auto-healing capacity, but 3D hydrogel needs extensive crosslinking and porous structure. To attain a stable porous interconnected structure, gellan gum was derivatized with divinyl sulfone. The composite of gellan gum–divinyl sulfone derivative and gellan gum was used to prepare 3D hydrogel and functionalized with laminin-derived peptide. The composite was encapsulated with skeletal muscle cells. The modification of gellan gum with divinyl sulfone improved cross-linking that stabilized the 3D framework of hydrogel [151].

4.4. Stability and Degradation Products of EPS Composites

Exopolysaccharide composites are comprised of EPS and natural or synthetic polymers. EPS mainly has polysaccharides as a major component, while natural polymers may also have proteins and lipids, etc. Each biopolymer, including sugar/carbohydrate, proteins, lipids, and nucleic acids, is prone to degradation and may endure only a few hours to a few weeks (depending upon the environment or type of polymer). The study conducted by McClatchy et al. [152] also proved that the stability of proteins depends upon the cellular environment. One of the advantages of the use of composite scaffolds made with natural polymers is the complete degradation of EPSs such as pullulan, dextran, etc., as well as secondary polymers, into CO_2 and H_2O. Some nitrogen-containing polymers, such as chitin and chitosan, release amino sugars which can be utilized by microorganisms and living cells [153,154]. To overcome the challenges associated with stability, synthetic polymers can be used instead of natural polymers. Some of the common synthetic polymers used in biomedical applications are polyarylsulfones, polysulfone, polyvinylpyrrolidone, polyamide, polycarbonate, polyacrylonitrile, PMMA, polyester polymer alloy, ethylene vinyl alcohol copolymers; and molecular-thin nanoporous silicon membranes are the common synthetic polymers used in hemodialysis membranes; polymethyl pentene in extracorporeal membrane oxygenation; polyester, polyether, and polycarbonate-based polyurethanes in catheters; nylon, polyurethanes with acrylate in wound dressing; fibrin glue in sealants and adhesive; expanded polypropylene, polytetrafluoroethylene, polyethylene terephthalate, polyvinylidene difluoride membrane in surgical meshes; and poly-tetrafluoroethylene, polyethylene terephthalate, dacron, nylon in scaffold and ligament repair [155]. It has been proven that the addition of synthetic polymer has improved the mechanical and woven strength of natural exopolysaccharides. Kenawy et al. [91] have shown that the addition of poly(vinyl alcohol) (PVA)to dextran has improved the mechanical stability of the composite for electrospinning. Further addition of citric acid as a cross-link had a conducive effect on the mechanical, as well as tensile strength and stability of the composite due to extensive cross-linking.

4.5. Side Effects of Synthetic Polymers

As discussed in the above section, synthetic polymers have positive results when blended with exopolysaccharides, but these synthetic polymers have shown side effects in a living system. Much literature has emphasized side effects and immunogenic responses with synthetic polymers. Acrylates are used in baby diapers, and prostheses [156], but acrylates may lead to serious allergic responses and dermatitis [157]. Similar kinds of side effects have also been observed with other synthetic polymers and adhesives, including anaphylaxis allergic reaction and arachnoid plasty in the case of fibrin glue [158,159], volitional swallowing, gastrointestinal obstruction, asthma, and allergy with polyethylene terephthalate [160] and oxidative stress from polypropylene used in mesh surgery [161]. On the other hand, some polymers have offered advantages over others, as their degradation product can be utilized in cells as in the case of poly-lactides. These kinds of polymers release lactic acids and respective oligomers upon degradation, which can be utilized by the cell during normal metabolism [162].

The major challenges associated with the commercialization of exopolysaccharide-based composites are its bulk-scale production, cost-effective recovery, and detailed analysis for optimum storage and transit conditions. However, there is no literature available that provides any information about the degradation of EPS composites, suitability of application, commercial cost of the products, and appropriate stability profile.

5. Conclusions

Exopolysaccharides have proven their candidature in the biomedical and healthcare sector mainly attributed to their biocompatibility, nontoxicity, and degradability. However, their limited mechanical and tensile strength, along with solubility in different solvents, obstructs their commercialization. Synthetic polymers have higher stability and strength, but are disadvantaged by side effects and compatibility issues. In order to achieve both compatibilities as well as tunability in strength, adaptability, and stability, composites are preferred over EPSs. The blending of natural and synthetic polymers might improve the physical and chemical characteristics. The composites have offered higher tensile and mechanical strength, along with water retention, slower degradation, drug-carrying capacity, and compatibility for biological applications, including 3D scaffold and wound dressing material fabrication, drug carrier properties, and biomedical sealant potential. Composites can support biological tissue and support healing due to improved adhesion and cell proliferation. The commercialization of composites needs in-depth study regarding stability in storage and transport, degradation behavior, and cost of the final product.

Author Contributions: V.A.: Conceptualization, methodology, writing—review and editing; A.K.B.: investigation, resources; J.R.B.: writing-review and editing; V.K.: writing—review and editing; G.K., G.K. and Y.-H.Y.: supervision, writing—original draft preparation; S.K.B.: Conceptualization, methodology, writing—review and editing, supervision. All authors have read and agreed to the published version of the manuscript.

Funding: This study was supported by the Research Program to Solve Social Issues with the National Research Foundation of Korea (NRF), funded by the Ministry of Science and ICT [Grant No. 2017M3A9E4077234], and by the National Research Foundation of Korea (NRF) (NRF-2022M3I3A1082545, NRF-2022R1A2C2003138 and NRF-2021R1F1A1050325). This study was also supported by the R&D Program of MOTIE/KEIT [grant number 20016324].

Institutional Review Board Statement: Not applicable.

Informed Consent Statement: Not applicable.

Acknowledgments: The authors acknowledge KU, the research professor the program at Konkuk University, Seoul, South Korea and Chandigarh University, Punjab India.

Conflicts of Interest: The authors declare no conflict of interest.

References

1. Kaur, N.; Dey, P. Bacterial Exopolysaccharides as Emerging Bioactive Macromolecules: From Fundamentals to Applications. *Res. Microbiol.* **2022**, *174*, 104024. [CrossRef]
2. Kuschmierz, L.; Meyer, M.; Bräsen, C.; Wingender, J.; Schmitz, O.J.; Siebers, B. Exopolysaccharide composition and size in Sulfolobus acidocaldarius biofilms. *Front. Microbiol.* **2022**, *13*, 982745. [CrossRef] [PubMed]
3. Qi, M.; Zheng, C.; Wu, W.; Yu, G.; Wang, P. Exopolysaccharides from Marine Microbes: Source, Structure and Application. *Mar. Drugs* **2022**, *20*, 512. [CrossRef] [PubMed]
4. Cázares-Vásquez, M.L.; Rodríguez-Herrera, R.; Aguilar-González, C.N.; Sáenz-Galindo, A.; Solanilla-Duque, J.F.; Contreras-Esquivel, J.C.; Flores-Gallegos, A.C. Microbial Exopolysaccharides in Traditional Mexican Fermented Beverages. *Fermentation* **2021**, *7*, 249. [CrossRef]
5. Jeong, J.; Kim, Y.; Hu, Y.; Jung, S. Bacterial Succinoglycans: Structure, Physical Properties, and Applications. *Polymers* **2022**, *14*, 276. [CrossRef]
6. Miao, K.H.; Guthmiller, K.B. *Dextran;* StatPearls Publishing: Treasure Island, FL, USA, 2022.
7. Nwodo, U.U.; Green, E.; Okoh, A.I. Bacterial Exopolysaccharides: Functionality and Prospects. *Int. J. Mol. Sci.* **2012**, *13*, 14002–14015. [CrossRef]
8. Ibrahim, H.A.H.; Abou Elhassayeb, H.E.; El-Sayed, W.M.M. Potential functions and applications of diverse microbial exopolysaccharides in marine environments. *J. Genet. Eng. Biotechnol.* **2022**, *20*, 151. [CrossRef] [PubMed]
9. Gupta, P.; Diwan, B. Bacterial Exopolysaccharide mediated heavy metal removal: A Review on biosynthesis, mechanism and remediation strategies. *Biotechnol. Rep.* **2017**, *13*, 58–71. [CrossRef]
10. Mohd Nadzir, M.; Nurhayati, R.W.; Idris, F.N.; Nguyen, M.H. Biomedical Applications of Bacterial Exopolysaccharides: A Review. *Polymers* **2021**, *13*, 530. [CrossRef]
11. Díaz-Montes, E. Dextran: Sources, Structures, and Properties. *Polysaccharides* **2021**, *2*, 554–565. [CrossRef]
12. Prasher, P.; Sharma, M.; Kumar Singh, S.; Gulati, M.; Kumar, D.; Gupta, G.; Kumar Chellappan, D.; Gregory George Oliver, B.; Wich, P.R.; Dua, K. Versatility of acetalated dextran in nanocarriers targeting respiratory diseases. *Mater. Lett.* **2022**, *323*, 132600. [CrossRef]
13. Yahoum, M.M.; Toumi, S.; Tahraoui, H.; Lefnaoui, S.; Kebir, M.; Amrane, A.; Assadi, A.A.; Zhang, J.; Mouni, L. Formulation and Evaluation of Xanthan Gum Microspheres for the Sustained Release of Metformin Hydrochloride. *Micromachines* **2023**, *14*, 609. [CrossRef] [PubMed]
14. Khare, P.; Chogale, M.; Kakade, P.; Patravale, V. Gellan gum-based in situ gelling ophthalmic nanosuspension of Posaconazole. *Drug Deliv. Transl. Res.* **2022**, *12*, 2920–2935. [CrossRef] [PubMed]
15. Patel, J.; Maji, B.; Narayana Moorthy, N.S.H.; Maiti, S. Xanthan gum derivatives: Review of synthesis, properties and diverse applications. *RSC Adv.* **2020**, *10*, 27103–27136. [CrossRef]
16. Dzionek, A.; Wojcieszyńska, D.; Guzik, U. Use of xanthan gum for whole cell immobilization and its impact in bioremediation—A review. *Bioresour. Technol.* **2022**, *351*, 126918. [CrossRef]
17. Chaudhari, V.; Buttar, H.S.; Bagwe-Parab, S.; Tuli, H.S.; Vora, A.; Kaur, G. Therapeutic and Industrial Applications of Curdlan with Overview on Its Recent Patents. *Front. Nutr.* **2021**, *8*, 646988. [CrossRef]
18. Di Mola, A.; Landi, M.R.; Massa, A.; D'Amora, U.; Guarino, V. Hyaluronic Acid in Biomedical Fields: New Trends from Chemistry to Biomaterial Applications. *Int. J. Mol. Sci.* **2022**, *23*, 14372. [CrossRef]
19. Guo, L.; Chai, Y.; Zhou, F.; Wang, P. Preparation and Properties of Hyaluronic Acid Hydrogel Modified by L-cysteine Hydrochloride. *IOP Conf. Ser. Earth Environ. Sci.* **2021**, *651*, 042022. [CrossRef]

20. Hussain, A.; Zia, K.M.; Tabasum, S.; Noreen, A.; Ali, M.; Iqbal, R.; Zuber, M. Blends and composites of exopolysaccharides; properties and applications: A review. *Int. J. Biol. Macromol.* **2017**, *94*, 10–27. [CrossRef]
21. Xu, W.; Qian, J.; Zhang, Y.; Suo, A.; Cui, N.; Wang, J.; Yao, Y.; Wang, H. A double-network poly(Nε-acryloyl L-lysine)/hyaluronic acid hydrogel as a mimic of the breast tumor microenvironment. *Acta Biomater.* **2016**, *33*, 131–141. [CrossRef]
22. Sieni, E.; Bazzolo, B.; Pieretti, F.; Zamuner, A.; Tasso, A.; Dettin, M.; Conconi, M.T. Breast cancer cells grown on hyaluronic acid-based scaffolds as 3D in vitro model for electroporation. *Bioelectrochemistry* **2020**, *136*, 107626. [CrossRef]
23. Buhome, O.; Wongwattanakul, M.; Daduang, J.; Limpaiboon, T. 3D Silk Fibroin-Gelatin/Hyaluronic Acid/Heparan Sulfate Scaffold Enhances Expression of Stemness and EMT Markers in Cholangiocarcinoma. *Vivo* **2022**, *36*, 1155–1167. [CrossRef] [PubMed]
24. Toullec, C.; Le Bideau, J.; Geoffroy, V.; Halgand, B.; Buchtova, N.; Molina-Peña, R.; Garcion, E.; Avril, S.; Sindji, L.; Dube, A.; et al. Curdlan–Chitosan Electrospun Fibers as Potential Scaffolds for Bone Regeneration. *Polymers* **2021**, *13*, 526. [CrossRef] [PubMed]
25. Angelin, J.; Kavitha, M. Exopolysaccharides from probiotic bacteria and their health potential. *Int. J. Biol. Macromol.* **2020**, *162*, 853–865. [CrossRef] [PubMed]
26. Zhang, Y.; Dai, X.; Jin, H.; Man, C.; Jiang, Y. The effect of optimized carbon source on the synthesis and composition of exopolysaccharides produced by *Lactobacillus paracasei*. *J. Dairy Sci.* **2021**, *104*, 4023–4032. [CrossRef] [PubMed]
27. Gientka, I.; Bzducha-Wróbel, A.; Stasiak-Różańska, L.; Bednarska, A.A.; Błażejak, S. The exopolysaccharides biosynthesis by Candida yeast depends on carbon sources. *Electron. J. Biotechnol.* **2016**, *22*, 31–37. [CrossRef]
28. Jin, H.; Jeong, Y.; Yoo, S.-H.; Johnston, T.V.; Ku, S.; Ji, G.E. Isolation and characterization of high exopolysaccharide-producing Weissella confusa VP30 from young children's feces. *Microb. Cell Factories* **2019**, *18*, 110. [CrossRef]
29. da Silva, J.A.; Cardoso, L.G.; de Jesus Assis, D.; Gomes, G.V.P.; Oliveira, M.B.P.P.; de Souza, C.O.; Druzian, J.I. Xanthan Gum Production by *Xanthomonas campestris* pv. *Campestris* IBSBF 1866 and 1867 from Lignocellulosic Agroindustrial Wastes. *Appl. Biochem. Biotechnol.* **2018**, *186*, 750–763. [CrossRef]
30. Choi, I.S.; Ko, S.H.; Lee, M.E.; Kim, H.M.; Yang, J.E.; Jeong, S.-G.; Lee, K.H.; Chang, J.Y.; Kim, J.-C.; Park, H.W. Production, Characterization, and Antioxidant Activities of an Exopolysaccharide Extracted from Spent Media Wastewater after Leuconostoc mesenteroides WiKim32 Fermentation. *ACS Omega* **2021**, *6*, 8171–8178. [CrossRef]
31. Pan, L.; Wang, Q.; Qu, L.; Liang, L.; Han, Y.; Wang, X.; Zhou, Z. Pilot-scale production of exopolysaccharide from Leuconostoc pseudomesenteroides XG5 and its application in set yogurt. *J. Dairy Sci.* **2022**, *105*, 1072–1083. [CrossRef]
32. Bhatia, S.K.; Gurav, R.; Kim, B.; Kim, S.; Cho, D.-H.; Jung, H.; Kim, Y.-G.; Kim, J.-S.; Yang, Y.-H. Coproduction of exopolysaccharide and polyhydroxyalkanoates from Sphingobium yanoikuyae BBL01 using biochar pretreated plant biomass hydrolysate. *Bioresour. Technol.* **2022**, *361*, 127753. [CrossRef] [PubMed]
33. Ahuja, V.; Kshirsagar, S.; Ghosh, P.; Sarkar, B.; Sutar, A.; More, S.; Dasgupta, D. Process development for detoxification of corncob hydrolysate using activated charcoal for xylitol production. *J. Environ. Chem. Eng.* **2022**, *10*, 107097. [CrossRef]
34. Ahuja, V.; Bhatt, A.K.; Mehta, S.; Sharma, V.; Rathour, R.K. Sheetal Xylitol production by *Pseudomonas gessardii* VXlt-16 from sugarcane bagasse hydrolysate and cost analysis. *Bioprocess Biosyst. Eng.* **2022**, *45*, 1019–1031. [CrossRef] [PubMed]
35. Yáñez-Fernández, J.; Herrera Ovando, M.G.; Patlán Ramírez, L.; Ramírez-Sotelo, G.; Guarin, C.A.; Castro-Rodríguez, D.C. Factorial Design to Optimize Dextran Production by the Native Strain *Leuconostoc mesenteroides* SF3. *ACS Omega* **2021**, *6*, 31203–31210. [CrossRef]
36. Jumma Kareem, A.; Abdul Sattar Salman, J. Production of Dextran from Locally *Lactobacillus* spp. Isolates. *Rep. Biochem. Mol. Biol.* **2019**, *8*, 287–300.
37. Koirala, P.; Maina, N.H.; Nihtilä, H.; Katina, K.; Coda, R. Brewers' spent grain as substrate for dextran biosynthesis by *Leuconostoc pseudomesenteroides* DSM20193 and Weissella confusa A16. *Microb. Cell Factories* **2021**, *20*, 23. [CrossRef] [PubMed]
38. Mangolim, C.S.; da Silva, T.T.; Fenelon, V.C.; Koga, L.N.; Ferreira, S.B.d.S.; Bruschi, M.L.; Matioli, G. Description of recovery method used for curdlan produced by *Agrobacterium* sp. IFO 13140 and its relation to the morphology and physicochemical and technological properties of the polysaccharide. *PLoS ONE* **2017**, *12*, e0171469. [CrossRef]
39. Prakash, S.; Rajeswari, K.; Divya, P.; Ferlin, M.; Rajeshwari, C.T.; Vanavil, B. Optimization and production of curdlan gum using *Bacillus cereus* PR3 isolated from rhizosphere of leguminous plant. *Prep. Biochem. Biotechnol.* **2018**, *48*, 408–418. [CrossRef]
40. Anane, R.F.; Sun, H.; Zhao, L.; Wang, L.; Lin, C.; Mao, Z. Improved curdlan production with discarded bottom parts of Asparagus spear. *Microb. Cell Factories* **2017**, *16*, 59. [CrossRef]
41. Nejadmansouri, M.; Razmjooei, M.; Safdarianghomsheh, R.; Shad, E.; Delvigne, F.; Khalesi, M. Semi-continuous production of xanthan in biofilm reactor using *Xanthomonas campestris*. *J. Biotechnol.* **2021**, *328*, 1–11. [CrossRef]
42. Raghunandan, K.; Kumar, A.; Kumar, S.; Permaul, K.; Singh, S. Production of gellan gum, an exopolysaccharide, from biodiesel-derived waste glycerol by *Sphingomonas* spp. *3 Biotech* **2018**, *8*, 71. [CrossRef] [PubMed]
43. Srivastava, N.; Kumari, S.; Kurmi, S.; Pinnaka, A.K.; Choudhury, A.R. Isolation, purification, and characterization of a novel exopolysaccharide isolated from marine bacteria *Brevibacillus borstelensis* M42. *Arch. Microbiol.* **2022**, *204*, 399. [CrossRef]
44. Roychowdhury, R.; Srivastava, N.; Kumari, S.; Pinnaka, A.K.; Roy Choudhury, A. Isolation of an exopolysaccharide from a novel marine bacterium *Neorhizobium urealyticum* sp. nov. and its utilization in nanoemulsion formation for encapsulation and stabilization of astaxanthin. *LWT* **2021**, *151*, 112105. [CrossRef]

45. Gangalla, R.; Sampath, G.; Beduru, S.; Sarika, K.; Kaveriyappan Govindarajan, R.; Ameen, F.; Alwakeel, S.; Thampu, R.K. Optimization and characterization of exopolysaccharide produced by *Bacillus aerophilus* rk1 and its in vitro antioxidant activities. *J. King Saud Univ. Sci.* **2021**, *33*, 101470. [CrossRef]
46. Miller, G.L. Use of Dinitrosalicylic Acid Reagent for Determination of Reducing Sugar. *Anal. Chem.* **1959**, *31*, 426–428. [CrossRef]
47. Bradford, M.M. A rapid and sensitive method for the quantitation of microgram quantities of protein utilizing the principle of protein-dye binding. *Anal. Biochem.* **1976**, *72*, 248–254. [CrossRef] [PubMed]
48. Waterborg, J.H. The Lowry Method for Protein Quantitation. In *The Protein Protocols Handbook*; Walker, J.M., Ed.; Humana Press: Totowa, NJ, USA, 2009; pp. 7–10. ISBN 978-1-60327-474-6.
49. Zaghloul, E.H.; Ibrahim, M.I.A. Production and Characterization of Exopolysaccharide from Newly Isolated Marine Probiotic *Lactiplantibacillus plantarum* EI6 with in vitro Wound Healing Activity. *Front. Microbiol.* **2022**, *13*, 903363. [CrossRef] [PubMed]
50. Säwén, E.; Huttunen, E.; Zhang, X.; Yang, Z.; Widmalm, G. Structural analysis of the exopolysaccharide produced by *Streptococcus thermophilus* ST1 solely by NMR spectroscopy. *J. Biomol. NMR* **2010**, *47*, 125–134. [CrossRef]
51. Piola, B.; Sabbatini, M.; Gino, S.; Invernizzi, M.; Renò, F. 3D Bioprinting of Gelatin–Xanthan Gum Composite Hydrogels for Growth of Human Skin Cells. *Int. J. Mol. Sci.* **2022**, *23*, 539. [CrossRef] [PubMed]
52. Alves, A.; Miguel, S.P.; Araujo, A.R.T.S.; de Jesús Valle, M.J.; Sánchez Navarro, A.; Correia, I.J.; Ribeiro, M.P.; Coutinho, P. Xanthan Gum–Konjac Glucomannan Blend Hydrogel for Wound Healing. *Polymers* **2020**, *12*, 99. [CrossRef] [PubMed]
53. Akkineni, A.R.; Ahlfeld, T.; Funk, A.; Waske, A.; Lode, A.; Gelinsky, M. Highly Concentrated Alginate-Gellan Gum Composites for 3D Plotting of Complex Tissue Engineering Scaffolds. *Polymers* **2016**, *8*, 170. [CrossRef] [PubMed]
54. D'Amora, U.; Ronca, A.; Scialla, S.; Soriente, A.; Manini, P.; Phua, J.W.; Ottenheim, C.; Pezzella, A.; Calabrese, G.; Raucci, M.G.; et al. Bioactive Composite Methacrylated Gellan Gum for 3D-Printed Bone Tissue-Engineered Scaffolds. *Nanomaterials* **2023**, *13*, 772. [CrossRef]
55. He, W.; Wang, X.; Hang, T.; Chen, J.; Wang, Z.; Mosselhy, D.A.; Xu, J.; Wang, S.; Zheng, Y. Fabrication of Cu2+-loaded phase-transited lysozyme nanofilm on bacterial cellulose: Antibacterial, anti-inflammatory, and pro-angiogenesis for bacteria-infected wound healing. *Carbohydr. Polym.* **2023**, *309*, 120681. [CrossRef] [PubMed]
56. Biswas, A.; Ahmed, T.; Rana, M.R.; Hoque, M.M.; Ahmed, M.F.; Sharma, M.; Sridhar, K.; Ara, R.; Stephen Inbaraj, B. Fabrication and Characterization of ZnO Nanoparticles-Based Biocomposite Films Prepared Using Carboxymethyl Cellulose, Taro Mucilage, and Black Cumin Seed Oil for Evaluation of Antioxidant and Antimicrobial Activities. *Agronomy* **2023**, *13*, 147. [CrossRef]
57. Ul-Islam, M.; Alhajaim, W.; Fatima, A.; Yasir, S.; Kamal, T.; Abbas, Y.; Khan, S.; Khan, A.H.; Manan, S.; Ullah, M.W.; et al. Development of low-cost bacterial cellulose-pomegranate peel extract-based antibacterial composite for potential biomedical applications. *Int. J. Biol. Macromol.* **2023**, *231*, 123269. [CrossRef] [PubMed]
58. Lahiri, D.; Nag, M.; Dutta, B.; Dey, A.; Sarkar, T.; Pati, S.; Edinur, H.A.; Abdul Kari, Z.; Mohd Noor, N.H.; Ray, R.R. Bacterial Cellulose: Production, Characterization, and Application as Antimicrobial Agent. *Int. J. Mol. Sci.* **2021**, *22*, 12984. [CrossRef]
59. Bai, F.-W.; Yang, S.; Ho, N.W.Y. 3.05—Fuel Ethanol Production from Lignocellulosic Biomass. In *Comprehensive Biotechnology*, 3rd ed.; Moo-Young, M., Ed.; Pergamon: Oxford, UK, 2019; pp. 49–65. ISBN 978-0-444-64047-5.
60. Orlando, I.; Basnett, P.; Nigmatullin, R.; Wang, W.; Knowles, J.C.; Roy, I. Chemical Modification of Bacterial Cellulose for the Development of an Antibacterial Wound Dressing. *Front. Bioeng. Biotechnol.* **2020**, *8*, 557885. [CrossRef]
61. Ojagh, S.M.A.; Vahabzadeh, F.; Karimi, A. Synthesis and characterization of bacterial cellulose-based composites for drug delivery. *Carbohydr. Polym.* **2021**, *273*, 118587. [CrossRef]
62. Ma, N.; Cheung, D.Y.; Butcher, J.T. Incorporating nanocrystalline cellulose into a multifunctional hydrogel for heart valve tissue engineering applications. *J. Biomed. Mater. Res. Part A* **2022**, *110*, 76–91. [CrossRef]
63. Zhu, Q.; Chen, X.; Liu, Z.; Li, Z.; Li, D.; Yan, H.; Lin, Q. Development of alginate-chitosan composite scaffold incorporation of bacterial cellulose for bone tissue engineering. *Int. J. Polym. Mater. Polym. Biomater.* **2023**, *72*, 296–307. [CrossRef]
64. Balakrishnan, B.; Soman, D.; Payanam, U.; Laurent, A.; Labarre, D.; Jayakrishnan, A. A novel injectable tissue adhesive based on oxidized dextran and chitosan. *Acta Biomater.* **2017**, *53*, 343–354. [CrossRef]
65. Andrabi, S.M.; Majumder, S.; Gupta, K.C.; Kumar, A. Dextran based amphiphilic nano-hybrid hydrogel system incorporated with curcumin and cerium oxide nanoparticles for wound healing. *Colloids Surf. B Biointerfaces* **2020**, *195*, 111263. [CrossRef]
66. Ma, C.; Zhao, J.; Zhu, C.; Jiang, M.; Ma, P.; Mi, Y.; Fan, D. Oxidized dextran crosslinked polysaccharide/protein/polydopamine composite cryogels with multiple hemostatic efficacies for noncompressible hemorrhage and wound healing. *Int. J. Biol. Macromol.* **2022**, *215*, 675–690. [CrossRef]
67. Iconaru, S.L.; Predoi, M.V.; Motelica-Heino, M.; Predoi, D.; Buton, N.; Megier, C.; Stan, G.E. Dextran-Thyme Magnesium-Doped Hydroxyapatite Composite Antimicrobial Coatings. *Coatings* **2020**, *10*, 57. [CrossRef]
68. Sabit, H.; Abdel-Hakeem, M.; Shoala, T.; Abdel-Ghany, S.; Abdel-Latif, M.M.; Almulhim, J.; Mansy, M. Nanocarriers: A Reliable Tool for the Delivery of Anticancer Drugs. *Pharmaceutics* **2022**, *14*, 1566. [CrossRef] [PubMed]
69. Rizvi, S.A.A.; Saleh, A.M. Applications of nanoparticle systems in drug delivery technology. *Saudi Pharm. J.* **2018**, *26*, 64–70. [CrossRef] [PubMed]
70. Kumar, A.; Rao, K.M.; Han, S.S. Show more Development of sodium alginate-xanthan gum based nanocomposite scaffolds reinforced with cellulose nanocrystals and halloysite nanotubes. *Polym. Test.* **2017**, *63*, 214–225. [CrossRef]
71. Feng, C.; Wang, F.; Xu, Z.; Sui, H.; Fang, Y.; Tang, X.; Shen, X. Characterization of Soybean Protein Adhesives Modified by Xanthan Gum. *Coatings* **2018**, *8*, 342. [CrossRef]

72. Zia, I.; Jolly, R.; Mirza, S.; Umar, M.S.; Owais, M.; Shakir, M. Hydroxyapatite Nanoparticles Fortified Xanthan Gum–Chitosan Based Polyelectrolyte Complex Scaffolds for Supporting the Osteo-Friendly Environment. *ACS Appl. Bio. Mater.* **2020**, *3*, 7133–7146. [CrossRef]
73. Neves, J.G.; Navarro da Rocha, D.; Lopes, C.C.; Barbosa, R.M.; Ferreira, L.F.; Westin, C.B.; Moraes, Â.M.; Calsa, B.; Santamaria, M., Jr.; Correr-Sobrinho, L.; et al. Calcium phosphates Chitosan-Xanthan composite scaffolds associated with mesenchymal stem cells for regenerative dentistry application. *Ceram. Int.* **2022**, *48*, 23088–23095. [CrossRef]
74. Lochhead, R.Y. The Use of Polymers in Cosmetic Products. In *Cosmetic Science and Technology*; Elsevier: Amsterdam, The Netherlands, 2017; pp. 171–221. ISBN 978-0-12-802005-0.
75. Nasrollahzadeh, M.; Sajjadi, M.; Nezafat, Z.; Shafiei, N. Chapter 3—Polysaccharide biopolymer chemistry. In *Biopolymer-Based Metal Nanoparticle Chemistry for Sustainable Applications*; Nasrollahzadeh, M., Ed.; Elsevier: Amsterdam, The Netherlands, 2021; pp. 45–105. ISBN 978-0-12-822108-2.
76. Coltelli, M.-B.; Danti, S.; De Clerck, K.; Lazzeri, A.; Morganti, P. Pullulan for Advanced Sustainable Body- and Skin-Contact Applications. *J. Funct. Biomater.* **2020**, *11*, 20. [CrossRef] [PubMed]
77. Chen, C.-T.; Chen, K.-I.; Chiang, H.-H.; Chen, Y.-K.; Cheng, K.-C. Improvement on Physical Properties of Pullulan Films by Novel Cross-Linking Strategy. *J. Food Sci.* **2017**, *82*, 108–117. [CrossRef] [PubMed]
78. Amrita; Arora, A.; Sharma, P.; Katti, D.S. Pullulan-based composite scaffolds for bone tissue engineering: Improved osteoconductivity by pore wall mineralization. *Carbohydr. Polym.* **2015**, *123*, 180–189. [CrossRef]
79. Cutiongco, M.F.A.; Tan, M.H.; Ng, M.Y.K.; Le Visage, C.; Yim, E.K.F. Composite pullulan–dextran polysaccharide scaffold with interfacial polyelectrolyte complexation fibers: A platform with enhanced cell interaction and spatial distribution. *Acta Biomater.* **2014**, *10*, 4410–4418. [CrossRef] [PubMed]
80. Kicková, E.; Salmaso, S.; Mastrotto, F.; Caliceti, P.; Urtti, A. Pullulan Based Bioconjugates for Ocular Dexamethasone Delivery. *Pharmaceutics* **2021**, *13*, 791. [CrossRef]
81. Chen, K.; Sivaraj, D.; Davitt, M.F.; Leeolou, M.C.; Henn, D.; Steele, S.R.; Huskins, S.L.; Trotsyuk, A.A.; Kussie, H.C.; Greco, A.H.; et al. Pullulan-Collagen hydrogel wound dressing promotes dermal remodelling and wound healing compared to commercially available collagen dressings. *Wound Repair Regen.* **2022**, *30*, 397–408. [CrossRef]
82. Baron, R.I.; Duceac, I.A.; Morariu, S.; Bostănaru-Iliescu, A.-C.; Coseri, S. Hemostatic Cryogels Based on Oxidized Pullulan/Dopamine with Potential Use as Wound Dressings. *Gels* **2022**, *8*, 726. [CrossRef]
83. Park, J.K.; Khan, T. 21—Other microbial polysaccharides: Pullulan, scleroglucan, elsinan, levan, alternant, dextran. In *Handbook of Hydrocolloids*, 2nd ed.; Phillips, G.O., Williams, P.A., Eds.; Woodhead Publishing Series in Food Science, Technology and Nutrition; Woodhead Publishing: Abington, PA, USA; Cambridge, UK, 2009; pp. 592–614. ISBN 978-1-84569-414-2.
84. Kırtel, O.; Avşar, G.; Erkorkmaz, B.A.; Öner, E.T. Microbial Polysaccharides as Food Ingredients. In *Microbial Production of Food Ingredients and Additives*; Elsevier: Amsterdam, The Netherlands, 2017; pp. 347–383. ISBN 978-0-12-811520-6.
85. Sezer, A.D.; Kazak, H.; Öner, E.T.; Akbuğa, J. Levan-based nanocarrier system for peptide and protein drug delivery: Optimization and influence of experimental parameters on the nanoparticle characteristics. *Carbohydr. Polym.* **2011**, *84*, 358–363. [CrossRef]
86. Choi, W.I.; Hwang, Y.; Sahu, A.; Min, K.; Sung, D.; Tae, G.; Chang, J.H. An injectable and physical levan-based hydrogel as a dermal filler for soft tissue augmentation. *Biomater. Sci.* **2018**, *6*, 2627–2638. [CrossRef]
87. Cinan, E.; Cesur, S.; Erginer Haskoylu, M.; Gunduz, O.; Toksoy Oner, E. Resveratrol-Loaded Levan Nanoparticles Produced by Electrohydrodynamic Atomization Technique. *Nanomaterials* **2021**, *11*, 2582. [CrossRef]
88. Osman, A.; Lin, E.; Hwang, D.S. A sticky carbohydrate meets a mussel adhesive: Catechol-conjugated levan for hemostatic and wound healing applications. *Carbohydr. Polym.* **2023**, *299*, 120172. [CrossRef] [PubMed]
89. Zheng, Y.; Liang, Y.; Zhang, D.; Sun, X.; Liang, L.; Li, J.; Liu, Y.-N. Gelatin-Based Hydrogels Blended with Gellan as an Injectable Wound Dressing. *ACS Omega* **2018**, *3*, 4766–4775. [CrossRef] [PubMed]
90. Kim, S.; Jeon, G.Y.; Kim, S.E.; Choe, S.H.; Kim, S.J.; Seo, J.S.; Kang, T.W.; Song, J.E.; Khang, G. Injectable Hydrogel Based on Gellan Gum/Silk Sericin for Application as a Retinal Pigment Epithelium Cell Carrier. *ACS Omega* **2022**, *7*, 41331–41340. [CrossRef]
91. Kenawy, E.-R.S.; Kamoun, E.A.; Eldin, M.S.; Soliman, H.M.A.; EL-Moslamy, S.H.; El-Fakharany, E.M.; Shokr, A.M. Electrospun PVA–Dextran Nanofibrous Scaffolds for Acceleration of Topical Wound Healing: Nanofiber Optimization, Characterization and In Vitro Assessment. *Arab. J. Sci. Eng.* **2023**, *48*, 205–222. [CrossRef]
92. Aki, D.; Ulag, S.; Unal, S.; Sengor, M.; Ekren, N.; Lin, C.-C.; Yılmazer, H.; Ustundag, C.B.; Kalaskar, D.M.; Gunduz, O. 3D printing of PVA/hexagonal boron nitride/bacterial cellulose composite scaffolds for bone tissue engineering. *Mater. Des.* **2020**, *196*, 109094. [CrossRef]
93. Zhang, M.; Huang, Y.; Pan, W.; Tong, X.; Zeng, Q.; Su, T.; Qi, X.; Shen, J. Polydopamine-incorporated dextran hydrogel drug carrier with tailorable structure for wound healing. *Carbohydr. Polym.* **2021**, *253*, 117213. [CrossRef]
94. Wang, H.; Xia, H.; Xu, Z.; Natsuki, T.; Ni, Q.-Q. Effect of surface structure on the antithrombogenicity performance of poly(-caprolactone)-cellulose acetate small-diameter tubular scaffolds. *Int. J. Biol. Macromol.* **2023**, *226*, 132–142. [CrossRef] [PubMed]
95. del Agua, I.; Marina, S.; Pitsalidis, C.; Mantione, D.; Ferro, M.; Iandolo, D.; Sanchez-Sanchez, A.; Malliaras, G.G.; Owens, R.M.; Mecerreyes, D. Conducting Polymer Scaffolds Based on Poly(3,4-ethylenedioxythiophene) and Xanthan Gum for Live-Cell Monitoring. *ACS Omega* **2018**, *3*, 7424–7431. [CrossRef] [PubMed]

96. Shawan, M.M.A.K.; Islam, N.; Aziz, S.; Khatun, N.; Sarker, S.R.; Hossain, M.; Hossan, T.; Morshed, M.; Sarkar, M.; Shakil, S.; et al. Fabrication of Xanthan gum: Gelatin (Xnt:Gel) Hybrid Composite Hydrogels for Evaluating Skin Wound Healing Efficacy. *Mod. Appl. Sci.* **2019**, *13*, 101–111. [CrossRef]
97. Malik, N.S.; Ahmad, M.; Minhas, M.U.; Tulain, R.; Barkat, K.; Khalid, I.; Khalid, Q. Chitosan/Xanthan Gum Based Hydrogels as Potential Carrier for an Antiviral Drug: Fabrication, Characterization, and Safety Evaluation. *Front. Chem.* **2020**, *8*, 50. [CrossRef]
98. Singh, S.; Nwabor, O.F.; Sukri, D.M.; Wunnoo, S.; Dumjun, K.; Lethongkam, S.; Kusolphat, P.; Hemtanon, N.; Klinprathum, K.; Sunghan, J.; et al. Poly (vinyl alcohol) copolymerized with xanthan gum/hypromellose/sodium carboxymethyl cellulose dermal dressings functionalized with biogenic nanostructured materials for antibacterial and wound healing application. *Int. J. Biol. Macromol.* **2022**, *216*, 235–250. [CrossRef] [PubMed]
99. Avsar, G.; Agirbasli, D.; Agirbasli, M.A.; Gunduz, O.; Oner, E.T. Levan based fibrous scaffolds electrospun via co-axial and single-needle techniques for tissue engineering applications. *Carbohydr. Polym.* **2018**, *193*, 316–325. [CrossRef] [PubMed]
100. Adrover, A.; Paolicelli, P.; Petralito, S.; Di Muzio, L.; Trilli, J.; Cesa, S.; Tho, I.; Casadei, M.A. Gellan Gum/Laponite Beads for the Modified Release of Drugs: Experimental and Modeling Study of Gastrointestinal Release. *Pharmaceutics* **2019**, *11*, 187. [CrossRef]
101. Lin, C.-C.; Chiu, J.-Y. A novel γ-PGA composite gellan membrane containing glycerol for guided bone regeneration. *Mater. Sci. Eng. C* **2021**, *118*, 111404. [CrossRef]
102. Yuan, H.; Zhong, W.; Wang, R.; Zhou, P.; Nie, Y.; Hu, W.; Tao, X.; Yang, P. Preparation of Cholesteryl-Modified Aminated Pullulan Nanoparticles to Evaluate Nanoparticle of Hydrophobic Degree on Drug Release and Cytotoxicity. *J. Nanomater.* **2020**, *2020*, e7171209. [CrossRef]
103. Mommer, S.; Gehlen, D.; Akagi, T.; Akashi, M.; Keul, H.; Möller, M. Thiolactone-Functional Pullulan for in situ Forming Biogels. *Biomacromolecules* **2021**, *22*, 4262–4273. [CrossRef] [PubMed]
104. Constantin, M.; Cosman, B.; Bercea, M.; Ailiesei, G.-L.; Fundueanu, G. Thermosensitive Poloxamer-graft-Carboxymethyl Pullulan: A Potential Injectable Hydrogel for Drug Delivery. *Polymers* **2021**, *13*, 3025. [CrossRef]
105. Shahriari-Khalaji, M.; Hu, G.; Chen, L.; Cao, Z.; Andreeva, T.; Xiong, X.; Krastev, R.; Hong, F.F. Functionalization of Aminoalkylsilane-Grafted Bacterial Nanocellulose with ZnO-NPs-Doped Pullulan Electrospun Nanofibers for Multifunctional Wound Dressing. *ACS Biomater. Sci. Eng.* **2021**, *7*, 3933–3946. [CrossRef]
106. Kushibiki, T.; Mayumi, Y.; Nakayama, E.; Azuma, R.; Ojima, K.; Horiguchi, A.; Ishihara, M. Photocrosslinked gelatin hydrogel improves wound healing and skin flap survival by the sustained release of basic fibroblast growth factor. *Sci. Rep.* **2021**, *11*, 23094. [CrossRef]
107. Zheng, C.; Liu, C.; Chen, H.; Wang, N.; Liu, X.; Sun, G.; Qiao, W. Effective wound dressing based on Poly (vinyl alcohol)/Dextran-aldehyde composite hydrogel. *Int. J. Biol. Macromol.* **2019**, *132*, 1098–1105. [CrossRef]
108. Wang, Y.; Guo, Z.; Qian, Y.; Zhang, Z.; Lyu, L.; Wang, Y.; Ye, F. Study on the Electrospinning of Gelatin/Pullulan Composite Nanofibers. *Polymers* **2019**, *11*, 1424. [CrossRef]
109. Ma, M.-X.; Liu, Q.; Ye, C.; Grottkau, B.; Guo, B.; Song, Y.-F. Preparation of P3HB4HB/(Gelatin + PVA) Composite Scaffolds by Coaxial Electrospinning and Its Biocompatibility Evaluation. *BioMed Res. Int.* **2017**, *2017*, e9251806. [CrossRef] [PubMed]
110. Navaei, T.; Milan, P.B.; Samadikuchaksaraei, A.; Davari, H.R.; Hardy, J.G.; Mozafari, M. Design and fabrication of polycaprolactone/gelatin composite scaffolds for diaphragmatic muscle reconstruction. *J. Tissue Eng. Regen. Med.* **2021**, *15*, 78–87. [CrossRef]
111. Choi, J.; Lee, J.; Shin, M.E.; Been, S.; Lee, D.H.; Khang, G. Eggshell Membrane/Gellan Gum Composite Hydrogels with Increased Degradability, Biocompatibility, and Anti-Swelling Properties for Effective Regeneration of Retinal Pigment Epithelium. *Polymers* **2020**, *12*, 2941. [CrossRef] [PubMed]
112. Hwang, Y.; Lee, J.S.; An, H.; Oh, H.; Sung, D.; Tae, G.; Choi, W.I. Hydroxyapatite-embedded levan composite hydrogel as an injectable dermal filler for considerable enhancement of biological efficacy. *J. Ind. Eng. Chem.* **2021**, *104*, 491–499. [CrossRef]
113. Wang, Q.-Q.; Liu, Y.; Zhang, C.-J.; Zhang, C.; Zhu, P. Alginate/gelatin blended hydrogel fibers cross-linked by Ca2+ and oxidized starch: Preparation and properties. *Mater. Sci. Eng. C* **2019**, *99*, 1469–1476. [CrossRef] [PubMed]
114. Heragh, B.K.; Javanshir, S.; Mahdavinia, G.R.; Naimi-Jamal, M.R. Development of pH-sensitive biomaterial-based nanocomposite for highly controlled drug release. *Results Mater.* **2022**, *16*, 100324. [CrossRef]
115. Rajput, I.B.; Tareen, F.K.; Khan, A.U.; Ahmed, N.; Khan, M.F.A.; Shah, K.U.; Rahdar, A.; Díez-Pascual, A.M. Fabrication and in vitro evaluation of chitosan-gelatin based aceclofenac loaded scaffold. *Int. J. Biol. Macromol.* **2023**, *224*, 223–232. [CrossRef]
116. Suflet, D.M.; Popescu, I.; Prisacaru, A.I.; Pelin, I.M. Synthesis and characterization of curdlan–phosphorylated curdlan based hydrogels for drug release. *Int. J. Polym. Mater. Polym. Biomater.* **2021**, *70*, 870–879. [CrossRef]
117. Sakai, S.; Ohi, H.; Taya, M. Gelatin/Hyaluronic Acid Content in Hydrogels Obtained through Blue Light-Induced Gelation Affects Hydrogel Properties and Adipose Stem Cell Behaviors. *Biomolecules* **2019**, *9*, 342. [CrossRef]
118. Velasco-Rodriguez, B.; Diaz-Vidal, T.; Rosales-Rivera, L.C.; García-González, C.A.; Alvarez-Lorenzo, C.; Al-Modlej, A.; Domínguez-Arca, V.; Prieto, G.; Barbosa, S.; Soltero Martínez, J.F.A.; et al. Hybrid Methacrylated Gelatin and Hyaluronic Acid Hydrogel Scaffolds. Preparation and Systematic Characterization for Prospective Tissue Engineering Applications. *Int. J. Mol. Sci.* **2021**, *22*, 6758. [CrossRef] [PubMed]
119. Dromel, P.C.; Singh, D.; Andres, E.; Likes, M.; Kurisawa, M.; Alexander-Katz, A.; Spector, M.; Young, M. A bioinspired gelatin-hyaluronic acid-based hybrid interpenetrating network for the enhancement of retinal ganglion cells replacement therapy. *Npj Regen. Med.* **2021**, *6*, 85. [CrossRef] [PubMed]

120. Reczyńska-Kolman, K.; Hartman, K.; Kwiecień, K.; Brzychczy-Włoch, M.; Pamuła, E. Composites Based on Gellan Gum, Alginate and Nisin-Enriched Lipid Nanoparticles for the Treatment of Infected Wounds. *Int. J. Mol. Sci.* **2022**, *23*, 321. [CrossRef]
121. Li, W.; Jian, X.; Zou, Y.; Wu, L.; Huang, H.; Li, H.; Hu, D.; Yu, B. The Fabrication of a Gellan Gum-Based Hydrogel Loaded with Magnesium Ions for the Synergistic Promotion of Skin Wound Healing. *Front. Bioeng. Biotechnol.* **2021**, *9*, 709679. [CrossRef]
122. Zhang, X.; Pan, Y.; Li, S.; Xing, L.; Du, S.; Yuan, G.; Li, J.; Zhou, T.; Xiong, D.; Tan, H.; et al. Doubly crosslinked biodegradable hydrogels based on gellan gum and chitosan for drug delivery and wound dressing. *Int. J. Biol. Macromol.* **2020**, *164*, 2204–2214. [CrossRef] [PubMed]
123. Xing, J.; Peng, X.; Li, A.; Chen, M.; Ding, Y.; Xu, X.; Yu, P.; Xie, J.; Li, J. Gellan gum/alginate-based Ca-enriched acellular bilayer hydrogel with robust interface bonding for effective osteochondral repair. *Carbohydr. Polym.* **2021**, *270*, 118382. [CrossRef] [PubMed]
124. May, T.; Herold, A.; Heinrich, D.C.; Heuschkel, I.; Sendrowski, H. Exopolysaccharide Production Microorganisms and Uses Thereof. WO2023036938A1, 16 March 2023.
125. Picken, S.J.; Zlopasa, J.; Binneveld, R.A.G.; Böttger, W.O.J. Modification of Biopolymers Using Polyols and Polyacids. WO2023038519A1, 16 March 2023.
126. Tsai, Y.-C.; Wu, C.-C.; Huang, C.-L.; Hsu, C.-C. Method for Treating Sleeping Disorders with Exopolysaccharides. WO2023025235, 2 March 2023.
127. May, T.; Heinrich, D.C.; Herold, A.; Stierl, R. Paenibacillus Strains Producing Low Amounts of Exopolysaccarides. WO2023020880A1, 23 February 2023.
128. a, C.B.; aaa, .A.; , B..; Vladimirovna, G.S.; Alekseevna, U.L.; Zhigzhitovich, T.V. *Paenibacillus polymyxa* Strain Essutm-2—Producer of Exopolysaccharides. RU2782953, 7 November 2022.
129. Colliec-Jouault, S.; Sinquin, C.; Garcia, J.M.; Heymann, D.; Zykwinska, A. Low-Molecular-Weight he800 Exopolysaccharide Derivatives with Anti-Cancer Properties and Uses Thereof. WO2023275343, 5 January 2023.
130. a, C.B.; aaa, .A.; , B..; Vladimirovna, G.S.; Alekseevna, U.L.; Zhigzhitovich, T.V. *Paenibacillus polymyxa* Vsgutu-1 Strain—Producer of Exopolysaccharides. RU2784088, 19 September 2022.
131. Domínguez, P.J.A.; Narváez, M.J.M.; Chowdhury, P.S. Bacterial Cellulose Biopolymer and Method of Obtainment. WO2023014213, 9 February 2023.
132. Peng, J.; Zhao, J.; Yang, F.; Liang, J.; Zhang, K.; Chen, G. Method for Separating and Screening Exopolysaccharide-Producing Lactic Acid Bacteria. CN115572690A, 6 January 2023.
133. Hu, J. Soothing and Repairing Composition and Application Thereof in Cosmetics. CN115554197A, 3 January 2023.
134. Fu, Y.; Mao, D.; Bian, X.; Wang, X.; Zhang, Y. Preparation Method of Chlorella Pyrenoidosa Exopolysaccharide with Antitumor Activity. CN115505610A, 23 December 2022.
135. Srivastava, N.; Choudhury, A.R. Microbial Polysaccharide-Based Nanoformulations for Nutraceutical Delivery. *ACS Omega* **2022**, *7*, 40724–40739. [CrossRef]
136. Theuer, R.; Augustine, T.; Kishter, L.; Lanspa, R. *Petition to Add Pullulan to the National List at §205.605(a) as an Allowed Nonsynthetic Ingredient in Tablets and Capsules for Dietary Supplements Labeled "Made with Organic (Specified Ingredients or Food Group(s))*; United States Department of Agriculture: Washington, DC, USA, 2013.
137. Costa, O.Y.A.; Raaijmakers, J.M.; Kuramae, E.E. Microbial Extracellular Polymeric Substances: Ecological Function and Impact on Soil Aggregation. *Front. Microbiol.* **2018**, *9*, 1636. [CrossRef] [PubMed]
138. Brandi, J.; Cheri, S.; Manfredi, M.; Di Carlo, C.; Vita Vanella, V.; Federici, F.; Bombiero, E.; Bazaj, A.; Rizzi, E.; Manna, L.; et al. Exploring the wound healing, anti-inflammatory, anti-pathogenic and proteomic effects of lactic acid bacteria on keratinocytes. *Sci. Rep.* **2020**, *10*, 11572. [CrossRef]
139. Becker, A. Challenges and perspectives in combinatorial assembly of novel exopolysaccharide biosynthesis pathways. *Front. Microbiol.* **2015**, *6*, 687. [CrossRef]
140. Schmid, J.; Sieber, V.; Rehm, B. Bacterial exopolysaccharides: Biosynthesis pathways and engineering strategies. *Front. Microbiol.* **2015**, *6*, 496. [CrossRef] [PubMed]
141. Mishra, A.; Jha, B. Microbial Exopolysaccharides. In *The Prokaryotes*; Rosenberg, E., DeLong, E.F., Lory, S., Stackebrandt, E., Thompson, F., Eds.; Springer: Berlin/Heidelberg, Germany, 2013; pp. 179–192. ISBN 978-3-642-31330-1.
142. Akdeniz Oktay, B.; Bozdemir, M.T.; Ozbas, Z.Y. Optimization of hazelnut husk medium for pullulan production by a domestic A. pullulans strain. *Prep. Biochem. Biotechnol.* **2022**, *53*, 317–330. [CrossRef] [PubMed]
143. Mirzaee, H.; Khodaiyan, F.; Kennedy, J.F.; Hosseini, S.S. *Production, Optimization and Characterization of Pullulan from Sesame Seed Oil Cake as a New Substrate by Aureobasidium pullulans*; Carbohydrate Polymer Technologies and Applications: Amsterdam, The Netherlands, 2020; Volume 1. [CrossRef]
144. He, C.; Zhang, X.; Zhang, Z.; Wang, C.; Wang, D.; Wei, G. Whole-crop biorefinery of corn biomass for pullulan production by Aureobasidium pullulans. *Bioresour. Technol.* **2023**, *370*, 128517. [CrossRef]
145. Aftab, M.N.; Iqbal, I.; Riaz, F.; Karadag, A.; Tabatabaei, M. *Different Pretreatment Methods of Lignocellulosic Biomass for Use in Biofuel Production*; IntechOpen: London, UK, 2019; ISBN 978-1-78923-988-1.
146. Kant Bhatia, S.; Ahuja, V.; Chandel, N.; Gurav, R.; Kant Bhatia, R.; Govarthanan, M.; Kumar Tyagi, V.; Kumar, V.; Pugazendhi, A.; Rajesh Banu, J.; et al. Advances in algal biomass pretreatment and its valorisation into biochemical and bioenergy by the microbial processes. *Bioresour. Technol.* **2022**, *358*, 127437. [CrossRef]

147. Santos, F.L.; de Amorim, G.M. Biotechnological challenges and perspectives of using exopolysaccharides. *J. Anal. Pharm. Res.* **2018**, *7*, 264–266. [CrossRef]
148. Singh, S.; Datta, S.; Narayanan, K.B.; Rajnish, K.N. Bacterial exo-polysaccharides in biofilms: Role in antimicrobial resistance and treatments. *J. Genet. Eng. Biotechnol.* **2021**, *19*, 140. [CrossRef] [PubMed]
149. Xu, R.-Z.; Cao, J.-S.; Feng, G.; Luo, J.-Y.; Wu, Y.; Ni, B.-J.; Fang, F. Modeling molecular structure and behavior of microbial extracellular polymeric substances through interacting-particle reaction dynamics. *Chem. Eng. J. Adv.* **2021**, *8*, 100154. [CrossRef]
150. Zou, P.; Yao, J.; Cui, Y.-N.; Zhao, T.; Che, J.; Yang, M.; Li, Z.; Gao, C. Advances in Cellulose-Based Hydrogels for Biomedical Engineering: A Review Summary. *Gels* **2022**, *8*, 364. [CrossRef]
151. Alheib, O.; da Silva, L.P.; da Silva Morais, A.; Mesquita, K.A.; Pirraco, R.P.; Reis, R.L.; Correlo, V.M. Injectable laminin-biofunctionalized gellan gum hydrogels loaded with myoblasts for skeletal muscle regeneration. *Acta Biomater.* **2022**, *143*, 282–294. [CrossRef]
152. McClatchy, D.B.; Martínez-Bartolomé, S.; Gao, Y.; Lavallée-Adam, M.; Yates, J.R. Quantitative analysis of global protein stability rates in tissues. *Sci. Rep.* **2020**, *10*, 15983. [CrossRef]
153. Hu, Y.; Zheng, Q.; Zhang, S.; Noll, L.; Wanek, W. Significant release and microbial utilization of amino sugars and d-amino acid enantiomers from microbial cell wall decomposition in soils. *Soil Biol. Biochem.* **2018**, *123*, 115–125. [CrossRef] [PubMed]
154. Beier, S.; Bertilsson, S. Bacterial chitin degradation—Mechanisms and ecophysiological strategies. *Front. Microbiol.* **2013**, *4*, 149. [CrossRef] [PubMed]
155. Maitz, M.F. Applications of synthetic polymers in clinical medicine. *Biosurface Biotribol.* **2015**, *1*, 161–176. [CrossRef]
156. Kostić, M.; Igić, M.; Gligorijević, N.; Nikolić, V.; Stošić, N.; Nikolić, L. The Use of Acrylate Polymers in Dentistry. *Polymers* **2022**, *14*, 4511. [CrossRef] [PubMed]
157. Kucharczyk, M.; Słowik-Rylska, M.; Cyran-Stemplewska, S.; Gieroń, M.; Nowak-Starz, G.; Kręcisz, B. Acrylates as a significant cause of allergic contact dermatitis: New sources of exposure. *Postep. Dermatol. Alergol.* **2021**, *38*, 555–560. [CrossRef]
158. Kanazawa, R.; Sato, S.; Iwamoto, N.; Teramoto, A. Allergic reaction following arachnoid plasty with a fibrin sealant. *Neurol. Med. Chir.* **2010**, *50*, 608–610. [CrossRef]
159. Orihara, M.; Takazawa, T.; Horiuchi, T.; Sakamoto, S.; Uchiyama, M.; Saito, S. Intraoperative anaphylaxis due to aprotinin after local application of fibrin sealant diagnosed by skin tests and basophil activation tests: A case report. *JA Clin. Rep.* **2021**, *7*, 68. [CrossRef]
160. Dhaka, V.; Singh, S.; Anil, A.G.; Sunil Kumar Naik, T.S.; Garg, S.; Samuel, J.; Kumar, M.; Ramamurthy, P.C.; Singh, J. Occurrence, toxicity and remediation of polyethylene terephthalate plastics. A review. *Environ. Chem. Lett.* **2022**, *20*, 1777–1800. [CrossRef]
161. Donati, M.; Brancato, G.; Grosso, G.; Li Volti, G.; La Camera, G.; Cardì, F.; Basile, F.; Donati, A. Immunological reaction and oxidative stress after light or heavy polypropylene mesh implantation in inguinal hernioplasty. *Medicine* **2016**, *95*, e3791. [CrossRef]
162. Casalini, T.; Rossi, F.; Castrovinci, A.; Perale, G. A Perspective on Polylactic Acid-Based Polymers Use for Nanoparticles Synthesis and Applications. *Front. Bioeng. Biotechnol.* **2019**, *7*, 259. [CrossRef] [PubMed]

MDPI
St. Alban-Anlage 66
4052 Basel
Switzerland
Tel. +41 61 683 77 34
Fax +41 61 302 89 18
www.mdpi.com

Polymers Editorial Office
E-mail: polymers@mdpi.com
www.mdpi.com/journal/polymers

www.ingramcontent.com/pod-product-compliance
Lightning Source LLC
Chambersburg PA
CBHW041017270926
51850CB00024B/1372
9783036574530